T0210222

Leprosy in Premodern Medicine

Leprosy in Premodern Medicine

A Malady of the Whole Body

Luke Demaitre

THE JOHNS HOPKINS UNIVERSITY PRESS
Baltimore

The Johns Hopkins University Press
2715 North Charles Street
Baltimore, Maryland 21218-4363
www.press.jhu.edu

Library of Congress Cataloging-in-Publication Data

Demaitre, Luke E., 1935–
    Leprosy in premodern medicine : a malady of the whole body / Luke Demaitre.
        p. ; cm.
    Includes bibliographical references and index.
    ISBN-13: 978-0-8018-8613-3 (hardcover : alk. paper)
    ISBN-10: 0-8018-8613-9 (hardcover : alk. paper)
        1. Leprosy—Europe—History—To 1500. 2. Leprosy—Europe—History—16th century.
3. Leprosy—Europe—History—17th century. I. Title.
        [DNLM: 1. Leprosy—history—Europe. 2. History, 17th Century—Europe. 3. History,
Early Modern 1451–1600—Europe. 4. History, Medieval—Europe.   WC 335 D369L 2007]
    RC154.4.D46 2007
    614.5'46094—dc22   2006026074

A catalog record for this book is available from the British Library.

Contents

Introduction

This study began as a history of an idea. An initial stimulus for the inquiry was the engaging monograph by Saul Nathaniel Brody titled *The Disease of the Soul: Leprosy in Medieval Literature* (1974). Explorations in early medical writings increasingly convinced me that physicians, unlike poets and preachers, viewed leprosy as a disease of the body rather than of the soul, and that they devoted far more attention to signs, causes, and cures than to metaphors and moralizations. This inchoate impression grew more precise when, in modest emulation of today's best historians of medicine, I began to connect medical concepts, conduct, and sociocultural settings. My insights gained a significant degree of concreteness when, after poring over dozens of treatises— and somewhat belatedly—I extended my research to archives. With this step came the considerable new challenge of integrating the conceptual and the concrete. This challenge made it appear wise not to overreach by speculating about the objective identity of "the malady called leprosy," and to stay with premodern responses to the disease.

The term *premodern* is understood somewhat idiosyncratically here, as preceding the modern history of leprosy, which began in the late eighteenth century. One milestone was the overseas expansion of the European awareness of the malady, first to its incidence in the West and East Indies. The definitive transition, however, came with the advance of clinical medicine and the advent of bacteriology. In Norway, Daniel Danielssen described the symptoms of leprosy with unprecedented precision in the 1840s, and his son-in-law G. Armauer Hansen identified the causative organism in 1873. *Mycobacterium leprae*, the cause of what is now known as Hansen's disease (HD), is a slowly multiplying microorganism. The bacterium is related to the tuberculosis bacillus, and it can survive in the soil for several months. It affects the peripheral nerves, the skin, and ultimately the bones. In people who are susceptible (less than 5 percent of most populations), the effects vary according to the individual immune response. After an initial and elusively indeterminate stage, from which spontaneous recovery may occur, and depending on how well the immune defense manages to check the multiplication of bacilli, the disease ranges between the poles of the milder tuberculoid (paucibacillary) and the disfiguring lepromatous (multibacillary) form. An intermediary or borderline form is possible. The lesions and progress of the lepromatous form are most directly devastating and dramatic. The tuberculoid form is manifested, often inconspicuously, in pale desensitized patches of the skin; eventually, however, it also cripples the person who has it, both physically and socially, with progressive nerve damage, loss of sensation, atrophies, secondary infections, and loss of extremities.

Long before the recognition of HD, a generally corresponding disease figured as a genuine medical entity. Neither the broad correspondence nor the appearance of specific parallels will be tested in these pages. In other words— and at the risk of disappointing some readers—I will not pursue the question, of long standing and great interest, of whether premodern leprosy was identical to Hansen's disease. For matching the written record, on which this inquiry is focused, with the physical evidence, I must defer to the expertise of paleopathologists. Nevertheless, it is reasonable to assume that some— or, according to recent research, many—of the people who were determined to have leprosy would today be identified as having Hansen's disease. This, of course, does not mean that the two diseases, or their defining criteria, were the same. In fact, when I seem to make an identification between other premodern diseases, such as morphea, and modern pathological entities, the reader should understand that the same reservation applies.

A parallel caution is in order for allusions to the prevalence of leprosy in Europe. This study is an examination of recorded responses to the disease, and the task of matching their frequency and tone with actual incidence will require the systematic assessment of epidemiologists. In the few instances in which I relate responses to epidemiological history, I rely on the insights of current scholarship. The most judicious scholars argue that the spread of leprosy through Europe was gradual, accelerating after the tenth century CE, and due to various causes among which—contrary to diehard simplifications—the Crusades played only a limited role. Much further quantification is needed to verify the appearance of a peak in the thirteenth and fourteenth centuries, which may be affected by changing social preoccupations. Negative evidence, mainly a combination of diminishing remains and waning concerns, leads historians to believe that the disease retreated from most of the Continent after the fifteenth century, while lingering in certain areas, particularly in Scandinavia. This decline, too, resulted from complex causes, rather than simply from recurrent waves of bubonic plague (from 1348 on), as had long been assumed.

Although I have collated the records in a chronological framework, their copiousness and diversity make the collation a preliminary survey rather than a comprehensive and sustained chronicle. A more definitive treatment must consider a host of factors. In particular, the establishment, organization, and transformation of leprosaria, which continue to generate a voluminous historiography, receive only passing attention here. My primary objective, then, is to explore the interaction between premodern medicine and a clearly exceptional malady: How did medical teachers and practitioners respond, in discourse and deed, to the disease? How did they affect the fate of patients? And how did changing contexts influence the interaction?

Our chronology will reach back to the first century CE, when the disease that became known as leprosy was first described in the West. We will concentrate, however, on the period between 1300 and 1700, when theoretical and practical responses to the disease were most intense. Our geographical focus is on France, Germany, and the Low Countries, with Italy and Scandinavia on the periphery. It would take a lifetime to explore all the riches in Continental repositories, to which trails have been blazed by pioneer historians Karl Sudhoff and Ernest Wickersheimer a century ago, and more recently by Albert Bourgeois, Griet Maréchal, François-Olivier Touati, and others. This abundance, together with a Belgian's relative facility in Dutch, French, and German, persuaded me to limit myself to the western Continent. The British Isles are beyond the scope of this examination, save for a passing reference. Eventually, the ongoing work of distinguished scholars in the United Kingdom, such as Keith Manchester, Piers Mitchell, and Charlotte Roberts, will allow for a comparative conspectus. Right now (while this book was in press), a comparative look has become possible, at least for the Middle Ages, thanks to Carole Rawcliffe, *Leprosy in Medieval England* (2006).

Within the defined geography and chronology, I have endeavored to cross some borders that remain visible to this day, namely, the frontier between Romance and Germanic idioms, and the usual boundary between the "medieval" and "modern" eras. The extended search, and the number of assembled documents, might reinforce the impression that leprosy was ubiquitous in premodern Europe. Their purpose, however, was to establish a solid basis for an assessment that is based on the exegesis of texts, with some attention for the iconography. Tables and diagrams, as well as selected images, will illustrate the exegesis. The flow of the discussion argued against the inclusion of statistics (such as the percentages of positive findings in examinations, gender ratios, or the chronological distributions of more than four hundred checked affidavits). One aspect, on the other hand, to which I have given special attention is the recognition and identification of ordinary people named in the sources: it seems worthwhile to lift the anonymity that has hung over people with leprosy as a result not only of stigma but also of history's "blind spot."

The first chapter introduces the types, authors, and contexts of the sources. The evidence consists mainly of two categories, namely, academic writings and public documents. Beyond this basic dichotomy, the texts vary widely, from head-to-toe encyclopedias to reminiscences of practitioners, and from diagnostic checklists to affidavits by health commissioners. By their diversity, the sources argue against facile generalizations; they also attest to exchanges between learned and common notions, links between theory and practice, and the coexistence of tradition and innovation. In Chapter 2, a wide range of textual—and pictorial—evidence supplies a script for the crucial encounter between medicine, the disease, and society. As genuine drama, the examination of people suspected of having leprosy featured distinct characters, expressed powerful ideas and emotions, and evoked deeper dynamics. The resulting judgment decided whether the suspect would go home free, be separated from the healthy, or receive a temporary but cautionary reprieve. In a stunning reversal, poverty induced indigents to fake rather than hide the disease. In either situation the role of judges fell to the physicians and surgeons, who gradually replaced nonmedical examiners. Verdicts, as well as "charges," often hinged on semantics.

The nomenclature of leprosy occupies the entire third chapter because it demonstrates the power of words, in both popular usage and learned discourse. The medical vocabulary had multiple roots in classical Greek, biblical Hebrew, early Christian Latin, Arabic, and vernacular languages. More names and predicates were attributed to leprosy than to any other ailment. The labels and attributes reflected emotional responses, terminological confusion, and attempts at rationalization. Chapter 4 reveals an additional element in the power of names, as the verbal definition of a disease, rather than microscopy or chemistry, held the key to pathology. The standard definition of leprosy adhered to the authoritative formulation by Avicenna. While not impervious to individual adaptation and changing emphasis, it was based on the physiology of the four humors and on Galen's idea of their combination, the construct of *complexio*. Humoral interpretations shaped and were shaped by other responses to the disease, so that they were not mere abstractions, even though they yielded to an empirical pathology at the end of the period under consideration.

Chapter 5 traces the development of two salient instances of linkage between attributes, concepts, and concrete implications. By definition and perception, if not by systematic observation, leprosy became identified as an essentially contagious and hereditary disease. The identification, which grew in breadth and intensity, emerged when age-old efforts in medicine—the efforts to understand the transmission of disease—merged with tightening social structures and intensifying apprehensions about "the other." Beyond the issues of contagion and heredity, however, medical authors organized an elaborate discussion of the causes and categories of the malady, which are surveyed in Chapter 6. Aristotelian natural philosophy, in addition to Galenic physiology, dictated the chief causes of leprosy, framed its division into stages and forms, and guided its allocation on the overall map of diseases. The explanations and classifications were of intensely practical interest for diagnosis and treatment.

Chapter 7 argues that the care in observing, describing, and differentiating the signs of leprosy culminated in the fourteenth and fifteenth centuries, when medical verdicts were most consequential. Also in other centuries, however, the visitation followed a set protocol for certain preliminaries, the physical examination, and diagnostic tests. Signs were differentiated according to the stage and form of the disease, itemized in checklists, and assessed for reliability. Notwithstanding the immediate importance of sense perception, diagnoses were colored by conceptual premises and popular assumptions, and even by the prospective consequences of a positive finding. The grim prognosis for confirmed leprosy, as Chapter 8 shows, suggested utter despair; paradoxically, however, it did not mean abandonment. The standard verdict of medicine was that the disease was totally incurable. The physicians' professed hopelessness and apparent fatalism, however, were contradicted by the pages they dedicated to prevention and therapeutics. They manifested their continued concern with patients in prescriptions for remedial and palliative treatment, memories of actual care, and pleas for prolonging life.

The chapters incorporate as many supporting and illustrative texts as the flow of the analysis allows. Most of these texts were not available in English until now. The translations are my own, unless noted. For the sake of space, original texts are not quoted in the notes, which provide full references to the printed and manuscript sources. In quotations of texts in English translation, the original spelling is retained for key words, such as *elephantia, elefantia,* or *elefancia.* When a term or phrase is particularly significant, evocative, or open to different interpretations, it is included in the quotation (in square brackets and in italics). The term *leprosy* is employed in the most generic sense in order to avoid the anachronism of "HD"; when the disease is mentioned in close association with the sources, I adopt the nomenclature of these sources—as consistently as possible. The word *leper* has been shunned consistently—except when it is part of a modern quotation—and consciously because it amounts to a slur (in English notably more than in other Western languages). After all, this book aims to show that premodern medicine considered the *leprosus* not as a reprobate, branded with a disease of the soul, but as a patient, burdened by a malady of the whole body.

Acknowledgments

Awealth of sources have allowed this study to grow in depth and range—and, I hope, in balance—beyond the initial analysis of a handful of treatises. Access to widely scattered documents was made possible by an Extramural Grant from the National Library of Medicine of the United States National Institutes of Health, Bethesda, Maryland. This aid was awarded, in large part, thanks to the powerful endorsements of my application. Therefore, this book would not have seen the light or, at best, would have been much poorer without the support of Ann Carmichael, Sheldon Cohen, Michael McVaugh, and John Wilhelm and Sally Squires. I am forever in their debt for their generous and brave recommendations and, as each one of them knows, for much more.

It is a pleasure to remember the kind attention I received at the Archives départementales de l'Aube (Troyes), the Archives départementales de la Côte d'Or (Dijon), the Archives départementales de l'Yonne (Auxerre), the Archives municipales de Nevers, the Stadtarchiv of Nördlingen, and the Universitätsarchiv in Tübingen. All the archivists and librarians, on both sides of the Atlantic, have been so unfailingly helpful that it seems unfair to single out any one. It would be even less fair, however, not to recognize those who extended extraordinary kindnesses. Two directors have gone beyond professional courtesy in facilitating my research: Mevrouw Hilde De Bruyne at the Archief en Kunstpatrimonium of the Openbaar Centrum voor Maatschappelijk Welzijn in Bruges, Belgium; and Frau Doctor Christine Sauer of the Handschriftenabteilung at the Stadtbibliothek, Nuremberg. At the National Library of Medicine, which remains an indispensable—if physically less and less accessible—resource, the History of Medicine Division has become a friendly haven as well as the ideal place for research.

While no reward could be greater than the knowledge that this book is deemed worth reading, I have enjoyed many blessings in the course of writing it. The journey, through unforgettable places and fascinating sources, opened new vistas of history and humanity. Most important, encounters with diverse people lifted my research beyond a bookish pursuit. I am particularly grateful to five persons who brought soul to my quest. Guenter Risse, who guided me to the riches of the Cologne archives, broadens the horizon of historians with his *philanthropeia*. Sally Squires of the *Washington Post*, who has championed the dignity of the last patients at the Gillis W. Long Hansen's Disease Center in Carville (Louisiana), showed me how to balance an interest in the past with passions about the present. Sister Veerle Ingelbeen and her group of recovered HD patients, Women United Through Handcrafted Lace and Embroidery, in Santa Barbara, Iloilo, Philippines, taught me, by their living more than with words, how tightly the medical aspects of a disease are interwoven with the social and cultural fabric. Richard Marks, former sheriff and now unforgettable guide in stunningly beautiful Kalaupapa, Molokai, Hawaii, not only brought to life the memory of being branded with "separation sickness": he also personifies the trials and triumphs of medicine and, above all, the resiliency of human nature. Arturo C. Cunanan Jr. enriched my visit to Culion Island, Palawan, Philippines, by exemplifying two themes that inspire this book, namely, the continuity of history and the grace of medicine. Uniquely bridging the past and the present (and defying the general trend of leprosaria to become mere relics or to vanish altogether), he geared the celebration of the centennial of the Culion Sanitarium (1906–2006) to the continued vitality of its health services. As head of the hospital's Leprosy Control and Rehabilitation Program, Dr. Cunanan devotes his life to the conquest of the disease and the full redemption of its victims.

Special gratitude goes to companions on my journey, especially to those who carried me over some rough spots. When illness intruded, Linda Voigts and Walt Schalick extended a network of caring prayers on both sides of the Atlantic; superbly skilled and attentive practitioners in the University of Virginia Health System kept cancer at bay while this book was nearing completion. Fourth-year medical students in my history elective, in the Humanities in Medicine Program at the University of Virginia, have taught me much about the physician's ideals. The lawyer's best sides shone forth in David Leive, who, out of sheer friendship—and at the risk of ending it—subjected himself to reading the early drafts of several chapters. A separate and elaborate acknowledgment would be needed for a full recognition of the extent to which this book was improved by the scholar who read the manuscript for the Johns Hopkins University Press; remaining flaws and errors are due to my personal limitations. The contribution of the anonymous reader was surpassed only by the blessing of a close-knit family. Our closeness, and especially the delight of our granddaughter Ava Sophia, provided an antidote of sanity against close-minded concentration. Above all, every mile and minute of these peregrinations—and of life in general—have drawn energy from the love of my wife, Dominique, to whom this book is dedicated.

Leprosy in Premodern Medicine

The Sources

Texts and Contexts

[I]ch frâgete vil gerne:
sô vil zuo Salerne
von arzenîen meister ist,
wie kumet das ir deheines list
ze iuwerm ungesunde
niht gerâten kunde?
herre, des wundert mich.

"I would like to ask you, if there are so many masters of healing in Salerno, how come that their art cannot succeed against your illness? My lord, this amazes me." With this inquiry of a concerned farmer, the poet Hartmann von Aue framed the topos of *Der arme Heinrich* (Poor Henry), his tale about a Job-like nobleman who fell into leprosy and would be cured only by submission to divine inscrutability. The poem, composed in the closing years of the twelfth century, stereotypically emphasizes the ineffectiveness of human healers. Nevertheless, it sheds light on reactions to leprosy and on their historical context. Above all, the poet cast the "masters of healing" (*von arzenîen meister*) in a dramatic role that stands in contrast with the invisibility of medical practitioners in early stories of miraculous cures of leprosy—for example, by Saint Martin of Tours.[1]

The poem also affords a glimpse into twelfth-century perceptions of leprosy. When Heinrich was seized by the disease ("miselsuht"), he "became repugnant to men and women, when they saw God's heavy chastisement in his body." The moral overtone of this description is undeniable. However, it should not make us oblivious to several concrete points that these lines convey. First, the name given to the disease is utterly unrelated to the term *lepra*, and, as we will see in Chapter 3, the etymology of *miselsuht* carries more social than nosological connotations. In addition, the essential effect of leprosy is reduced to Heinrich's repulsive appearance, rather than to his pain or, more significant, to people's fear of contagion. "He now became so ugly that nobody could stand to look at him." Medical teaching, filtered through popular rumor, surfaces in the explanation that Heinrich did not despair immediately but sought help because he had often heard "that this disease is very variable, and curable in many cases." He "hastened to Montpellier for the counsel of physicians," only to receive the prompt and blunt news "that he would never be saved."[2]

IN THE PUBLIC FORUM
Distant Echoes: Physician as Healer

It is significant that Heinrich sought medical intervention when his leprosy was unmistakable. He did not need the physicians for identifying his disease but for curing it. This perspective was apparently not unusual at the time, if we consider an episode in a chronicle by William of Tyre, Hartmann von Aue's contemporary. As the tutor of Baldwin IV, the young son of the crusading king Amalric, in the 1170s, William was the first to discover that the boy had lost sensation in his arm, an alarming but still unclear sign. When reminiscing, about a decade later, the chronicler suggested that the courtiers ("we") eventually reached their own diagnosis, while expecting different help from the *medici*.

> I reported all this to his father. Physicians were consulted and prescribed repeated fomentations, anointings, and even poisonous drugs to improve his condition, but in vain. For, as we later understood more fully as time passed, and as we made more comprehensive observations, this was the beginning of an incurable disease. I cannot keep my eyes dry while speaking of it. For as he began to reach the age of puberty it became apparent that he was suffering from that most terrible disease, leprosy. Each day he grew more ill. The extremities and the face were most affected, so that the hearts of his faithful men were touched by compassion when they looked at him.[3]

In both cases, of poor Heinrich and the future "Leper King," the physicians were consulted for a cure. Less than a century later, their primary role would begin to shift from the treatment to the diagnosis of leprosy.

Poor Heinrich's last resort was to seek "the art of the wise healers" in Salerno. There, "the best master" immediately told him "a strange story: that he was curable [*genislîch*] yet would forever be unhealed [*ungenesen*]." Aside from the punning on the difference between curing and healing, both the ultimate recourse to Salernitan masters and their reputation as the most advanced in the West had been part of a topos for some time. Montpellier occasionally shared that reputation, but in second place. Indeed, the only extant discussion of *lepra* that might have some ties to late twelfth-century Montpellier was by a physician who went there from Salerno but did not stay long before moving on to Paris. Gilles de Corbeil (ca. 1140–1224) may have composed his versified handbook on the signs and symptoms of diseases while in Montpellier. He devoted 89 of its 2,358 lines to leprosy, which,

> *destroying Nature's temple by contagion,*
> *results from several causes. The diet produces*
> *a surplus of unclean blood when an earthy humor,*
> *produced by a coarse diet, corrupts the nourishment*
> *of the organs. First to be mentioned are the meats*
> *of donkeys and goats; cabbages, milo, beans, lentils.*
> *[The disease] arises from food that blackens the burnt blood.*

The passage ended with the warning

> *with unwavering and unyielding reason,*
> *consider the place, time, lifestyle,*
> *strength, complexion, and appearance of the patient.*

While Gilles's verses aptly summarized much of medical lore on the signs and causes of leprosy, they apparently represented what he had learned in Salerno—and his own idiosyncrasies—rather than Montpellier teaching. A Salernitan corpus was well established by the end of the twelfth century. It reached back to a brief treatise on leprosy by Constantine the African (d. before 1098/9); and it included passages in the practical manuals compiled by masters such as Gariopontus (fl. 1050), Johannes Platearius (fl. 1130), and Johannes de Sancto Paulo (fl. 1190).[4]

The situation of learned physicians between the twelfth and mid-fourteenth centuries presents a paradox. Even if they personified the highest human ability to heal, and while they were compiling a rich medical literature on leprosy, they were hardly visible in actual testimony about patients. When physicians were named in records of the proliferating foundations for *leprosi*, they figured as witnesses, landholders, or donors rather than caregivers; when they appeared in chronicles, they left mostly indistinct traces. Official documents regarding hospitals preserved scant information on matters of health, as their chief objective was to secure ownerships and incomes. Typical of this priority was the general silence on the medical criteria for admission, which stood in contrast with the precision on financial requirements at entry. In November 1220, for example, the *scabini*, or syndics, of the town of Ieper (Ypres) in Flanders mentioned no procedure for determining that someone was "stricken with leprosy and should therefore not be allowed to circulate among people," whereas they dictated precise terms for the person's hospitalization.

> If he wishes and requests to be received in the leprosarium of the Blessed Magdalen, if he has no children and legal spouse, he will be accepted in the said house with all his movable and immovable possessions that can be verified. If he has a wife and children, their entire property will be divided into three shares, and the leprous patient will bring his third to the said house. If he has children and no wife . . . If he has a wife and no children . . . If he comes into some property at the death of one of his own parents, the said leprosarium is to receive and hold it entirely and without claims. If a suit or claim arises about these provisions, the *scabini* of Ieper will settle it in accordance with their best and most discerning knowledge.

Not only the decrees of rising urban powers but also the twelfth- or thirteenth-century cartularies of leprosaria reveal little about medical concerns. When charters included rules for the inmates, their chief aim was, at best, to secure a monastic lifestyle and, at worst, to prevent chaos and dissolution. This is evident in an elaborate statute drawn up in 1301 for the leprosarium of Ieper. The charter regulated in detail—with stipulated fines for most of the transgressions—matters of conduct ranging from marital visits to gambling and quarreling, but it did not mention the communal care of health, or even health-related subjects such as nutrition or cleanliness. The silence of the sources does not warrant the inference that the leprous inmates of almshouses were utterly bereft of care. Many of the foundations included "brothers" and "sisters" who dedicated their lives to caring for the residents. These attendants apparently acquired valuable experience in the diagnosis as well as the care of leprosy, and they passed their knowledge on to apprentices. Nevertheless, the absence of medical practitioners from hospital documents into the fourteenth century remains worth noting.[5]

In chronicles and analogous sources of the same period, the presence of physicians was at most casual and shadowy. This is particularly striking in the reports of the most atrocious episode. In 1321, after spasms of popular unrest, King Philip V of France first condoned, and then ordered, the burning of *lépreux* in their shelters across the country. Rumors, spreading like wildfire, had accused them of conspiring, together with Jews and Muslims, to poison wells and rivers and to destroy Christendom. Within weeks, the panic crossed into the kingdom of Aragon, where King Jaume II also ordered leprous subjects as well as strangers to be rounded up as suspects, tortured, and burned if judged guilty. Several chroniclers reported the outburst and responses in vivid terms, but with omissions and passing references that trigger questions about certain medical aspects. In general, one wonders whether physicians, who at that time were becoming more visible in princely retinues and populous communes, played any role in the sequence of events. Did they voice support or condemnation of the harsh royal measures, at court, at faculty gatherings, in the afflicted towns? If so, were they heard? Eventually, the excesses of the episode and the irrationality of the responses may have contributed, as Michael McVaugh suggests, to "a gradual recognition that leprosy should be established as a specifically medical condition." During the turmoil, however, how were learned medical practitioners perceived: as protecting the healthy by identifying the dangerously sick, or as favoring the sick to the extent of aiding their conspiracies? The latter possibility is insinuated, for example, by the claim of one witness that "the advice of physicians" had guided the preparation of poisonous powders; also, a physician had translated, from Arabic into French, two letters from Muslim kings in support of the conspiracy.[6]

More indirectly, the episode of 1321 raises the possibility that medical teaching, which will be examined in the following chapters, contributed to mass hysteria by characterizing the disease in qualitative terms, by intensifying the fear of contagion, and by portraying *leprosi* as threats to the social order. While a royal order of August 18, 1321, to purge the land of the "rottenness" of the "stinking lepers" no doubt echoed popular repulsion, it was

phrased in terms ("putriditas" and "fetidus") that harked back to the qualitative descriptions by physicians. Furthermore, although the popular association of leprosy with wickedness was fomented most intensely by the
metaphors and moralizations of preachers, it may have been reinforced by
the emphasis in medical textbooks on "evil" in the causes, essence, and consequences of the disease. The linking of patients and Jews as conspirators
was natural for rabble fanaticism, but it could find support in at least one
medical opinion. In 1306, the learned surgeon Henri de Mondeville asserted
that, "because Jews seldom have intercourse during menstruation, few Jews
are leprous." The perception that a minority enjoyed exceptional immunity
was likely to fuel the envy and suspicions of a prejudiced populace.[7]

An Emerging Profile: Physician as Judge

For the sake of perspective, it is useful to compare Mondeville's assertion, as
well as the riot of 1321 and the setting of the early 1300s, with the later case
of a leprous Jew which is documented in the archives of Bourg-en-Bresse
(now in the French Département de l'Ain). On July 21, 1462, the syndics of
Bourg reported to the town council,

> [T]he people are clamoring about the Jew Eliannus de Montel, why he is not
> being expelled from this town even though he is and has long been leprous
> [*lazarus*] and many, Christians as well as his own folk, have dealings with him. Be-
> cause of this constant interaction [*conversacione*], several have allegedly been
> affected and stricken by the said disease and many more can expect to be unless
> remedial action is taken. Furthermore, while Christians are expelled from the
> town when they become leprous, this Jew is allowed to stay here in town, thanks
> to numerous irregular favors.

In a petition to the governor for the Duke of Savoy, the syndics requested
"that said individual be examined [*visitetur*] in order to know whether or not
he is leprous, and that the physicians Jacobus and Michael and the barbers
Johannes Paradi and Johannes de Viry be commissioned for this." Several
prominent burghers raised the stakes by demanding that also the son of
Eliannus, Master Sansuinus, be summoned and checked "because it is publicly known that he, too, is seized by the disease of leprosy." Unyielding,
Eliannus rejected all pleas to "avoid expenses" by voluntarily admitting his
affliction and accepting isolation.

Eliannus maintained that he was healthy, did not need an examination, and would pursue his own appeal to the governor. The syndics insisted "that his person should be examined in order to know the truth and to remove all doubt," and that, if the examiners "declared him to be infected with leprosy," a petition for his expulsion should be sent to the Duke. If he stayed in town, it would be inevitable that

> many, Christians as well as Jews, would communicate with him all the time, espe-
> cially his wife, children and household, and the town's other Jews who trade and
> deal with Christians on a daily basis throughout the town, not only in the mar-
> ket and streets but everywhere. Furthermore, it should be kept in mind that these
> Jews also bake bread in the town's ovens; draw and use water from its springs,
> wells, and streams; buy meat, eggs, cheeses, fruits and other vittles in its food-
> market, squares and shopping quarters; and that they touch all these things with
> their hands. They breathe in the face of the Christians when they come close,
> walk or stand around, talk, and so on, and they infect the air.

Armed with these arguments, "and many other considerations too long to write down," the syndics brought their complaint directly to the residence of Amedeus, heir apparent of Savoy. Further litigation was prevented by a compromise. With regard to Sansuinus, who claimed that he was wronged by the allegations, and that "he had only the beginning of this disease, with signs in his eyes, eyelashes, face, and speech," an agreement was reached

> to insist on the demand that he be examined in order to remove all doubt and to
> calm the people's clamor. Further, if the said Eliannus and Sansuinus should not
> be found infected with the said disease, the town would pay the expenses and the
> examiners' fee; on the other hand, if they were found infected, the said Jews
> would pay the expenses and the examiners' fee.

Without informing us about the procedural steps that followed, the council's minutes tersely report their results. In short order, Eliannus de Montel left town. "He was brought [*portatus*] to the grange of the Jewish cemetery of Bourg, and with his own mouth he admitted and confessed that he was leprous." After his removal to a more distant place, the castle of the lord of Romans, his son Sansuinus agreed with the syndics "to add up the payment for the town's expenses."[8]

The story of Eliannus and Sansuinus is tantalizingly unique in the archives, so that it cannot answer questions about the incidence of leprosy or about the treatment of leprous patients among Jews in medieval Europe. Nevertheless, the report is keenly illustrative of the continuity and changes over one and a half centuries. In Bourg-en-Bresse in 1462, as in the French kingdom in 1321, street rumors carried considerable momentum. When they swirled around Jewish neighbors, they boiled with suspicious envy of the seeming immunity, either from leprosy or from its social consequences. On the other hand, there is a marked difference in the settings of both episodes. The brutal eruptions of prejudice across France occurred at the edge of towns and throughout the countryside, after having been stirred by marauding mobs that were lumped together as Pastoureaux, or "Little Shepherds." The rumors in the urban neighborhoods of Bourg may have sounded ominous, but mob violence does not appear to have been imminent, thanks to increased civility—one likes to think—but above all to civic safeguards. Even so, it seems fair to suppose that popular pressure hastened the expulsion of Eliannus and to surmise that tensions may have affected the medical verdict.

A crucial contrast with the outbursts of 1321 lies in the intervention by the municipal government of Bourg in 1462 and, most of all, in the decisive power it granted to the medical practitioners for settling the issue. While we may wish that the council minutes contained an account of the examination, their silence suggests that the syndics took the work and decision of the commissioned practitioners for granted. Moreover, the syndics evidently did not need to search for candidates when they identified Jacobus and Michael as the physicians, and Johannes Paradi and Johannes de Viry as the barbers, who would carry out the official examination. These examiners personified the diagnostic phase of medical intervention, which the earlier popular perception had largely ignored. Moreover, their ability to recognize leprosy was trusted (though not absolutely, as we will see), whereas earlier practitioners were rated as able—or unable—to heal. Most significant, the Bourg commissioners were typical of the "professional" group that took over and eventually monopolized the inspection of suspected individuals in the fifteenth century. The *visite*, or *Schau*, had been the prerogative of town councillors, hospital regents, and even leprous patients themselves until the late 1300s. Between 1400 and 1600 the number of certificates by medical examiners surged— literally hundreds of them constitute a mosaic for this study—in one of the clearest manifestations of the expanding presence, prestige, and power of organized medicine. As might be expected, however, the takeover did not proceed without resistance and conflict.

A Documentary Showcase, 1456–1460

Two years before the Bourg incident, the medical practitioners commissioned by Frankfurt-am-Main protested a new town ordinance. The disagreement reminds the reader of the frequent contentions—modern as well as medieval—between the professions of medicine and law. The Frankfurt measure, ostensibly promoted by the judicial authorities but conceivably occasioned by contested diagnoses, required the participation of a judge in each examination. On May 16, 1460, the municipal physician and his two associates addressed a formal complaint

> To the honorable men of the Council of Frankfurt . . .
>
> In your wisdom you have commissioned us to examine people for leprosy under the oath, which we have sworn to your Honors: to speak the exact truth in accordance with our best understanding; to prepare a letter about the examination; and to present the verdict in proper form in that letter. Until now, people have been satisfied with this kind of letter and verdict without the presence and oath-taking of a judge, and we the undersigned believe that we are charged and sworn adequately.

When the medical men came to the crux of their complaint, the strokes of their quill became somewhat bolder.

> Now, the chief justice is of the opinion that we should be charged and sworn for every individual examination. We, however, believe that it is sufficient to keep the verdict under the oath of our sworn commission. This has been done from old, and people have found it quite satisfactory. If the judge himself assigns us someone of Frankfurt, or if someone else wishes to have the judge present for administering a particular oath, then it is customary to add the judge and the oath. In this case, the judge charges two Taler, and for sealing the letter he charges two more Taler.

With some irony, the medical practitioners then summoned the past for contradicting the judge, whose normal role it was to uphold precedent.

> It is not the custom to take the judge for every examination, to have us swear an oath each time, and to charge an additional fee. It should be clearly understood that a standing oath is stronger than an oath for each single verdict. We have

sworn the former to our lords, and "the pomp and circumstance of the occasion put weight upon the one who swears an oath [*circumstantie enim et sollempnitates adhibite grauant jurantem*]."

From this legal tone, which was reminiscent of the court, the physician and his associates then switched to another urban chord, voicing sympathy with the people.

> Until now, people have been satisfied without a judge present. Many do not want to be examined in his presence or with his knowledge, and the judge cannot be of any use there. People do not like to be burdened by more than the fee for our work. To impose on them an unnecessary and useless burden, in addition to their great pain of being judged, strikes us as an injustice. Something useless should not be forced upon people, nor should we be forced to burden them by collecting that money for the judge.
>
> Furthermore, whereas the judge should not be sealing a letter for a fee of two Taler, we wish that, in your wisdom, you allow us to seal the letter with the small seal of our lords, since it has been the custom, and since it is more credible and honorable; or, that the sworn physician of the town be allowed to apply the seal for us.
>
> We hereby humbly request you to keep us under the good custom; not to burden us with the oath proposed by the chief judge; and not to tax poor people, without ground and need, beyond what has been laid down in writing.[9]

The letter was signed by Heinrich Losser, sworn town physician; Johann Augsburg, master barber; and Johann Harp, barber and wound healer. The town council apparently did not reverse course. The Frankfurt archives say little about the remainder of Losser's commission until his death in 1468, and then they make no mention of examinations for ten years. We should keep in mind, however, the haphazard survival of the evidence, since most of it was in the form of affidavits that wound up in the hands of the examined subjects. Nevertheless, the gap in the archives from 1460 to 1478 stands in contrast with the eloquent testimony about the four years that had culminated in Losser's public stand of 1460.

Several of the documents about these four years of a practitioner's career make for fascinating reading. In addition to their inherent drama, the eloquent Losser documents highlight the immense value of archival sources in general, and they illuminate several facets of leprosy as a medical issue. Heinrich

Losser, also known as Heinrich Glyperch of Koblenz, was *licentiatus in medi-cina* and possibly an alumnus of the University of Cologne. He had been physician to the archbishop of Trier until 1450, when the council of Frankfurt-am-Main appointed him sworn physician (*phisicus iuratus*). In November 1456, he presented an affidavit on the examination of Rupprecht Snyder, in which Johann Augsburg and Johann Harp had assisted. He reported finding "a number of apparent signs that pertain to leprosy," but not the complete set of signs "that bear on the fundamentals of the permanent disease, which is called *lepra confirmata*, and which calls for total and speedy separation from people."[10] The report concluded with the tentative judgment that Rupprecht did not need to be isolated immediately, but it also enjoined him to avoid gatherings and to take special care of his health until he would return for a second examination. The qualified judgment, with diagnostic details and thoughts about the implications, will supply illustrations for various points in the following chapters, in which the customary caution of examiners should also become increasingly apparent.

The circumspection of Heinrich Losser and his coexaminers adds irony to a complaint against them, which the Frankfurt burgomaster and councillors received on March 3, 1457. The plaintiff, Gerusse, wife of Peter Gul-denlewe, alleged that, when the council sent her to the sworn medical men, they had "looked at me in my womanly sickness." She claimed that they had proceeded,

> even after I gave them to understand that, at such a time, every woman is heavily burdened with this womanly sickness in all her veins and members, and I pleaded with them to skip their order to undress, for the sake of womanly honor, so that it would not happen that they might possibly discover it in the last of my clothes.

We may infer that the examiners found her to be leprous and in need of removal from the community, for her letter continued with the plea,

> by the Lord God and his Blessed Mother Mary, we humbly beg from your wisdom that, through the mercy of God and his Blessed Mother Mary, you graciously grant me poor woman permission to remain hidden in my house for one month, for my husband and I have debts that we wish to settle, and we are not able to come out without great shame after everything that happened ...
>
> But now I that have entrusted everything to the will of Almighty God and his Blessed Mother Mary, I appeal to your soul and honor. I beg your provident wis-

dom, by the will of God and Mary his Blessed Mother, not to deny this plea from a poor woman.[11]

This excerpt is quoted here with all its redundancy because the reiterated pious invocations underscore a folksy tone not often heard in preserved formal documents. We will encounter (in Chap. 6) more evidence of lurid associations between leprosy and menstruation, which were burdened by hoary and heavy taboos. In more immediate terms, the pathos invites us to consider one or both of two implications. It probably complemented a melodramatic—and conceivably fraudulent—charge, in an outcast's cry of despair at being condemned. However, we should not overlook the possibility that the theatrical tone underscored a patient's pained and valid complaint against the insensitivity of the medical officers.

In any event, the council immediately granted the couple a fourteen-day respite and forwarded the complaint to the examiners. On the same day, Heinrich Losser fired off a pained and obviously hasty rebuttal to the "unfair and untrue words." He made it clear, first, that on the morning in question *he* had wished to defer the examination, but Gerusse did not agree; second, she had not told them about her period; and third,

> in no way would we have examined her if we had known this. We treated her with such propriety and dignity that, when uncomfortable, Gerusse could have covered herself at her convenience and wish, sufficiently that neither we nor anyone else would have noticed anything. Taking great pity on her, we were very compassionate and friendly, and we did not wish to cause her fright and grief.
>
> If we had suspected anything about her indisposition when proceeding with the examination, out of compassion we would not have made a pronouncement about separation at that time.[12]

Losser assured that, in spite of the plaintiff's unjust charges, the examiners would do whatever the case still required. At the same time, he urged the councilmen "to hold Gerusse to it, that her fables, lies, and untruths will not with impunity harm our honor and standing."[13]

The honor and credibility of organized medicine were at stake in most examinations, as we will see in more detail in Chapters 2 and 7. Heinrich Losser faced this challenge again the very day after the disturbing incident with Gerusse Guldenlewe. On March 4, 1457, the council sent him and his fellow commissioners a widow, Jan Fransbergerinn, with mixed signs that called for an ambiguous judgment and, almost as urgently, for a letter that was carefully worded yet authoritative. Losser declared,

> [W]e have found that said widow has a stubborn ailment in the nose and lips, which is called the cancer, coming from a rheum of the head. Avicenna says in the fourth book, with regard to leprosy, that "cancer is leprosy of one member," and that a cancer has something in common with true general leprosy [*mit der rechten gemeyn maledie*]. The widow does not have this *maladie*, which originates in a weakness of the liver, is specifically called "*lepra universalis*" or "*dye gemeyn maladie*," can infect the entire body, and is a just reason for separating people.

After this apparent exoneration, however, the examiner reversed course by stating that the ailment,

> on account of its predisposition to leprosy, is likely to result in unhealthy and noxious breath, which may cause harm. Therefore, reasonable and thoughtful consideration makes us conclude that it is just to separate this widow from the people, with the wise permission of our lords, and that she may make use of the privileges and alms of such separated persons. We have made this kind of decision because this matter is of grave and exceptional consequence.[14]

A letter like this enables us to gauge at least something of the patient's agony and the examiner's burden, and to glimpse part of the social setting and the conceptual framework of the encounter between medicine and leprosy. The richest documentation grew out of a drawn-out case that evidently tested Doctor Losser's patience as much as it challenged his credibility. On September 29, 1458, Hans Maderus addressed a petition to the mayor and council of Frankfurt. He complained that, after "meyster Heyrich" and his associates had "charged" (*geschuldiget*) him with leprosy, he had been forced to travel to Cologne for a second examination. Two implications of this complaint deserve notice, at least in passing. There is no extant affidavit of Losser's examination of Maderus, which reminds us of the haphazard survival of evidence. Second, by 1458 Cologne was indeed emerging as a court of appeals; it was staffed by university masters, who eventually produced the greatest number of extant certificates.[15] For Maderus, the visit to Cologne was a mixed blessing, as he claimed:

Now almighty God has given me the grace to be declared free of leprosy by the masters there, and to obtain a sealed letter to this effect. But, dear Sirs, as you can no doubt understand, this matter has brought me heavy expenses, grief, and shame. Therefore, I ask you kindly to compensate me properly or to appear before the *Vice-Dominus* of Neustadt or the Burggrave of Alzey, both officials of my lord the Count Palatine.[16]

The council forwarded the petition to Losser, from whom it drew a response that affords exceptional insights, not only into the unique place of leprosy among diseases, but also into the many implications for the relations between practitioner and patient.

Within ten days, Losser submitted a pointed reply. First of all, Maderus had requested the examination himself. What is more, he had insisted on an instant result and refused the conditional delay (*Frist*) that examiners usually granted in equivocal situations.

For this reason I made an exception and, in my house in Saint Antony street, I explained to Hans that, if his infection was incipient, it would be treatable. However, for two or three days he came to me three times, with crying eyes and bent knees, in the presence of my wife Elisabeth, to obtain a decision. Moved, I said that, by God, I would give a result in a year and a day, and told him that we have helped many people this way. Hans offered to give me a reward, but I told him that I did not wish to take a reward, in view of his poverty, but that I would do it purely to serve by the will of God. I also informed him that, once he was healed, he should be examined once more and at that time, if he was pronounced clean, he could give me what he could afford.[17]

This is only a brief excerpt from the elaborate rebuttal that, as soon as he had written it, Losser seems to have found insufficiently forceful.

Two days later, Losser sent an additional letter to the council. In this substantial addendum, he made it clear that, when Maderus,

quite disfigured in the face, was suspected of the disease and shunned by people, he came to us of his own accord, and he begged us humbly and meekly to inspect and test him thoroughly. Considering his plea, we took it upon ourselves to act in keeping with our custom and due process. We found that, at the time, he was in the onset of leprosy and infected to a considerable degree. This is what we declared, according to our best sense and understanding, and as we are paid and sworn to do.

The physician's explanations did not satisfy Maderus, who further besieged the council with two rather blustering missives. First, he professed being amazed ("das mich befremdet") at not having received an answer, and he set a two-week ultimatum before suing. Next, dismissing the examiners' letters, "which contained a lot of words, and which I read to my good friends," he asserted,

> your men may write about me what they want, but there is no doubt that you have the letter concerning leprosy, which I obtained in Cologne . . . and it should be understood that, perhaps out of hatred, your men may have done me wrong and not dealt with me truthfully.

The council did not deign a response, but Maderus apparently persisted. Five months later, Heinrich Losser proposed mediation by two members of the council and two representatives of the plaintiff, "so that, after complaint and answer, justice may be done to us and Hans."[18]

The intense correspondence over this case brings to life various circumstantial aspects, such as the consequences and complexity of medical pronouncements, the official deference to university physicians, and the combination of an urban and feudal context. More poignantly, the letters of Maderus, like the complaint of Gerusse Guldenlewe, bring us closer than most sources to the rarely voiced reactions of people found to be leprous. These and other "subjects" in examinations, however, were still at large before *entering* the official status of *leprosus*. Once they *crossed* this threshold, and although they may have retained legal rights and even personalities, they largely ceased to testify about their own fate. The imbalance between our knowledge of the disease and our information about its victims has persisted into modern times. Only recently have concerted efforts been made to present the perspective of the person who was classified as "leprous"—regardless of medical definitions and official name changes. The most notable of these efforts include eye-opening interviews with former patients in Kalaupapa in Molokai, Hawaii; a sensitive and lucid chapter in Guenter Risse's history of hospitals; and a sympathetic documentary on the last residents of the Hansen's Disease Center in Carville, Louisiana. Before such endeavors and broader cultural changes, leprosy was arguably the one disease that was most completely borne by the people who had it and most exclusively discussed by those who did not, ranging from preachers in their moral exhortations to physicians in their learned treatises.[19]

Academic Writings

The centerpiece of the Hans Maderus affair, namely, Losser's elaborate rejoinder, points from the forum to the *studium,* and it will indeed lead us from the municipal archive to the faculty library.[20] His defense of the examiners' actions tied the individual case to conceptual foundations, and it hinged on a crucial distinction. In order to have a fundamental understanding, the physician argued, "it is necessary to know that there are two kinds of examination." One relates to the separation of individuals who are found to be completely infected. Their illness is permanent, "called *lepra confirmata* by the authoritative masters, and this is what Gordonius [Bernard de Gordon] writes in the special chapter of his *Lilium.* These persons should not receive a respite." The other examination is for leprosy in an early stage, when it is still "curable according to the honored masters, Avicenna, [Bernard de Gordon] the *liliator,* and Galen." Patients with this stage of the disease "should receive a respite: and we found Hans to be among these."[21]

Authoritative Sources

Losser underscored the seriousness of his brief by citing Galen, Avicenna, and Bernard de Gordon, who were the three sovereign authorities on the subject. Galen of Pergamum (ca. 129–200 CE), whose unparalleled influence on premodern medicine is well known, mentioned *lepra* in several of his numerous works. However, as will be seen in the following chapter, he also left a legacy of confusion about the identity of the disease. His pervasive influence was intensified by a surge of translations of his works from Greek into Latin—to which historians now, following the pathbreaking lead of Luis García-Ballester, refer as "the new Galen"—in the late-thirteenth-century.[22] Around the same time, an author who systematized much of Galenic teaching was competing for the mantle of "Prince of Physicians." Avicenna, or ibn Sina (980–1037 CE), compiled the Arabic *Qanun* (*Canon of Medicine*), which, in the Latin translation by Gerard of Cremona (ca. 1114–1187), became the standard medical encyclopedia in the West. Avicenna devoted three substantial chapters of the *Canon* to the nature and causes, signs, and treatment of leprosy (Canon IV, bk. 3, *fen* iii, chaps. 1–3). The juxtaposition between leprosy and cancer, though sometimes attributed to the seventh-century Byzantine compiler Paulus of Aegina, became axiomatic—as in Heinrich Losser's report on the widow Fransbergerinn—and commonplace owing to Avicenna's authority. For more than four centuries, his chapters shaped the discourse on the disease in comprehensive works, selective commentaries, and specialized treatises.

No Latin author was cited more often as the spokesman for Galen and Avicenna on the subject of leprosy than Bernard de Gordon, who taught medicine at the University of Montpellier from 1283 to sometime after 1308. He wrote at least a dozen treatises, expounding on various theoretical and practical aspects of the curriculum, from the basics of humoral physiology (on which more below) to the diagnostic techniques of reading pulses and urines, and from the *regimen,* or management of health, to the treatment of disease with complex medicines and the application of phlebotomy, or bloodletting. In 1299, in a synopsis of criteria for choosing appropriate therapeutic methods, Bernard observed that great experience was needed for the eradication of morbid material from deep within the body: therefore, the physician should carefully "consider which medicine is suitable for confirmed and unconfirmed *lepra.*" The challenge prompted Bernard to promise, "[W]ith God's grace we will devote a special treatise to this subject." This proposed essay became a chapter in his most copied and quoted book, a *practica,* or medical encyclopedia, that he called *Lilium medicine* (The Lily of Medicine). Bernard completed the work in 1305, exactly when leprosy appears to have reached the height of its visibility in a wide range of documents, and when the disease was becoming increasingly "medicalized" in a process that we have already seen manifested and which will run as a thread through the rest of this study.

In Plate 1, we see Bernard portrayed in an early manuscript copy (1322) of his *Lilium:* seated in his magisterial chair (*cathedra*) and lecturing to attentive students, he is pointing at the authoritative textbook, perhaps Avicenna's *Canon,* while he seems to be enumerating—as he was wont to do—the salient points with his left hand. Medical practice is illustrated in the opposite corner of Plate 1, in a scene from a group examination in Nuremberg in 1493 (discussed in Chap. 2, where the entire woodcut is illustrated in Plate 2). I have superimposed the illuminations on a page of *Fleur de lys,* the French translation of Bernard's *Lilium* which was made in Rome in 1377 and printed in

PLATE 1. (*Clockwise, from bottom left*) Bernard de Gordon teaching: illumination from the *Lilium medicine,* Vienna Österreichische Nationalbibliothek MS 2338 (dated 1322), fol. 1r; courtesy of the ÖNB/Wien. The chapter on leprosy in the French translation (Rome, 1377) of the *Lilium medicine,* titled *La pratique de tres excellent docteur et maistre en medecine Maistre Bernard de Gordon qui sappelle fleur de lys en medecine* (Lyon, 1495), p. [68]; courtesy of The College of Physicians of Philadelphia. Physicians examine presumed leprosy patients, Nuremberg, 1493: detail from the woodcut reproduced in Plate 2.

Lyon in 1495: the page shows the opening of the chapter on leprosy, with a title indicating the chapter's practical relevance by stating that it "speaks of the treatment of *lepre*," although it covers a great deal more.[23]

Bernard's chapter on *lepra*, one of the heftiest of the 163 chapters in the *Lilium*, soon gained special recognition. The papal physician and surgeon Guy de Chauliac declared in 1363, "Gordonius, following Avicenna, treated this subject very well."[24] This citation by Chauliac, however laudatory, is somewhat misleading, for it overstates the derivative character of the treatment of leprosy in the *Lilium*. Even though Bernard adopted much of Avicenna's teaching, he barely mentioned him in this chapter. On the other hand, he named Galen eleven times, citing the newly available writings (the "new Galen") on complexions, natural faculties, internal organs, diseases and symptoms, prognostication, method of healing, and hygiene. Bernard also drew on more recent authors, including Theodoric Borgognoni of Cervia (1205–1298), although he did not identify them. He also added a considerable number of his own interpretations and observations. There is ample evidence that teachers and practitioners consulted Bernard not merely as an interpreter of Avicenna but as an authority in his own right. Physicians cited him as the Liliator, author of a canonical book; and they quoted his anecdotes as well as his theses, in notarial depositions and in scholarly treatises, in support of individual verdicts and fundamental theories. Moreover, his coverage of the disease has drawn the attention of modern historians, beginning a century ago with Karl Sudhoff—whose publication of copious source materials from archives and libraries has been of immense value for this study. Anyone who wades through a substantial amount of premodern writings on leprosy is likely to be struck by the comprehensiveness, lucidity, and structure of Bernard's treatment. In about 7,200 words, it covers the topic in the usual encyclopedic form, but more methodically than in any of its counterparts.

The *Lilium* chapter on *lepra* opens with the definition, nature, and categories of the disease. These fundamental notions then support the identification and classification of remote and immediate causes. The center of the chapter deals with the signs, their differentiation, their assessment for certainty and gravity, and their correspondence with a specific offending humor. The palpable urgency of this section reflects the existential importance of issuing the right judgment over someone suspected of leprosy—we have already seen ample evidence of this importance, and the *iudicium* will continue to occupy a central place throughout this study. Bernard's discussion of the differential diagnosis opens the way to prognostication, which entails not only the prediction of the outcome of the disease but also an estimation of its curability. The last section of the encyclopedic summary deals with treatment (*cura*). This part accounts for almost one-third of the entire chapter, and thus it contradicts the assumption that medicine generally gave up on leprosy patients. On the other hand, the eleven therapeutic approaches are arranged in a strange sequence, beginning with phlebotomy and ending with dietetics, which reverses Bernard's normally conservative progress from hygiene to medication and then to manual intervention. On the whole, however, the logical structure of the chapter provides an ideal blueprint for an examination of the many facets of *lepra*—and it has shaped the organization of much of my book.

A large part, almost the entire second half, of the *Lilium*'s chapter on leprosy consists of a didactic elucidation of twelve (numbered) issues that were open to question or "doubt" (*dubium*). Bernard was conscious of this disproportionate length. He warned the reader, "[T]his subject matter is so messy [*sordida*] that it calls for much clarification, and therefore no one should hold it against us if we dwell on it somewhat longer." At first sight, this clarificatory section may strike a modern reader as abstract and "typically Scholastic" exercises in dialectic. The very first *dubium*, however, illuminates the link between speculation and actuality. It expresses the author's frustrating struggle to reconcile his own experience with the received definition of "confirmed leprosy" as necessarily involving the face. After admitting that "the doubt occurs to me above all on account of what I saw in a certain man," Bernard makes the quandary more concrete by narrating the anecdote.

> This is what happened. There was someone whose toes and fingers were so deformed, disfigured, and falling apart that they had only one joint left, without nails; they were completely abnormal and ugly. And yet, in spite of such great deformity nothing at all was apparent in the face. From this example, it seems that one may be confirmed leprous after hands and feet become deformed and fall off without any sign of leprosy appearing in the face.

After trying, largely without success, to reason away the contradiction between this personal observation and conventional doctrine, Bernard intensified the poignancy of his reminiscence.

> I assure you that I wished to absolve him, and I questioned him repeatedly whether any sign had appeared in his face. He had remained like this for nearly

twenty years and he still lives with this ugliness of the extremes yet nothing is visible in the face. Therefore, I guess, with conjecture in approximation of the truth, that it was not leprosy and that it could not have lasted for so long without disfigurement of the face. Even though I had once believed differently, now, after having labored diligently in this work, I am of another opinion and now I would not declare someone leprous. However, God knows the truth, I do not know.[25]

In contrast with this unresolved situation, anecdotal experience served Bernard, earlier in the chapter, to underscore a conventional theory. After dwelling rather insistently on the risk of contracting leprosy by having intercourse with a diseased woman, Bernard added,

[L]et me tell you what happened: a certain leprous countess came to Montpellier and wound up in my care, and some bachelor in medicine attended to her, lay with her and impregnated her, and became completely leprous. Lucky the person who is made cautious by the perils of others.[26]

It is important to note, as we did with regard to the prominence of the section on *cura,* that the anecdote contradicts, at least implicitly, the facile generalization that physicians abandoned patients with confirmed leprosy. The vignette is of additional interest because it implies, both in the common tradition and in the cautionary anecdote, connotations about sex which had as much to do with moral tenets as with natural philosophy—as we shall see in Chapter 6.

Bernard's direct and evocative references to personal memories, together with his attempts to integrate them into his synthesis of conventional wisdom, were the first stimuli for my curiosity about the implications and context of his views on leprosy. To the extent that he appeared representative of learned medicine, his *Lilium* chapter inevitably triggered broad questions. How did university-trained physicians in general view *lepra* before they had an inkling of what Armauer Hansen discovered in 1873? How did practitioners deal with the disease or diseases covered by that label? How much change was there in medical theory and practice? How did the science and profession of medicine interact with cultural and social responses to leprosy? Answers to such sweeping questions, however, are bound to be tentative and incomplete. They also depend on a systematic exploration of as many sources as possible, along the distinct avenues opened by Bernard. His remarkable chapter makes one wonder, in particular, about the circumstances in which this physician and his colleagues encountered leprosy, the correlation between their practice and abstraction, the foundations and objectives of their positions, and the impact of their teaching.

No matter how much Bernard may have resorted to sophistry in his endeavor to interpret the standard concept of "confirmed leprosy," he left no doubt about the paramount objectives of his speculation. His dialectic led to the emphatic conclusion that

> no one should be judged leprous unless disfigurement is manifest. And let me tell you why I repeat myself so many times about disfigurement: because it appears to me that *leprosi* are judged very poorly [*pessime iudicantur*] these days; therefore, let anyone pay attention who has ears and is willing.[27]

An erroneous diagnosis of full-blown leprosy would condemn someone wrongfully to isolation and, in the best of cases, to treatment that was merely palliative. Therefore, most of the learned practitioners, echoing Bernard's intellectual humility, professed painful awareness of the social and medical consequences. They displayed a general reluctance toward issuing categorical verdicts. Paradoxically, they also laid claim to privileged practical knowledge by virtue of their learning.

Pleas for Learned Knowledge

In his teaching on *lepra*, Bernard cultivated learning, particularly the recourse to rationalization and written tradition, because he believed that it served practical purposes (an objective overlooked by some critics of "Scholastic" medicine). Inversely, a growing and increasingly vocal group of physicians claimed that book learning was a prerequisite for dealing with leprosy. These claims, which were still incipient and diffuse at the time of Bernard de Gordon, coalesced in the course of the fourteenth century. They came to full fruition across French and German regions after 1400, when more and more feudal and municipal authorities deferred to the judgment of learned practitioners. This convergence shaped situations such as those in Frankfurt in 1460 and in Bourg-en-Bresse in 1462, which we have seen unfold in previous pages. It was a complex process, with a combination of continuity and change, with fluid relations between town councils and faculties of medicine, and even with shifting priorities in medical teaching on *lepra*. By compressing the developments of several centuries, we can place this process—and the testimony of Bernard at a turning point—in sharper relief. Before examining the major changes, it is useful to dwell on one constant, namely, the conviction of physicians that their learning was indispensable to proper intervention and that laypeople, regardless of their experience with the disease as well as patients, lacked the necessary "understanding."

This conviction found a formal and forceful expression in 1574 at an assembly of the Faculty of Medicine in Cologne. On July 16, Bernard Cronenburch and Dietrich Birckman, doctors of medicine ("der arznei doctorn") read to the faculty senate a letter that they were addressing to the town fathers. At issue was their rejection of a judgment by the matrons of the leprosarium, who had declared the poor and very elderly Tryn von Vrechen to be leprous. The physicians found, after careful inspection and consultation with other doctors, that

> the good woman is burdened by old age rather than the disease, for her problems are relatively minor and do not add up to such gravity that they could be taken for leprosy. . . . There are so many innocent people who, for some reason such as the lack of a limb, are condemned, in utter rashness, by the uninformed masters of the leprosarium [*unverstendigen Melatzmeister*]. Therefore, it is most urgent to design a better policy for these affairs.[28]

Perhaps not in direct connection, but hardly by sheer coincidence, only four years earlier the famous anatomist Volcher Coiter, who was sworn town physician of Nuremberg since 1569, had drafted a policy for responding to rampant abuses in such examinations. In this draft, to which we will return in a later chapter, Coiter emphasized the need for collegial consultation among the commissioned physicians. His policy proposal, together with actual manifestations of medical solidarity, may have aggravated the friction between the seven municipal physicians and the town council in Nuremberg, which is revealed by other documents.[29]

There can be little doubt that Coiter would have agreed wholeheartedly with his Cologne colleagues, who declared,

> A matter of this high importance, which may decide the well-being of a person or the community, should more justly and in accordance with old custom be entrusted to the medical profession [*professioni medicae*] than to such wretched illiterates [*slechten Idioten*]. [Professionals] know sufficiently that this exceptional dis-

ease, which calls for separation, manifests itself not only in a changed complexion but, much more, in the destruction of the shape and appearance, particularly of the face. Those leprous people [*Melaten*], however, seem wickedly to stick to their opinion.

The doctors may have betrayed some deeper prejudice, if not by referring to the matrons as "leprous" (and thus echoing the professional disapproval of patients managing leprosaria), at least by associating their obstinacy with wickedness. Indeed, the situation raised moral, ethical, and judicial implications, and it was difficult to disentangle these from the issue of professional self-interest. At the heart of the protest was the demand, by the leprosarium regents, for a joint examination by both parties. This request, the learned protesters claimed—with the added flourish of a Latin phrase—was "entirely against the dignity of the medical doctors [*contra omnem medicine doctorum dignitatem*]." Moreover,

> to our knowledge, it has never been done before. It is also contrary to all justice and fairness that the higher judge in his sentence should take counsel from the lower one. The lepermasters have neither the knowledge nor the aptitude to discuss our recommendations [*consilia*]. Therefore, we hope that your Lordships will give more credence to the doctors of medicine than to those wretched persons in this case, which reflects on the honor and reputation of our faculty. We stand by our prior sentence, which has also been confirmed by other doctors.

One of these other doctors signed the protest and added his endorsement: "[F]or many reasons, I cannot expel this woman on the pretext of leprosy, as has been done by those who lack the understanding of our profession [*den unverstendigen unser profession*]."[30]

It is clear that, from Bernard de Gordon to Bernard Cronenburch and Dietrich Birckman, university-trained physicians maintained that correct decisions required not only a readily acquired ability to recognize numerous signs and symptoms but also literacy, a grounding in medical theory, and even a differential knowledge of categories such as "changed complexion" and "destruction." The reader will object that the shared positions of these physicians should not blur the considerable distance that separates 1305 Montpellier from 1574 Cologne. To be sure, the *Lilium medicine* and the protest letter belong to two different classes of written evidence. This duality might seem to confirm the stereotype of separation between town and gown. Universities that produced writings on leprosy, from Montpellier to Marburg, left few traces of participation by professors in actual responses. Towns that kept copious records of medical commissions, from Brugge to Nördlingen, showed little awareness of academic centers. However, we will come across many points at which academic treatises and official documents intersected, and we will find even more abundant indications of interaction between teaching and life. The most remarkable evidence lies in German repositories: the municipal archives of Frankfurt and Nuremberg, as well as the university archives of Cologne and Tübingen, bridge the documental divide by preserving a wealth of information on interactions between medical faculties and public officials—which will be more fully examined in the chapters that follow. In any event, I would argue that any dichotomy in the sources does not neutralize the distance between 1305 Montpellier and 1574 Cologne but, rather, results from the same causes. By the sixteenth century, the faculty secretary and the municipal notary were upstaging the scholastic scribe, and public records were piling up faster than scholarly folios. Archives attest both to the expanding involvement of government in health matters and to the matching assertiveness of a medical "profession" as distinct from the lay populace.

Behind these obvious developments lay broader and deeper changes, in the confrontation between medicine and leprosy as well as in the historical context. Regional divergences—for example, between French and German areas—led to different responses to the disease, which, in turn, may suggest differences in its incidence. Events with a direct impact ranged from wars to epidemics, from pious movements to material innovations, and from economic fluctuations to scientific advances. The "Great Mortality" of pestilence and the ravages of syphilis colored evaluations of leprosy. Spasms of anarchy and fear, dovetailing with calls for conformity and authority, added weight to the judgment of people rumored to be infected. Waves of dire poverty, appeals to charity by religious reformers, and tensions between town and country affected the prospects of confirmed patients—and of cheaters. The effects of these and other external factors will be an integral part of our panorama. They may recede into the background when we first concentrate on internal developments in the medical discourse. Our sources, however, will lead us back to the outside world by converging paths, when academic doctrine is joined by public involvement, and conceptual medicine by clinical observation. While fifteenth- and sixteenth-century archives have already afforded us a glimpse of this convergence, it is most fully documented in libraries.

Evidence of Changing Emphases

An overview of the medical literature reveals a gradual, though not linear, shift in the discourse on leprosy. Proportions changed in the attention that was allotted to each of the four conventional areas, namely, the definition, causes, signs, and treatment of the disease. It is not feasible to reconstruct this trend from the nearly one hundred treatises that were examined in an effort to have a reliably representative basis for general observations. A simple reconstruction is precluded not merely by the volume of these writings but also by the numerous variables in their formats, their authors' idiosyncrasies, and their intended audiences. The only manageable way is to draw up a comparative sample, with due awareness of the bias inherent in selecting, no matter how carefully it was done. Word counts yield a graph (Figure 1.1) that, by comparing twelve authors, illustrates the changing proportions of the coverage. Proceeding chronologically from the eleventh to the eighteenth century, the graph spans an admittedly long period. Nevertheless, it conveys an image that is largely confirmed when the collation is extended to more authors.[31]

The most visible shift, highlighted by the proportions in Bernard de Gordon's discussion, lies in the widening of the theoretical foundations (left side) and the shrinking of their practical application (right side). Before we concentrate on this shift, the two middle segments in the graph call for a brief comment. The sections on *causae* show the greatest irregularity, at least in part because they depended most on the individual authors and their milieus; in any event, their course is too erratic in this graph to allow for any generalizations. The opposite is true for the sections on *signa*, which grew rather steadily. The preoccupation of physicians with the presentation and outlook of leprous people intensified after 1300, fueled by the social implications, as we have seen in vivid testimony. However, this preoccupation also drew on internal dynamics of medical teaching, which progressively elaborated the Hippocratic and Galenic primacy of diagnosis.

Whereas the fourteenth century produced numerous lists of the signs of leprosy without express reference to diagnostic principles, by the sixteenth century there were authors who first concentrated on these principles and then applied them to their discussion of leprosy. Thus, the Basel town physician Johann Jakob Huggelin (d. 1564) presented semiology as a principal part of medicine in his book *De semeiotica medicinae parte.* When he wrote a shorter tract on leprosy, *Von dem Aussatz* (which was almost certainly the first to be originally composed in German), he focused on "the signs by which to recognize an infected person." It is worth mentioning that Huggelin registered

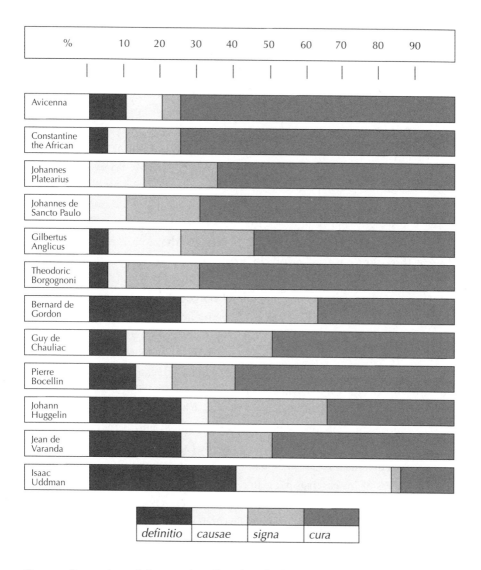

Figure 1.1. Proportions of the attention allotted to the four conventional areas of medicine

his personal observation of the loss of sensation, "as I saw in Pforzheim, in a boy of fifteen," who, as long as he looked away, was unaware of being pricked "by the barber, with a needle about half a finger in length." A contemporary and kindred spirit, Jodocus Lommius (Jodocus Lommius Buranus, Josse van Lom, ca. 1500–ca. 1564), town physician of Brussels, composed a medical manual that concentrated on observation, *Medicinalium observationum libri tres*— a manual that found wide circulation among physicians for more than two centuries. In his discussion of leprosy, he provided an almost "clinical" synopsis of the signs, which was stripped of their ubiquitous humoral interpretations and virtually silent on their social impact. We will be able to draw further insight from comparisons of Lommius, Huggelin, and many other authors in separate chapters on the causes and on the signs of leprosy.[32]

Although the definition and the treatment, too, will each have their own chapter, their prominently changing proportions in Figure 1.1 call for further comment. The expansion of the *definitio* corresponds to the growing emphasis on understanding and rationalization, which we have observed in both scholastic and judicial sources. The receding share of *cura* parallels the diminishing concentration, by poets as well as physicians, on healing—and, as we will see, a clearer position on the issue of curability. It is useful to point out that, in this graph, the relative exceptions to the reduction in *cura* represent three surgeons who, ex officio, were likely to display a more than average interest in treatment. Morever, there are significant differences among these three learned surgeons. Theodoric Borgognoni was a member of the Order of Preachers and bishop of Bitonto (near Bari) when he wrote his *Cyrurgia*; later, when he became bishop of Cervia (near Ravenna), he spent most of his days in the bustling university town of Bologna. His chapter on leprosy gives the impression of a busy and practical man, and its theoretical part relies mainly on Avicenna.[33]

Two generations after Theodoric, Guy de Chauliac (ca. 1300–1368), medical courtier in Avignon, achieved a virtual equilibrium between practical concerns and theoretical justifications in the chapter on *lepra* in his surgical encyclopedia, the *Inventarium sive chirurgia magna*. The equilibrium and comprehensiveness of his coverage made him one of the most influential authors on the subject for at least two centuries. This influence is evident, for example, in the first work on leprosy originally written in French, *Practique sur la matiere de la contagieuse et infective maladie de lepre* (1540). The author, Pierre Bocellin, was the town surgeon of Belley in Savoy, and he compiled the book for "those who are unable by themselves, and without writings, to know this disease."

While he wrote recognizably as a public practitioner and a popular educator rather than a scholastic don, he cited statements from all the standard authorities. An alumnus of Montpellier, Bocellin gave pride of place to the past authors of his alma mater. Paying special homage to Bernard de Gordon, "excellent docteur,"[34] he repeatedly quoted other Montpellier masters, including Arnau de Vilanova (d. 1311), Guy de Chauliac, Velasco de Tharanta (fl. 1418), and Jean Falco (fl. 1498–1541). Around the time that Bocellin dedicated his humble booklet to "the very learned doctor Dionysius Fontaneus, Regius Professor at Montpellier," academic counterparts, at Paris as well as Montpellier, wrote more ambitious treatises, not only in Latin but occasionally also in French. The illustrious and controversial Parisian professor Jean Fernel (1497–1558), physician of Henri II and Diane de Poitiers, discussed *lepra* in his *Pathologiae libri VII*. The even more renowned author Ambroise Paré (ca. 1509–1590), physician and surgeon to several kings, discussed the disease in *Traicté de la peste . . . avec une briefve description de la lepre*.

While Bocellin as well as his university colleagues consistently cited the teachings of Galen and Avicenna as the principal sources of their knowledge (as will be seen in Chap. 3), Figure 1.1 suggests that there may have been an interruption in this theoretical legacy. The definition of leprosy received little attention from Constantine the African, and none from the Salernitan masters Johannes Platearius and Johannes de Sancto Paulo. This is all the more striking because the schools of Salerno cultivated the connection between practical medicine and natural science in the notion of *physica*. The section on *definitio*, which extended from the names to the nature of leprosy, began to grow larger after 1300. The growth was characteristic of academic writings, though rarely as impressive as the leap in the *Lilium* of Bernard de Gordon, who was exceptionally conscious of the need to understand a very "messy" subject. Three centuries later, Jean de Varanda (also known as Varandaeus, 1563–1617), dean of the Faculty of Medicine and regius professor in Montpellier, compiled a treatise in which leprosy was the first subject, syphilis the second, and "weakness of the liver" the third. He began by carefully spelling out not only the terms for leprosy but also their etymology, even while repeating the hoary maxim that there was little need to worry about names and their origins as long as it was clear what they really signified ("quae res significetur"). Then, as he invited the reader to advance with him "toward knowing the reality," his key to this knowledge lay not in observing leprosy but in defining the concept. After criticizing the formulas of most of his predecessors for "not having satisfactorily expressed the proper nature and impact, kinds, and causes" of the disease, he constructed his own definition with the conventional building blocks (qualities, humors, and balances); for particulars, he proceeded by logical distinctions. An equally conceptual and dialectical approach characterized the *Dissertatio* about leprosy which was part of the "debates," or *Agonismata*, of Heinrich Petraeus (1589–1620) at the university of Marburg.[35]

While always complementary—but usually secondary—to the conceptual quest for reality, the empirical approach eventually gained the ascendant in the various branches of medical science. For leprosy, as for other subjects, this process was both a cause and an effect of the merging of academic and public medicine. Few medical authors personify this stage more eminently than Samuel Hafenreffer (1587–1660), professor at the Faculty of Medicine in Tübingen (see Plate 5, in Chap. 4). In his textbook on diseases of the skin, *Nosodochium* (which is historic for its description of the mechanism of itch), Hafenreffer dispensed with speculation when he discussed *lepra*—in a chapter that also covered the "signs and treatment" of scabies and syphilis. He combined humoral premises, and rather credulous deference to other authors, with a remarkably sharp focus on observable data and diagnostic precision. He included detailed instructions for the *examen leprosorum*, which we will consider in detail later. Hafenreffer stands out because, in addition to a noteworthy academic treatise, he produced at least one certificate of a medical examination (reproduced in Plate 5) and other official documents that attest to his involvement in public affairs—and which are pristinely preserved in the Tübingen Universitätsarchiv.[36]

The complex and lengthy transition from a primarily conceptual to a more experiential response to leprosy becomes richly evident if we consider the contrast between the last author in Figure 1.1 and his predecessors. Isaac Uddman, a Finnish physician, represents the most pronounced difference in the distribution of the four areas. Indeed, he made a point of being different when, in 1765, he presented his doctoral dissertation on *lepra* to the Swedish Royal Academy in Uppsala, under the stellar aegis of Carolus Linnaeus (Carl von Linné, 1707–1778). Uddman did not wish to come up with "cabbage that had been recooked twice or a hundred times," even though this seemed inevitable.

One can find almost nothing that has not been compiled and percolated from many authors, so that the repeated and recopied works can only emerge, just like a brook meandering far from the source, turbid and impure. Add to this the extreme inconstancy of the authors as, regardless of the considerable disparity of experience, they borrowed from others in good faith where all of their own observations had ceased.[37]

Readers will find this criticism borne out by quite a few passages in the chapters that follow. Nevertheless, it was a rather sweeping and youthful rejection of age-old and multiple efforts to get a handle on a baffling disease. With their limited successes, and even with their failures, these efforts acted as powerful catalysts in a variety of circumstances.

The setting of Uddman's discussion undoubtedly shaped the contrast, reflected in Figure 1.1, between preceding positions and his own, with his marked concern with the nature of leprosy, his disproportionate attention to causes, and his relative neglect of diagnosis and therapeutics. Nevertheless, the seemingly idiosyncratic deviations in his distribution also reflected changes in physicians' approach to leprosy. Like many of his predecessors, Uddman felt challenged by a disease "so obscure in origin, variable in form, and difficult to treat." His chief departure from hoary routine was that he did not build a formulaic definition. He deemed terminology far less important than precise descriptions of the disease, the geography of its incidence (not limited to Scandinavia), and, above all, the collected inferences from personal observations. In line with the clinical tendencies of his age, he cited Boerhaave and Needham rather than Hippocrates or Galen. He showed no interest in teleology or intangible principles when he inquired into causes. His inquisitive empiricism led him, intriguingly, to speculate that leprosy was spread by specific agents, animalcules still smaller than those already discovered under the microscope. It might be worthwhile, but it would lead us too far afield, to inquire whether this speculation was directly influenced by Linnaeus, who in 1757 had published a dissertation on "skin diseases and the organisms that cause them, based on Leuvenhök's microscope observations." Unfortunately, Uddman does not appear to have carried his burst of genius into lifelong research or further discovery on the transmission of the disease.[38]

Only two years after Uddman's dissertation, and across the continent, similarly new trends surfaced, side by side with traditional coverage, in a more voluminous work by François Raymond (fl. 1756–1769). An alumnus of the medical faculty of Montpellier and member of the College of Physicians in Marseille, Raymond evinced an academic taste for names, etymologies, and classical sources. However, he also extended his definition to a detailed description, the "countries of both hemispheres where the disease still teems," and reports of observed cases. An excerpt of Raymond's first case may illustrate the clinical slant in his approach to leprosy—as well as his continued indebtedness to an age-old legacy of humors and complexions.

> In 1746, the wife of Etienne Menager, thirty-two years old, of a sanguine-melancholic temperament and of a vigorous constitution, came to consult me on the horrible state she had found herself in for several years. Her face was covered by crusty tubercules, very thick, protruding, and cracked. The base of her nose was somewhat depressed, her nostrils stopped up, her voice hoarse and dark.

After noting that medical treatments did not save her from dying two years later and that, in fact, "her death was hastened by the rubbings with mercury, which a surgeon gave her," Raymond further observed that both her children were killed by the "virus" before puberty but that "her husband, who never ceased knowing her, did not contract any infection [vice], even though he was of a melancholic temperament." The report concluded with specifics on the patient's environment:

> This family lived in the hamlet of Peypin, four leagues north-northeast of Marseille, in an area with high pine-covered mountains that form very narrow and humid valleys. The inhabitants feed on dried or salted fish that are not of the best quality.[39]

Isaac Uddman and François Raymond bring us to the threshold between premodern and modern medicine, not only with regard to leprosy but even, to some extent, in general. Their inquisitive writings constitute an antipode to the literary sources in which, at the beginning of this chapter, poets and chroniclers evoked physicians as shadowy sages and famed but ultimately failing healers. Medical treatises and municipal archives show that changing preoccupations with leprosy intersected with the rising visibility of learned practitioners. From the twelfth to the eighteenth century, from Salerno and Montpellier to Uppsala, there was a growing conviction that patients, potential patients, and society benefited most from decisions and care that were based on literate learning. The claim was not unchallenged, and some disputes afforded rare glimpses of individuality. The content of learning was neither static nor homogeneous, and by the 1300s a profusion of texts competed with the authoritative teachings of Galen and Avicenna. From 1400 on, the academic faculty became more involved in the public forum. The distance between the classroom and the clinic narrowed steadily, to be bridged in the 1700s by Uddman, Raymond, and others. François Raymond, the last author in the period under our consideration, was the first to call his discourse on leprosy a *histoire*. His personal interests, which reflected the culmination of changes in emphases, lay with description rather than definition, observation rather than speculation, and diagnosis rather than treatment. His "history" encapsulated the notarial and didactic evidence of three centuries, in which medical intervention revolved far more around the judging than around the healing of *leprosi*. Indeed, the profile of the physician as a judge of patients is so pronounced, not only in written but also in pictorial sources, that it calls for a closer look before we consider medical responses to the disease itself.

CHAPTER TWO

Iudicium leprosorum
Medical Judgment

To the Venerable College of Sworn Masters of Surgery in the City of Toulouse.

Not long ago, in the company of some of this city's Medical Doctors, and together with one of you, I was officially assigned by the Court of Parliament to the process of visiting an itinerant, accused of being leprous. The visitation followed the praiseworthy custom of well-governed cities, and it was performed so solemnly that nothing better could be expected for so serious and important a judgment. The Court, nevertheless, found the report, which contained our complete advice, too general in its conclusion. Therefore the Court summoned us to reassemble, and to declare whether leprous patients are contagious when they have only a predisposition and not the actual disease, or whether this disposition is contagious. This was done, and the doubts were suspended in favor of the accused.

Now, the terms of "disposition" and "act" in leprosy, as Mr. Guy de Chauliac interprets them, are full of ambiguity and reasonable doubt. Therefore, for the benefit of all the ordinary surgeons who are interested in reading only this author, I wish to show not only what is leprosy in disposition and act, but also the method of testing and judging leprous patients. In doing so, I hope to present something that will please you and profit the Commonwealth of Toulouse, if—in emulation of a great physician of our time who has favored his country by writing a compendium on the examination of leprous patients—I let our Tolosains see, in French, the extraordinary care that must be observed in such visitations.

I n this prologue to his spirited little book *Examen des éléphantiques ou lépreux,* "compiled from numerous good and renowned authors, Greek, Latin, Arabic, and French," the surgeon Guillaume des Innocens touched on several implications of the formal process of identifying leprous patients. He alluded to the relationships between physicians and surgeons, between theory and practice, and between medicine and justice. In addition, his prologue offers a glimpse into the adoption of judicial terminology by medical examiners, the possibility that their pronouncement would be challenged, and their awareness of the public context. The treatise, which was published in 1595, also illuminates some of the changes and continuity in the contribution of medicine to the overall response to leprosy in the three centuries since Theodoric Borgognoni and Bernard de Gordon.[1]

"Due Process"

The prologue of Guillaume des Innocens introduced the principal dramatis personae of a *iudicium leprosorum.* A reported but unnamed person with symptoms of leprosy has exited; sworn medical practitioners and the municipal court and magistrates occupy center stage; two learned authors are solemnly posted at each side; the less learned ordinary surgeons may be imagined in front, seated on the floor; and the townspeople crowd the background. We note the absence of two types of characters, however, namely, priests and nonmedical examiners. This absence is due to time and place. Priests had initially played the leading role in initiating, conducting, and enacting the examination and judgment of anyone who was suspected to have leprosy. From the second half of the thirteenth century, their involvement was eclipsed by the expanding power of towns and the rising status of learned physicians. The ecclesiastical legacy lingered into the sixteenth century, in biblical echoes and in parts of the judging procedure. It is hardly surprising that references to the Bible, particularly to the directly pertinent passages in Leviticus 13 (which we will examine in Chap. 3), persisted in the language of reporting,

characterizing, inspecting, and sentencing people suspected of having leprosy. The crucial difference, however, was that the medical discourse omitted
the biblical implications of moral and ritual purity, concentrated on physical factors, and adopted only the most general religious concerns. This adoption of ecclesiastical procedure was most visible in guidelines for the examination. A typical medical *ordo examinandi* began with a prayer for divine
assistance. Next, in anticipation of the worst, the physician was instructed
to reassure the subject, with "comforting words," that "this disease is salvation for the soul" and that "leprous patients, while shunned by people, are
dear to God, as we see that poor Lazarus was borne by angels into Abraham's
bosom while the spoiled rich man was buried in hell."[2]

The milieu in which des Innocens was writing was responsible for the
absence of a second major character from the cast in his prologue: the lay
examiner. In many places, sworn juries consisted partly or entirely of people
who were not medical practitioners. In the orbit of a university (Montpellier in the case of des Innocens), physicians joined and gradually took control of the examining process. This shift was neither universal nor straightforward. Although the transition began in Italy, extant records of medical
examinations for leprosy are more profuse when we move northward. In addition, there is a difference between the French archives, which generally hold
less information on examinations than on leprosaria, and German repositories, where the reverse is true; archival holdings in the Low Countries present an intermediate profile. It would lead us too far here to pursue the social,
cultural, and historiographical—and, perhaps, epidemiological—ramifications. Suffice it to be aware that, from place to place, juries varied greatly in
their commission and composition, until they disappeared in the early eighteenth century. Nonclerical juries might be commissioned by the court of a
diocese or county, a town council, or regents of a leprosarium. They consisted
of leprosarium residents or regents, appointed citizens, town councillors, barbers, surgeons, physicians, or a combination of these.

In two instances, which seem to have been exceptional, leprous outsiders
("forains" and "malades de dehors") examined potential patients, at Lille in
1378 and 1388; even more surprisingly, these examinations took place *in* the
leprosarium, where they were normally conducted by the residents, staff, or
regents. Allowing for odd situations such as this one, and for more likely
exceptions—for example, when high dignitaries were visited in their own
mansions—we may note that changes of venue often reflected shifts in the
composition and control of juries. Examinations migrated, at different paces

in different milieus, from a church or chapel to the leprosarium, to the local lord's castle courtyard, to a town park or square, and eventually to the physician's house. In 1380, Doctor Conrad de Dannstet examined Petrus Pirchfelder "in a room [*stuba*] of my house" in Vienna. Out of 179 official judgments recorded in Cologne between 1491 and 1664, 137 were pronounced in the house of the dean of the Faculty of Medicine. Indeed, while there were many variables, it is possible to recognize certain trends and tensions with regard to the participation of medical practitioners in the *iudicium leprosorum*.[3]

Medical Practitioners Move to Center Stage

The first medical representation at a judgment for leprosy appears to have been recorded at Siena in 1250. The record mentions that four examining physicians "judged and sentenced" Pierzivallus to be infected, although the document concerned primarily the "remuneration for the service that they performed at the command of the podesta and the court." In 1261, the podesta and bishop's council of Castiglione del Lago (on Lake Trasimene) chose two "good men" to examine, with the advice of physicians ("cum conscilio [*sic*] medicorum"), whether a person was infected by leprosy. At Pistoia in 1288, Master Laurentius, *medicus*, "formally pronounced a sentence [*sententialiter pronuntiavit*]," under oath, in the chapel of a priest and in the presence of three witnesses. One year later, he dictated a judgment to a notary, "in the house of the heirs of Spina Philippi" (presumably the family of the examined countess) and in the presence of four witnesses. Without reducing the change to a linear progression, we may see physicians moving from the position of performing a service and providing advice to a more prominent role in the judgment.[4]

The physicians' rise to prominence became more clearly defined in the context of emerging professional identities. This emergence was aptly outlined in 1324 by Marsiglio (Marsilius) of Padua (1275/80–1342) in his seminal treatise on government, *Defensor pacis*. Marsiglio, who once studied medicine, captured the diverging prerogatives of priests and physicians with regard to leprosy and judgment. He set the stage by defining the function of the clergy. According to the biblical commentary of Saint Jerome, with a characteristic penchant for metaphor, "priests have that right and duty which rabbis once had under the law in the treatment of lepers. They cancel or retain sins when they judge and show that these sins have been canceled or retained by God." A distinction between healing and judging was foreshadowed in the clarification "[I]n Leviticus lepers are commanded to show themselves to the priests, who do not make them leprous or clean, but discern who are clean and who are unclean."[5]

When Marsiglio turned to the connection between judgment and punishment, particularly by excommunication, he built an analogy between ecclesiastical and medical verdicts. He alluded to several developments that were accelerating at that time and which affected the culminating notoriety of leprosy. The collegiality of the clergy, which was being adopted by members of medical colleges and faculties, was central in Marsiglio's thesis that an examination, which might lead to excommunication, "ought to be made by a judge with the help of a group of priests or of a determinate number of the more experienced from among them." By juxtaposing priests and physicians, Marsiglio implicitly acknowledged the degree of autonomy which medicine was gaining. He argued that it was up to the priests to decide when

> a person must be cut off from the company of the faithful lest he infect others; just as a physician or a group of physicians has to judge about the bodily diseases for which a diseased person, such as a *leprosus*, must be separated from the company of others lest he infect them.

In his elaboration of the parallel, Marsiglio's focus was on the power of the commonwealth, but he also underscored the contrast between literal and metaphorical disease.

> Just as it does not pertain to any physician or group of physicians to make a judgment or to appoint a judge having coercive power to expel *leprosi*, but rather this pertains to the whole body of citizens or to the weightier part of them, . . . so too in the community of the faithful it pertains to no one priest or group of priests to make a judgment . . . to expel persons from the company of the community because of a disease of the soul.

A second layer in this analogy pointed to two contrasts: between a mandate from the "body of citizens" and the corporate "community of the faithful," and between views of leprosy as "a disease of the soul" and a disease of the body. This double dichotomy determined the fundamental differences between ecclesiastical and medical judgments—and sequestrations.[6]

In Italy, the assertion of medical competence presented itself mainly in juxtaposition and succession to clerical prerogative; north of the Alps, as an alternative to nonuniversity intervention. In the case of Montpellier, it is instructive first to go back in time and to turn to the university statutes of 1240, which were drawn up by two friars with the approval of the bishop of Maguelonne. One of the many articles ordered the bachelors and masters "to swear that they will not accept the care of a leprous patient in Montpellier, for longer than eight days, between the two rivers, except with the permission of the royal court, if the patient is in the king's territory, or of the episcopal court, if in the bishop's territory." By their focus on the therapeutic role of the physician, this statutory provision typically harked back to the earlier age that was illustrated at the opening of Chapter 1. Montpellier appears to have reached a turning point, toward an emphasis on diagnostic competence, in the second half of the thirteenth century. While faculty members were still treating leprous patients, they drew a contrast, at least implicitly, between their own ability to identify infected individuals and the limited knowledge of lay examiners. In the orbit of Montpellier, the Crown of Aragon may have been the first to manifest, in the apt phrase of Michael McVaugh, "the medicalization of leprosy"—a phrase that I would venture to fine-tune to "the medicalization of the diagnosis of leprosy." With his usual perceptiveness, McVaugh notes "a steady change in public attitude towards both the disease and the authority over it of the physician."[7]

Around 1300, Bernard de Gordon still had a leprous countess in his care in Montpellier, as we may recall from the misfortune of his libidinous assistant. In 1305, Bernard de Gordon was berating lay examiners—rather than fellow physicians or priests—when he observed that "*leprosi* are judged very poorly in our day." His observation suggested a growing readiness among physicians to remedy the inadequate judging. This readiness no doubt fueled the efforts of his colleagues and successors at Montpellier, including Jordanus de Turre, to formalize the diagnostic protocols. Their efforts were not confined to academic compilations. In 1327, the lieutenants of the royal court and the consuls of Nîmes requested three members of the Faculty of Medicine, assisted by "some barbers who, as they claim, have participated many times in such procedures," to examine six inhabitants who had been "denounced as infected by leprosy." The physicians prefaced their report with the assurance that they had diligently observed "all the essential signs that the great philosophers and the masters in medicine have taught us on this matter. Before God, and under the oath that we have sworn to the University of Montpellier, we present our unanimous conclusions and opinions."[8]

It is not clear whether the physicians were referring to their pledge to the statutes in general or to a special oath that pertained to their participation in juries (the latter case is more plausible), but that would not have been recorded in the faculty constitution. Not much later, however, a new constitution attested to the importance of this aspect. The statutes, drawn up by lawyers in 1340, contained an article, "regarding *leprosi*," which ordained

> that no Montpellier master of medicine shall give, to anyone who is leprous or suspected of leprosy, a testimonial letter stating that he was cured of leprosy or not infected by it, unless this person has first been judged, by two masters, to be healthy or not infected, and unless the granted letter is sealed with the seals of the two masters while said judgment is in effect. Should one single master act to the contrary, however, let him be deprived of the community and honor of the university of masters.

The severe sanction of expulsion left no doubt that infractions were taken seriously—and were frequent enough to cause concern. The efforts to secure the examinations, the increasing involvement of physicians, and the seriousness of their responsibility were exemplified most eloquently in Guy de Chauliac's methodical discussion of the *iudicia*. Guy introduced his instructions with the warning "[G]reat attention should be paid to the examination and judgment of *leprosi*, for it is extremely unjust to sequester those who should not be sequestered and to send the infected among the people."[9]

Physicians at the University of Paris were relatively late in entering the stage of the *iudicium leprosorum*. In 1411 they formally cleared a university colleague of suspicions of leprosy, as we will see in Chapter 3: their exoneration proved that they knew the procedure, but this was an intramural affair. When the Paris masters first expressed an interest in public examinations, they reached beyond leprosy, at least in part because the stakes had been raised by the Great Mortality of 1348–1350 and subsequent epidemics. In 1426, the Faculty of Medicine requested Pope Martin V to grant that "any persons who are infected with or suspected of a contagious disease, be obliged and held to undergo an examination and judgment of this Faculty, as often as they are summoned by the same Faculty." There may have been a correlation between the political setting of Paris, the direction of the faculty's epidemiological interests, and the low profile of the university in records of leprosy examinations. The recorded involvement of one Parisian doctor seems to have been atypical. Roland L'Escripvain, a onetime dean of the medical faculty (and famous for sitting at the trial of Joan of Arc), was called in 1455 and 1460 to examine patients in the leprosarium Saint-Pierre in Brussels: this service, in the employ of the Duke of Burgundy, was hardly comparable to the task of medical commissioners. It is further striking that, in the orbit of the University of Paris, juries continued to include lay members on equal footing with medical practitioners, even if the latter were municipal health commissioners. In 1547, for example, "the sick brothers and sisters, residents in the hospital of Saint-Ladre in Valenciennes," certified that they found Loys Fourment not infected after having "seen, palpated, and examined him in mature consultation with" the sworn doctor—inferentially, a Paris graduate—and surgeon of the city.[10]

Tensions between Lay and Learned Examiners

The cooperation between lay and medical examiners was not always as harmonious as at Valenciennes in 1547, whether they served on the same panel or constituted separate juries. Tensions were most frequently voiced in the German lands, where the reach of university power and the involvement of faculty members became also most clearly delineated. The earliest recorded verdict, or *sentencia*, was issued in Bavaria, around 1335, by a churchman: the subject was found clean "in an examination by physicians in our presence"; he was ordered, however, to return in one year, for a reexamination by the unnamed physicians and, if necessary, a new judgment by the churchman. In 1357, three "physicians of the city of Cologne," two of whom held clerical rank, found Johannes Junghen, parish priest of Kruppach, clean of leprosy. There is no indication of where—or even whether—these "phisici et medici" received an advanced education in medicine; Cologne did not have a university until 1388. The Faculty of Medicine remained proportionately small, but it acquired a stature beyond its size and became a court of appeals for a wide area. In 1419, three doctors in Cologne cleared a man who had "submitted himself to our examination, which, as is known, pertains to us by full right"; unfortunately, although two of the doctors affixed their seals, no one signed or indicated any relation to the university.[11]

The first official testimony to an assertion by the University of Cologne in this area was a 1455 concession by the city fathers that the decision of the Faculty of Medicine would henceforth be conclusive in judging reported leprosy patients. We saw (in Chap. 1) this appellate authority in action in 1458, in the disagreement between one persistent plaintiff and the Frankfurt examiners: Hans Maderus appealed Doctor Heinrich Losser's judgment by going to Cologne, where, he claimed, he was vindicated. Supporting evidence may once have existed in the faculty archives, but all the Cologne records predating 1490 were lost. For the years between 1491 and 1712, however, Lepra-Untersuchungsprotokolle in the Deanery Books preserve a cornucopia of information. Quite a few of these minutes shed light on the primary objective of the 1455 concession, which was to provide a counterbalance to the judgments made by the examiners of Melaten, Cologne's chief leprosarium. As late as 1658, the regents swore in three female and three male inmates as examiners (*Probemeisters*). Melaten differed from most of its counterparts by taking in noncitizens and citizens of Cologne alike. Shrewdly managed admissions could enrich the institution and enhance the power of the regents. With this prospect, it was difficult to avoid slanted judgments: there was a natural bias toward finding someone leprous when the symptoms were ambiguous. Such judgments were repeatedly overruled by faculty members, often with a terse angry comment. In 1533, for example, they found Elsgina Smyt "simply clean," in contradiction with "the brazen judgment" of the Melaten residents who had, "against all fairness, pronounced her infected with leprosy." In 1608, the Cologne physicians cleared an adolescent boy from Antwerp who had "brazenly and falsely been held for unclean" by the *leprosi*, "whom it does not behoove to render judgment on this infection."[12]

Municipal authorities, not only of Cologne but also of nearby cities, long preferred to send their suspects to Melaten rather than to the physicians. I can only surmise that inertia was the major reason, but other factors may have played a role, including fiscal considerations, political connections, and perhaps even the impression (to which we will return shortly) that too many medical verdicts were inconclusive. In any event, municipal preferences for Melaten stirred a few heated protests from the Faculty of Medicine. In a memorandum of 1544, Friedrich Broich (or Grevenbroich), the dean of the Faculty of Medicine at Cologne, and his colleagues reported,

> When Arnaldus Dalem, after being reexamined by us, returned home with our testimonial letter, the councilmen of the city of Wesel asserted that this judgment belongs only to the *leprosi* outside the walls of Cologne, not to the doctors of medicine. After much ado, they replied at last that they would be willing to abide by our verdict if the senate of Cologne would testify that, by right and ancient custom, the judgment on leprosy belongs to the medical faculty. Then we requested a testimonial letter from the regents [of Melaten].

One of these regents, "forgetting" that he also had ties to the university, insisted that this judgment should be assigned to the leprous examiners. This riled the doctors. Before noting their qualified victory, and with a display of their superior Aristotelian learning, the university men huffed that *leprosi* "have obtuse senses and hence a dense intellect—because nothing is in the intellect if it was not previously in the senses." In the end, under pressure from "certain powerful members of the senate," the Melaten regents conceded that a certificate from the doctors could be sent, as long as it bore the seal of the city of Cologne: "[O]nce the councilmen of Wesel received the letter, they allowed the examined subject to circulate freely among their citizens."[13]

From 1547 to 1551, Doctor Broich was embroiled in a legal battle with the city council of Bonn, after he had cleared one of its citizens, Stina Berchs. The contest was settled only when the rector of the university intervened as referee or, to be more precise, as *iudex ordinarius*. Broich's memorandum on the Wesel councilmen was tame in comparison with the scathing letter (examined in Chap. 1) that the faculty members addressed to their own town fathers of Cologne in 1574, to protest the condemnation of elderly Tryn von Vrechen by the "wretched illiterates" of Melaten. The relative importance of such disagreements becomes clearer when we know how frequently the occasion arose. Friedrich Broich (d. 1557) participated in at least twenty official examinations, and his elder brother Johann (d. 1537) in at least twenty-seven. Their participation paled in comparison with that of their predecessor as dean, Heinrich Andree von Sittard (d. 1551), who attended fifty-eight judgments and had himself brought—in a sedan chair—to the dean's house for two of these, when he was eighty years old and ailing. Examinations by the faculty continued, largely regardless of the incidence of leprosy. In 1712, Dean Thomas Steinhaus reported,

> I was called by the full senate to examine the *leprosos* in the house called "Melaten." They were twelve in number, and together with my most expert senior colleague, the permanent examiner Thurn, we examined them and did not find a single one infected, except one married woman, named Momburs. Except for her, therefore, they have all been released and provided with a clean bill of health [*testimonio purificativo*].

Doctor Steinhaus seems to have been unaware of the historic weight of his terse epitaph.[14]

Cologne was hardly the only German city where learned medical practitioners exerted a growing and often far-ranging authority on official examinations for leprosy. Among the many other cities, three in the same general southwestern sector of Germany (with copious extant records in spite of the recurrent waves of war) stand out as illustrating not only the emerging visibility of medicine but also the substantial differences from place to place. Frankfurt, which lacked an academic corps of physicians, affords a glimpse into the emergence of commissioned examiners, and into the personal challenges they faced. Tübingen, one of the two revered university towns on the Neckar, exemplifies the impact that political powers could exert on the reach of a medical faculty's influence. In Nuremberg, without a university but with a distinguished college of physicians, medical and social developments converged in the most dramatic fashion.

The first sworn examiners of Frankfurt presented rather blurry profiles in the documents. From 1428 to 1439, they appeared obliquely, in letters from neighboring towns which requested the mayor and council of Frankfurt to have the bearer examined because "we have learned that you have several people who understand the disease and are able to inspect and test [*besehen und proberen*] patients," and who, in addition, "are commissioned with the inspection [*zu besichtigen verordnet*]." From 1447 on, the requests identified the examiners with greater precision, as "your health commissioners" (*uwer siechen Meister*). The mayor of Aschaffenburg was still vague in 1490 when he sent Peter Spengeler to Frankfurt with a plea to the council, "[P]lease have him checked by those among you who are able to understand the things to be inspected." The mayor's letter continued, quite remarkably, with the request to report either outcome because, if Peter was infected, "we wish to take care of him," presumably in a town leprosarium; should he not be infected, "we also wish to help him with our alms like the other poor folk, since he has lived among us for a long time, has lived piously and honestly, and is really poor." Although the identity of the Frankfurt examiners was still obscure to Aschaffenburgers in 1490, the haze of anonymity had lifted in 1450 with the swearing in of Doctor Heinrich Losser, who served until his death in 1468. We have become familiar, in the preceding pages, with Losser's restless career and with the challenges he faced, not only from patients but also from his employers on the town council.[15]

A different kind of power came to hold sway over Tübingen and to challenge the outreach of medical authority. The University of Tübingen, founded in 1477, became famous—and vibrant with zeal for the common good—after the establishment of a Reformed seminary in 1534. From 1538 on, the Faculty of Medicine recorded requests for inspections, though not the results. Then, however, repeated interventions by the dukes of Württemberg kept the range of influence from expanding. From 1568 to 1586 they protected the local monopoly of the *medici* of Stuttgart, thirty kilometers away, after "a dispute had arisen with regard to the *iudicium leprosorum*" between these physicians and the faculty. From 1599 to 1613, the dukes repeatedly ordered the "highly learned and our beloved faithful, the dean and doctors" of the medical faculty, to limit their examinations to Tübingen and to leave the rest to the "commissioned court physicians" in Stuttgart.[16]

After the disruption of the Thirty Years' War (1618–1648), the Tübingen faculty resumed examinations, which were recorded until 1666 and requested until at least 1689. Samuel Hafenreffer, whom we have encountered as the author of academic instructions for diagnosing leprosy, in his textbook on skin diseases, was also a participant in *iudicia*. Of all the authors cited in this study, he bequeathed the most concrete and official evidence of his participation. On September 26, 1656, as dean of the faculty, Hafenreffer presided over the inspection of Maria Jacoba Maunesin of Mengen, age fifty-three, who had been brought before her town council as "afflicted with the horrible [*abscheulichen*] disease of leprosy." Before the beadle or notary drew up the certificate, one of the doctors—perhaps the aging Hafenreffer himself, as one is tempted to speculate—wrote a succinct report, in a tremulous hand. Revealing their reliance on questioning as part of the procedure, the examiners reported that one of two large lesions, which Jacoba "has had for more than twenty years, took its origin in erysipelas." Evoking a vast human drama in one phrase, they added that the lesion was made worse by "cold and poor eating, in the troubles of war." As the Tübingen doctors found no further signs, they concluded that Jacoba was "completely free and clear of leprosy."[17]

A Showcase: Nuremberg

While the human drama loomed in the background of many an individual who appeared under suspicion of leprosy, it dominated another type of *iudicium*. Several cities held periodic group inspections of mendicants who were suspected of having leprosy. No annual examination is better documented, or more immediately relevant to the interaction between medicine and society in reactions to leprosy, than the annual *Schau* at Nuremberg. For a frame of reference, we should survey the general situation of *leprosi* around Nuremberg and many other western European towns. Since the fourteenth century, most leprosaria right outside the town gates received municipal subsidies. However, they admitted only citizens of the town, with the Cologne Melaten as a notorious exception, as we have seen. In addition, the town leprosaria usually set fiscal conditions for admission, and they imposed exacting rules on the inmates. As a result, the total number of sick residents for all the "lazar houses" of one city, such as the four leprosaria of Nuremberg, did not exceed a few dozen at any given time. Far more *leprosi*, who did not reside in these institutions for a variety of reasons, found shelter in more remote chapels, smaller shelters at some crossroads, or huts in the fields. Across Gallic as well as Germanic areas, these were the "outsiders" (*forains*), itinerant "external patients" (*Sondersiechen*), or "patients in the fields" (*veldzieken, akkerzieken, hoge zieken*). They depended mostly on alms for their livelihood and were in the news far more often than their counterparts in the city leprosaria. As local governments were concerned with preventing contagion and, more immediately, disorder and deception, they tried to control begging by issuing licenses. One way to obtain or renew a permit was to be certified as leprous by sworn examiners, usually at one of the periodic inspections. Thus, in startling contrast with the stereotypical suspected patients who desperately sought to be cleared, there were applicants who wished to be judged leprous. For some of these, the objective was admission to a town leprosarium, but most of them wanted to enter the ranks of licensed beggars.

In the fifteenth century, vagrant *leprosi* drew attention because of their growing numbers, occasional unruliness, and appeal to charity. Although the resurgence of charitable activism, which paralleled the twelfth-century spread of leprosaria, intensified with the Reformation, it preceded sixteenth-century developments, as a graphic document from Nuremberg will demonstrate. Kindly disposed burghers, viewing the group inspection (*Schau*) as an ideal opportunity for practicing and showing Christian charity, turned it into a feast for the attending poor. Thus, they promoted an event that was already attractive to destitute patients and aspiring mendicants alike and which now drew increasing numbers of out-of-town visitors. Every year, for four days during Holy Week, leprous paupers flocked to Nuremberg, where they were welcomed, seen by doctors, and treated to spiritual and bodily succor. By 1446, a hospital was built within the city walls (where today we find the picturesque Weinstadel), exclusively for accommodating the most ill during those four days. The thriving celebration was captured in 1493 in a broadsheet, a single-page woodcut that may have served as a poster and which is reproduced in Plate 2. This picture brilliantly, and quite artistically, illuminates a wide range of medical facets of the *Sondersiechenschau*.[18]

In this crowded drama, religion and charity are upstaging medicine, not only because they give the impetus to this manifestation of mercy, launched by "the hardworking people of Nuremberg" (as the caption in the lower right corner explains), but also because they offer the only remedy once leprosy has been confirmed. The medical examiners risk being ignored in this crowded tableau. The crowd is festively gathered on a lawn inside the city wall that runs along the Pegnitz and which (shown as an upper border behind the crowd) is surpassed in height by the pulpit. The preacher encourages the suffering audience to "be patient" and reminds them of the innocent sufferings of Jesus, their duty of gratitude, and their chance of gaining God's grace by offering their "great pains." His words are balm for the spirit, judging from the meek smiles of three people in his audience (one person, who may not be smiling, has covered the lower face to hide erosive sores, in a prototypical image of leprous patients—which has endured). Confession and communion provide medicine for the soul. Food and drink are being served to comfort the body, at least momentarily, and a cartload of bales of cloth awaits distribution to protect the poor from the elements. The physicians, who, we may recall, had once been viewed as healers, are cast here as gatekeepers. The good folk of Nuremberg have engaged "two doctors for inspecting these men and women diligently, to see who may be sick or healthy."

In the upper right corner, four or five patients are filing past two youthful doctors, magisterially seated and in full academic garb. Neither of the examiners seems to be a surgeon, unless the artist intended to show the one on the far left as touching the patient. Their inspection seems to be quite perfunctory, especially when compared with a depiction that was drawn in Basel a little more than two decades later (Plate 8 and discussion in Chap. 7). The most obvious reason for the contrast is that, while the Basel woodcut portrays a private examination, the Nuremberg broadsheet advertises a public celebration. This distinction, of the procedure itself as, respectively, private and public, should not be confused with the distinction between personal and official initiative. Jean de Varanda mixed the two distinctions in his "model of a judgment, which is performed in well organized commonwealths when *elephantici* are examined publicly, or also privately when they wish to be more certain about their present condition." In fact, the differentiation also blurs for some examinations that were scheduled in cities. In Brugge, for example, the visitation room of the Magdalen leprosarium was open twice a week for

PLATE 2. *Lepraschau*, public examination of people reported or claiming to have leprosy: woodcut broadsheet, Nuremberg, 1493; reproduced in Casimir Tollet, *Les édi-
fices hospitaliers depuis leur origine jusqu'à nos jours* (Paris, 1892), p. 45. At the National
Library of Medicine, Bethesda, Maryland.

seeing *veldzieken* since at least 1309 (it is worth noting that, for this entire year, only seventeen names were recorded in total).[19]

This perspective adds relief to the eminently public character of the *Schau* at Nuremberg. Furthermore, while the presence of examiners lent the title and pretext to this event, it was not the focus but almost a ceremonial part of the celebration. In principle, their judgment determined who received the spiritual and physical solace and who was sent away as insufficiently sick or, worse, an impostor. Two or three of the examinees in the scene are likely to be invited in because they present the telltale facial nodules. Similar symptoms are visible in other scenes, on the faces of several indigents who have already been admitted, such as the confessing person in the center (enlarged in Plate 6, in Chap. 5). The leprous patients, not only in the examination scene but in the entire sheet, are further identifiable by the leggings inserted into their boots, which may be contrasted with the more elegant footwear, for example, of the waiter down below. The first patient in line for examination is having his breath checked by the physician, who, given the closeness, does not seem overly concerned with becoming infected.

Two readings make sense for the next set in this scene. The second examinee seems to have a cavity where the tip of his nose should be, and to have an ear missing or severely shrunk on the right side of his face, which is exposed to the physician's gaze. This physician may be pointing at the face, or gesturing as someone about to state an opinion. In an alternate and preferable reading, however, he is already pronouncing a judgment, which is captioned above the scene, "Friend, you are not an outcast patient [*sundersiech*]; you have only neglected yourself. . . . Let someone else come here, and you go in back there." This would mean that the physician is unmasking a pretender, who is dejectedly turning away after trying hard to masquerade as a leprous indigent. The wide-brimmed hat in the man's left hand was an obligatory item for itinerant *leprosi*. The staff in his right hand was also a standard part of the outfit in many areas, leading Ambroise Paré to refer to "a clapper and a stick" as the distinctive mark of leprous beggars. A symbol of crippled patients in much medieval iconography, the staff became indispensable with the advance of neurological impairment and damage to the extremities.

Completing his "trademark" profile, and opening his cape as if on purpose, the pretender at the Nuremberg *Schau* prominently displays his purse and clapper. At least five more clappers are visible in the drawing as a whole, and we will look more closely at their symbolic significance later in this chapter. Behind this second subject, a man or woman is holding his or her nose against the stench, if not against infection: this bystander is more likely a Nuremberg notable than a medical practitioner. The examinee at the far right displays more advanced facial lesions than his companions. He is less dis-figured, on the other hand, than other individuals in the woodcut, including the person at the far right in the preacher's audience, the man who is receiv-ing communion in the scene below, and several of the well-served guests at the table.

The growing popularity of this event may not be surprising, but it grew to an alarming level in the course of the sixteenth century, owing to a com-bination of pious fervor among the burghers and dire misery for marginal people. Little more than a generation after the broadsheet was printed, the number of medical examiners was doubled, and within one or two decades it was expanded to six. More than three thousand beggars streamed to the *Schau* in 1570, posing a grave threat to the possessions, security, and health of the burghers. Of 3,240 supplicants who came in 1574, the examiners judged 709 to be fraudulent—leaving us to wonder whether these, then, were de-clared healthy, and whether the others were certified as truly leprous. It would appear, however, that by this time medicine had become largely irrelevant to the yearly event. In 1575, the city's welcome was suspended, and, as a memo-randum of the council noted tersely, "the *Sundersiechen* were not let in, but all were inspected and given pocket money [*Zehrpfennig*]." The following year, the event was permanently moved outside the town walls, to the grounds of Saint John's leprosarium (where the historic cemetery the Johannisfriedhof is lo-cated today). The annual celebration at Nuremberg faded as fast as it had exploded, and a simple explanation is that sheer success was the cause of its undoing. Further research is needed to assess the respective roles played by the apparent decline in the incidence of leprosy, a heightened concern with public order, and the dulling of charitable impulses. By 1623 only one hun-dred indigents were seen. The upheavals of the Thirty Years' War hastened the demise of a memorable tradition.[20]

While the Nuremberg city council seems to have lost interest in the med-ical aspects of the *Schau* after 1574, it could not ignore the indigent but pos-sibly contagious crowds at the gates. The influx of 1570 prompted the city fathers to assess the balance between public health and public order. The coun-cil demanded a written consultation (*Gutachten*) from the health commission-ers. The five sworn physicians of Nuremberg obliged. Fortunately for us, the originals of their replies, and of four follow-ups by colleagues, are pristinely preserved. The nine consultations deserve to be published, parsed, and compared in a separate study. We can only glance at them in this space, with focus on the place of medicine in the *iudicium leprosorum*. A stellar member of the medical team, the Dutch-born Volcher Coiter (1534–1590), who had been professor of anatomy and surgery at Perugia and Bologna, wrote the most methodical report. He began with an elaborate statement of the basic problems that made it difficult, in the disorder and insufficient time of the *Schau*, to distinguish leprosy from "common syphilis [*gemeinen frantzosen*] and other foreign diseases." He then itemized six points to be observed for improving the situation. His preoccupation throughout was that people "do not make themselves sick for the sake of large alms. We find quite many who know how to make themselves up, with glues and plasters and other trickery, to look as if they were really sick." Coiter stressed that such deceit was practiced "to the disadvantage of the doctors and the community, and to the harm of the sick themselves." In his final point, he returned to the combined threat of chaos and deception by urging that the municipal officials make sure "that the sick are let in one at a time, in orderly fashion, and that not all kinds of fraudulent beggars sneak in."[21]

Two other commissioners, Justinus Müller and Johannes Schenck, supplemented Coiter's concern with security by suggesting that the note issued by the examiners, the "Zettl," or certificate, serve as an admission ticket for the following year, or even as a sort of identification. The two remaining sworn physicians, Georg Rücker and Erasmus Flock, claimed that they were too young and inexperienced to offer worthwhile advice, and they deferred, with reverent "obedience," to their "learned" seniors. Georg Palma (1543–1591), a distinguished Nuremberg physician, was far less timid when he wrote a follow-up in 1572. His recommendations to the city council seemed to vent some pique, for he noted that the situation had caused "us to be accused, as if it were due to our negligence and imprudence." Indeed, in defense of his profession, Palma emphasized that leprosy was extremely difficult to diagnose accurately. At the onset, before becoming "confirmed" and producing "a horrible look" (*abscheulichen anblick*), the disease not only was almost imperceptible but might "be cured and removed with the help of medicines" or even "by nature alone." The physician's difficulty was compounded by the fact that the symptoms were so "multiform" (*ungleichförmig*). Worst, however, was the widespread "deception, particularly by itinerant beggars who, burned in the sun, are sometimes as hideous to behold as those truly infected with evil leprosy, especially when they make themselves up with ointments and other materials to create an appearance greatly resembling the miserable plague."[22]

BEING JUDGED: PATIENTS AND EXAMINERS

Taken together, the reports of the Nuremberg physicians on the *Schau* high-light a dual change in outlook, which manifested itself in various places over the course of the sixteenth century. The credibility of the examined patients and of the medical examiners eclipsed leprosy itself as the paramount issue in judgments, and a new kind of suspect tried to counterfeit rather than to conceal the disease. The shift in emphasis, from the menace of leprosy to the crisis of credibility, reverberated in echoes of the courtroom, which were spreading through examinations—to be surpassed only by the pervasiveness of legal concerns in today's medicine. As in judicial procedures, the stakes ranged from money to honor, and from freedom to authority. Affidavits and protocols evinced a search for greater specificity and permanence. The presence of witnesses, however foreign to the idea of medical confidentiality, secured at least an image of objectivity. The proliferation of oaths, required from the judged and the judges, signaled growing apprehensions about deception. The challenge of deception by hiding symptoms or, in a remarkable inversion, simulating them suggested that "truth" was legally ambivalent. Appeals were pursued not only by those who received an adverse judgment but, often, also by the authorities. Among the examiners themselves, claims to trustworthiness were contested between lay juries and medical practitioners, and occasionally between surgeons and physicians. In the contest, the doctors strove to safeguard their credibility with empirical tests, rational explanations, collegial solidarity, and circumspect judgments. The circumspection had almost as much to do with jurisprudence as with medicine, and a juridical tone entered the language and construction of the *examen leprosorum*. Indeed, to the extent that a preoccupation with credibility colored the medical response to leprosy, it reflected a dilemma for the physician, between judging and healing.

Concealing and Counterfeiting Illness

The patient's truthfulness, while important for many medical diagnoses, was deemed critical in examinations for leprosy. This perceived urgency was due to the ambiguity of most symptoms, as we will see later (in Chap. 7), but also to the convergence of the particular ramifications of the disease with the natural inclinations of patients and the circumstances of the time and place. Some people who are ill tend to minimize their condition, out of self-delusion, a wish to calm worried bystanders, or a fear of diminished self-hood; other people, on the contrary, affect signs of illness, out of sheer hypochondria, need for attention, or fear of some obligation. Such impulses were greatly compounded by the exceptional perception and consequences of leprosy. Terror tempted many an examinee to prevent or belie a positive diagnosis by any means possible.

Poignantly documented attempts to conceal the symptoms of leprosy include the case of Grette Swynnen Thielens of Diedenhofen, who twice failed to be judged clean. After her parish sent her to Metz, where she was declared infected, Grette appealed for a second opinion in Cologne, where, it was believed, "everyone usually received due justice." The parish sent her to Cologne in 1492, to be seen by the *Probemeister* of Melaten—rather than by members of the faculty, we may note. In the accompanying letter, the parish council requested that she be investigated with special care, "not only in her body, but also with inspection of her drawn blood." They explained, "[W]e understand that she is suppressing and hiding the disease in her body with medicines." Within a week, a reply from Cologne assured that, after seeing Grette "in the customary and traditional manner," the examiners "found her completely unclean and heavily infected with the disease of leprosy." Their silence on procedures and criteria deprives us of all clues to the examiners' credibility and the patient's truthfulness; we can imagine, however, the tragic impact of the "sentence" on Grette's life.[23]

As an effective deterrent to deceit, one could solemnly formalize a statement—and invoke the omniscient and just God—by imposing an oath. While already part of examinations before medical practitioners participated regularly, oaths gained weight and prominence in proportion to growing concerns with power and order. In 1363, compiling instructions for the *iudicium leprosorum,* Guy de Chauliac ordered that the examining physician, before questioning suspected patients, "make them swear to speak the truth about the things to be asked." This oath, however, like the invocation and pious consolation of the patient which were to precede it, undoubtedly harked back to earlier ecclesiastical procedures. In the course of the fifteenth century, a second oath was added, at least in some places. In 1511, the Cologne Faculty of Medicine recorded a formula that was to *follow* the pledge "to tell the truth about all the points of interrogation" and which, the faculty asserted, "has always been observed for ages." The Faculty of Medicine formula required the examinee to swear that, if he receives a different sentence from the faculty than he would have preferred to hear, he will on that account not inflict trouble, in person or through friends, directly or indirectly, on members of the faculty of medicine as a group, individually, or on some dependency of the faculty. If it should happen, perish the thought, that members of the faculty of medicine, as a group or individually, or one of its dependencies, are harassed by anyone, or forced into expenses because of the sentence, whether already issued or to be issued, he will absolve and indemnify them. If a priest is to be examined, he will swear this oath saying, with the hand on the chest, "So I swear, and I promise to observe my oath." If the examinee is a layman or secular, he will swear the oath while touching sacred objects, a crucifix, the gospels, and the like, or in the hands of the notary public and of the university beadle.

The formulation, far more precise than for the first oath, suggested that the examiners were more concerned with shielding themselves than with guaranteeing truthful answers.[24]

The oath to tell the truth became so rooted in the diagnostic routine that it was expected even from worried individuals who, of their own accord, sought a private—and, it would seem, informal—inspection. This was the case when, in the third quarter of the sixteenth century, a man came to the Dutch physician Pieter van Foreest (Petrus Forestus, 1522–1597) and "a certain surgeon" in the city of Pithiviers, south of Paris, "for the purpose of a check [*experiendi*], to know from us whether or not he was afflicted by true leprosy." Foreest recalled that, while proceeding rather smoothly ("iuxta ordinem satis facilem"), they followed the usual order, and "we proposed to him that he should swear to us to answer our questions." By this time, however, the conventional oath was applicable to a dwindling number of cases, for a variety of reasons. The concern with tracking down people with leprosy had waned, while other diseases such as plague and syphilis drew more attention. Most important, circumstances such as widespread poverty inverted the truth that might have been guaranteed by an oath. Aside from the fearful suspects who, by swearing, were deterred from hiding their illness, more and more examinees, driven by despondency or indolence, would stop at nothing to appear leprous. This problem, to which we were introduced in the Nuremberg *Schau*, became increasingly marked during the sixteenth century.[25]

Caution should be applied, as Robert Jütte reminds us, to generalizations about fraudulent *leprosi*. The simulation of disease was not limited to leprosy, and a heightened occurrence cannot be automatically inferred from its increased presence in documents of the sixteenth and early seventeenth centuries. Jütte argues that the attention to this counterfeiting was due, at least in part, to a "heightened collective sensitivity to social deviance" from the end of the Middle Ages on. This sensitivity also affected the physician's approach to diagnosis, albeit more subtly in assessing a patient's health than in judging a claimant's authenticity. In addition, while the practitioner's credibility was seriously challenged by the task of identifying a deceptive disease, it was almost comically—but, nevertheless, threateningly—tested by the appearance of a cunning subject. The potential erosion of the doctors' authority was no doubt one reason that Volcher Coiter worried about the presence of impostors among the masses to be examined. On the other hand, the impostor gave medical practitioners an opportunity for proving their privileged knowledge by exposing trickery, either in public, as at the Nuremberg *Schau*, or in private but still with official authority, as in the following story.[26]

With characteristic verve, Ambroise Paré related an encounter, in Vitré (Brittany) in the mid-1520s, of his brother Jehan with "a big knave of a beggar who counterfeited a leper." The beggar sat in front of a church, with several coins strewn in a handkerchief at his feet. He was "holding in his right hand some clappers, making them click rather loudly: his face covered with large pustules, made of a certain strong glue and painted in a livid, reddish fashion, approximating the color of lepers, and he was very hideous to see; thus, out of compassion, everyone gave him alms." When Jehan Paré inquired, the man, fully cognizant of popular stereotypes about leprosy as a disease, "answered in a broken and rusty voice that he had been a leper from his mother's womb and that his father and mother had died of it, and that their limbs had fallen off of them piecemeal because of it." The artful impostor tightened, from beneath his cloak, a rag that he had wrapped around his neck, "so as to make the blood mount to his face, in order to render it still more hideous and disfigured, and also to make his voice husky." Jehan, however, chatted with him until he needed air and was forced to loosen the rag. With his suspicion thus confirmed, Jehan, who was a surgeon, "went directly to the Magistrate, asking for his kind help to find out the truth about this." The man was ordered to be brought to Jehan's house for an examination. After removing the rag, the surgeon "washed his face with warm water, which caused all his pustules to become detached and to fall off; and his face remained full of life and natural, without any sign of disease." Echoing the diagnostic guidelines of the textbooks, Jehan then "had him strip naked, and he found on his body no sign of leprosy, either univocal or equivocal." Jailed

PLATE 3. Clapper from the sixteenth or seventeenth century: preserved at the Onze-Lieve-Vrouw-ter-Potterie Hospitaalmuseum, Brugge; copyright Stedelijke Musea van Brugge.

as a "thief of the people," the beggar was interrogated by the magistrate. He confessed "that he knew how to counterfeit several illnesses, and that he had never found greater profit in it than when he counterfeited the leper."[27]

Bypassing Ambroise Paré's shocking conclusion (in which the punishment of the beggar degenerates into something like a mob lynching), we should not ignore several significant ramifications of the story. In the perspective of this chapter, the most salient points are the lure and trademarks of begging, the drastic joint response of the surgeon and the magistrate to a solitary leprosy simulator, and the implicit contrast between the perspicacity of the medical practitioner and the gullibility of the populace. First, begging as a *leprosus*, whether genuine or counterfeit, could be lucrative. The lure was so tempting that, in early sixteenth-century Flanders, some people were said to seek contact with leprous patients in the hope of becoming lightly infected, so that they would be "allowed to beg freely and to ask for their bread, like other leprous folk." In 1531, when waves of mendicants swept through several of the Habsburg territories, Emperor Charles V prohibited begging by "all those who were able to earn their bread," and only leprous subjects were allowed to collect alms. When this ordinance caused an apparent surge in the number of infected people, the emperor decreed in 1547 that every leprous person had to carry a certificate of an examination.[28]

While a certificate provided official legitimacy, the popular mark of the leprous beggar's privilege, as well as a symbol of the counterfeiting vagabond, was the rattle or, to be more accurate, "clapper." The instrument, illustrated in Plate 3, was shaken so that slats of wood, hitting each other, produced a sound like clapping hands (rather than a rattling sound made by, for instance, a baby rattle or the tail of a rattlesnake). The clapper reproduced here is a small but prized possession of a museum in Brugge. Yet the item, probably dating from the early seventeenth century, awaits a careful investigation into its provenance and iconography. Carved on one side, a rather well dressed burgher is identified as a leprous mendicant by a large clapper in the right hand, a beggar's basket in the left, and a wide-brimmed hat, while the dog may refer to Poor Lazarus. It is interesting to compare this burgher with the obviously more rustic and poorer beggar who was portrayed—plausibly around the same time—shaking a far simpler clapper and accompanied by a dog (and, for added effect, holding a crucifix) on the title page of a book on skin diseases by Samuel Hafenreffer (Plate 5, in Chap. 4).

On the reverse of the Brugge clapper, a woman is holding the martyr's palm and a tower, the symbols of Saint Barbara, who, to the best of my knowledge, had no connection to leprosy (Saint Magdalen, patroness of people with leprosy, would be holding an ointment jar rather than a tower). If one assumes that this clapper was once owned and used by a certified *leprosus*, the artistic execution might reinforce the impression that, in some bourgeois settings and at a relatively late stage, begging as a victim of leprosy became a "gentrified" occupation. Some cities in Flanders indeed issued marked "standard clappers," in order to prevent fraud. On further thought, however, and in view of its pristine condition, it seems more likely that the Brugge piece was, rather, a ceremonial object, perhaps one of the insignia of a leprosarium or of a fraternity of *akkerzieken*.[29]

Heinrich Losser spotlighted the trademark of the leprous wanderer in 1458, while arguing that the Cologne physicians were too ready to exonerate suspected *leprosi*: "The respectable doctor, Master Bartholomeus, found the valet of our colleagues in Mainz infected, while the doctors in Cologne pronounced him clean; at this time, he is going around near Aschaffenburg with a clapper [*cleppirchen*], as Master Bartholomeus told me recently." A century later, Ambroise Paré contrasted the biblical injunction, that the expelled "unclean" wear torn clothes, with his own time, when leprosy sufferers were "given a clapper and a stick, so that they may be recognized." Guillaume des Innocens characterized the "cliquettes," rather strangely, as "invented by the school of physicians (in my opinion), for safeguarding the public interest, by preventing conversation of leprous patients with healthy and clean people." When warned by the clapping, people could avoid "the breath of these poor folk, giving them a wide berth—and an occasional alm." While des Innocens related the clapper only casually to begging, Jean de Varanda made a direct connection. Varanda referred to patients under the protection of Saint Lazarus who "dwell in hospitals far removed from the edge of towns, and beg for alms with wooden clappers." The vital connection between the claim of illness and the noisy solicitation of succor was most evident in actual situations. In 1647, when Grietgen Meurers, resident of the leprosarium near Elberfeld (Lower Rhine), was found clean, the council ordered her to undergo another examination and, meanwhile, to refrain from using the clapper. When another resident, Arndt von der Finkenlandwehr, who had reported Grietgen's remission (and was himself found clean, three years later), was accused of brawling with neighbors and indulging in drink and tobacco, the judge warned him that, "unless he mended his ways, the clapper would be taken away from him."[30]

The clapper had become a privilege, indispensable to the livelihood of some and, apparently, a source of income for others. As a measure of the profit that could be gained by being recognized as leprous, we have concrete information about a kind of fraud that, while different, paralleled the case exposed by Jehan Paré. At the end of the seventeenth century, a worker at Melaten was caught selling false leprosy certificates to healthy customers, at a price of thirteen to fifteen silver coins, which equaled a week's wages for a building artisan. It is not clear whether the certificates were intended for obtaining the permission to beg or for gaining admission into a leprosarium. A similar ambiguity marks the prosecution, in 1628, of two residents of the leprosarium of Étampes and two associates for "having written and fabricated false letters, falsely signed" by the grand almoner. It is possible that buyers of these fake certificates were trying to gain admission to a town leprosarium rather than a begging license. This possibility cautions us against oversimplifying distinctions between resident and itinerant patients, and between those who were sustained by urban welfare and those who depended on their ability to stir mercy. In all these patterns, however, there were economic incentives for fraud, as well as extraneous challenges for medical examiners.[31]

Challenges to Competence

The intervention of Jehan Paré pertained more to an officer of the law than to a health official. It illustrated the practitioner's exercise of forensic authority, deputized by the magistrate. This exercise, while normal in individual inspections for leprosy, became problematic when false patients were more numerous *and* perceived more as a public threat to order than to health. If deception by a single beggar was of such concern to the magistrate of Vitré (and to the populace, as the lynching would suggest), it is easy to imagine the general consternation over a proliferation of swindlers. Medical participation, however, in response to collective counterfeiting was bound to be ceremonial rather than effective, as developments in Nuremberg proved. The sworn city doctors shared Coiter's complaint that, "amid so much confusion and in such a short time, it is not quite possible to perform an orderly and intelligent inspection." Medical and official concerns diverged when authorities were more concerned with maintaining order than with controlling disease. In 1560, Duke Wilhelm of Cleves was alarmed by the "many beggars in our land who give the appearance of being leprous [*uthsettich*] while they are found not to be, with the result that the truly sick are barred from and robbed of their alms."

Therefore, the duke issued stern measures against "such uninfected beggars in leprous disguise" (*sulche unbefleckte betler in melatischem scheine*). He did not mention examiners, but, preferring that the beggars police themselves, he ordered them to be members of the fraternity or guild (a *bruderschaft* led by a *gildtmeister*) and to convene once a year, in order to prevent abuses by means of communal measures: it stands to reason that these measures would have included a *Schau*. If guild members were entrusted with the inspections, however, their authority was not likely to rest on medical experience or knowledge, and their decisions probably had little to do with the disease of leprosy.[32]

The question of credibility was less acute when examinations were not entrusted to itinerant patients but to residents of leprosaria, as was common practice. Nevertheless, verdicts by inmates were the most frequent targets of appeals, whether brought by protesting subjects or by dissatisfied authorities. The populace accused leprous examiners of wishing to bring others into their misery, and physicians harshly belittled their "dense intellect," as we have seen. Nonleprous residents, most of whom were under religious vows and directed by a mother superior, were also challenged as lacking the required knowledge and judging "brazenly" or without due deliberation. In 1582, Philippus Schopff, who had served as municipal physician of Bad Kreuznach for seven years, found that suspects constantly came to appeal to "us, the commissioned medical inspectors" (*Medicis inspectoribus*), when they contested the judgment by nonmedical examiners. This persuaded him to publish a compendium on leprosy (*Aussatz*) that he had originally compiled for his own "daily use and recollection." Not much later, the Swiss physician Felix Platter (1536–1614) recalled that, in his forty-three years of practice, he had seen many erroneous judgments by "inexperienced individuals." He observed that these errors caused harmless people to be unjustly "expelled by the magistrate" or, inversely, infected patients to remain in the community "until they can be recognized even by the common people, with a collapsed nose, hoarse voice, and horrible appearance."[33]

It is difficult to minimize the distance between popular and medical knowledge, or the benefits of more careful examinations, even if we have reservations about the notion of progress. Furthermore, even though criticisms often reflected an effort to retain or gain control, physicians and people had reasonable grounds for being wary about the reliability of verdicts issued by leprosarium staffers. These examiners might be swayed by the consideration of the suspect's economic status, and of their own comfort in the shelter. In addition, their understanding relied on anecdotal experience. Such experience, which was termed "fallacious" in the Hippocratic tradition, lacked the presumably objective foundations of logic and physiology. However, neither ulterior motives nor limited insight was an insurmountable obstacle to credible diagnoses. Bias could be remedied by an appeal to conscience, secured by an oath. The members of the commission at the Brugge Magdalen hospital swore

> by Almighty God, to inspect properly and faithfully, to palpate, and to test all those who come to be examined; to state truthfully whether they are healthy or sick; not to be swayed by hatred or love, fear or money, or any reason in the world; to maintain the vigor and honor of the *prouve;* and to reveal no secrets of the inspection, by word or any sign, to friends or enemies, women or men.

The lay examiners' intellectual liability, on the other hand, was mitigated by an unequaled closeness to the disease. Their lack of theoretical foundation could be remedied, as it was in Brugge and Ghent, by vocational solidarity, cumulative empirical knowledge, and, ultimately, reference to learned medicine.[34]

Dissatisfied with being pronounced clean at two separate examinations at the Brugge Magdalen in 1542, Ursula Balcaen sought a different verdict in Ghent. The "mistress and her company of Saint Lazarus hospital" there found her sick and would have admitted her if she were a Ghent citizen. However, they withheld judgment, as they wrote to their Brugge colleagues, "out of respect for your community and against gossip from people out there." The Brugge panel took notice and duly found Ursula leprous at another examination. Longtime relations between Ghent and Brugge indicated that the solidarity reached beyond honor, to the sharing of information. In 1477, after forty-five years of service in the Ghent leprosarium, Our Lady of Lazerie, Sister Barbara Sagers recorded her diagnostic experience. She distinguished three kinds of leprosy by color, as "brown," "white," and "green"; her label for a fourth type, "pakerie," awaits a cogent interpretation. These four kinds corresponded to categories of other nonmedical examiners, but the distinction had no exact counterpart in the taxonomies of physicians (see Chap. 6). Sister Barbara's guidelines were informal yet authoritative: they were copied and expanded, more than a century later, by the examiners of the Brugge Magdalen. It is possible that, in the intervening years, some medical influence seeped in. Eight numbered signs for the "brown sickness," which included nodules, brown and yellow macules, and collapsed nose, pointed toward the form that learned authors called "leonine"; ten signs for the "white sickness" included symptoms that the physicians emphasized, particularly the inability to blink, longitudinal sores and pale spots, spasms and paralysis of the extremes, and susceptibility to getting burned. The Brugge version concluded with the reassurance that the signs were not only clarified in court records, of an actual appeal by one Christoffel Pychaut at Ghent in 1445, but also "confirmed by forty doctors."[35]

Competence: Contested—and Costly

Doctors may have wielded the highest authority for the recognition of leprosy, but their credibility was far from invulnerable. Challenges to the fairness of their judgments inevitably nudge us to a question that, because of the built-in trap of presentism, I have intentionally avoided in most of this book. When premodern practitioners declared someone leprous, how often were they "right"? It is wise to heed the caveat by Piers Mitchell, "There is evidence for other diseases to have been grouped under the umbrella term of 'leprosy' in the past." Allowing for the deceptiveness of the disease, which has troubled diagnosticians into modern times, I believe that, if and when premodern diagnoses were erroneous, this was due to social pressure as often as to medical error. The consequences of their verdict placed practitioners under a heavy and acknowledged onus, as we have seen, so that their efforts to differentiate leprosy surpassed their care in recognizing any other disease— or, for that matter, even impending death. This exceptional diligence, together with the identification of signs that were universally recognized as "essential" (particularly macules and loss of sensation), reduced the likelihood of gross misidentification by reasonably conscientious and informed examiners. Instead, a trend that increased after the fourteenth century (at the same rate as that at which the notoriety of leprosy decreased), was to *admit* other diseases "under the umbrella" of leprosy.[36]

According to a dual and diehard stereotype, numerous people were wrongfully labeled leprous and expelled, and this was at least partially due to the primitive state of medicine. The mere numbers argue against this stereotype. After tabulating hundreds of records, I arrive at the admittedly broad estimate that medical examiners found fewer than 20 percent infected among the people who were brought to them—even though all of these, obviously, were displaying signs that pointed to leprosy. Even the most careful historians are liable to contribute to the assumption of widespread errors. One historian, for example, infers too readily that if verdicts of leprosy were overturned on appeal, they *must* have been based on "erreurs de diagnostic"; another suggests that patients who were discharged from leprosaria, after stays that ranged from one and a half to twenty years, would have been admitted on the basis of an initial misdiagnosis. These interpretations overlook not only the possibility of an illness going into remission but also, far more important, the many extraneous factors that affected verdicts and appeals, as well as leprosarium admissions and expulsions.[37]

The growing influence of extrinsic considerations on medical judgments after the fourteenth century, and the accompanying change in parameters, are signaled by two disparate witnesses. The pious Tolosain surgeon Guillaume des Innocens concluded his *Examen des éléphantiques* (1595) on a note that was positive but not intrinsically related to leprosy. People who were found healthy should receive an affidavit,

> so that they may go and find the rector, pastor, or vicar of their parish who (following the example of the priests of the Old Testament) publishes and reads aloud in his Sunday sermon the total absolution delivered by the physicians and surgeons. Thus they may, without fear and doubt, dwell and work among the healthy, attend church and assemblies, and enjoy all the privileges of a person who is found completely healthy and clean. Praise God.

Looking in the opposite direction, but under an equally broad umbrella, the Tübingen dean Samuel Hafenreffer taught in his *Nosodochium* (1660) that all

> examinations are instituted primarily with this objective, that these wretched persons may obtain life sustenance when they are not in an established leprosarium. For this reason, the customary declaration is that those persons who do not know from what to live, much less where to seek treatment, should be separated from the healthy and sustained in a leprosarium, as they are afflicted with a malignant and evil scabies in the same way as leprous patients [*elephantici*].

The striking title page of Hafenreffer's remarkable work is reproduced in Plate 5 (Chap. 4) together with an affidavit of one of his own examinations.[38]

The role of social considerations in diagnosing leprosy is further documented in the archives. For example, it is brought to life, with echoes of the broader human drama, in a series of terse notes extant in Auxerre. In 1661, when Jeanne Minee, the wife of Jacques Collotte Savetier, was suspected of leprosy, the authorities ordered that she "be examined *and declared leprous* by a

physician or surgeon *in order to receive a pension* of 150 livres." A surgeon, Henri
Foulon, indeed certified that he "found several signs of *lepre*," whereupon
Jeanne Minee, meanwhile widowed, "received the order to withdraw to an
alcove of the leprosarium chapel of Saint Margaret." We do not know
whether she ever received any "pension," but subsequent accounts mention
disbursements for "a morsel of bread" and "a bundle of straw" for Jeanne
Minee, "poor leprous woman"; in the end, she appears simply as "the poor
leprous Minee." In sum, the treatises and archives indicate the importance of
social frameworks, and the need to take these into account if one wishes to
assess the reliability of premodern diagnoses by modern criteria.[39]

From my vantage point as a historian, I am primarily interested in understanding which challenges to their credibility medical examiners faced in
their own time, and how they responded. A few preliminary avowals are in
order. Some positions that are presented here as defensive may just as well be
viewed as integral to standard procedure; it is occasionally difficult to separate an issue from the intended solution; and some defensive measures were
two-edged swords. The truthfulness of the medical practitioners was secured
first by an oath, as was the case with lay examiners, except that the former
had sworn their pledge to the municipal authorities or a faculty of medicine.
Increased pressures triggered attempts by the authorities to reinforce this security with a second oath (these attempts, ironically, paralleled the demands
made by physicians of suspects to be examined). We may recall that Doctor
Losser and his fellow commissioners vigorously protested a new requirement
that they swear an oath and have a court judge present at each examination.
The Frankfurt city council ignored their protest. Losser lacked the corporate support of a university or a *collegium*, unlike his counterparts in Cologne
and Nuremberg. In these cities, the persuasive power of association was evident. In a direct application of collegiality to credibility, Volcher Coiter advised that, as

> there is so much disagreement among the very learned in this domain, the sworn
> physicians should assemble a few days before the *Schau*, and agree what they wish
> to consider true signs of *Aussatz* at the time of the inspection, so that they are
> able to focus the inspection on two or three.[40]

Solidarity could adversely affect credibility by reinforcing the perception
that medical practitioners were cliquish. In economic terms, this perception
overlapped with the perennial complaint about the price of health-related

services. Even when examiners delayed a decision and demanded a repeat visit, ordering medical supervision for the intervening period, they might be accused of exploiting a case. When a physician found Geert ten Starte of Kampen "predisposed" and in need of preventive treatment, his wife called medical practitioners "thieves." The fees for the examinations themselves must often have seemed exorbitant or burdensome, especially since they produced no inherent therapeutic benefit. In Frankfurt, Hans Maderus complained bitterly about the "heavy expenses" he had incurred for seeking a second opinion in Cologne. The documents contain numerous clues to the monetary aspects of examinations, which merit a systematic study. Even the oldest extant record of a medical examination, in Siena in 1250, informs us that each of the four attending physicians received 20 solidi. In the setting of this chapter, a precise and comprehensive appraisal of the level of examiners' fees is precluded by the chronological and geographical range, variable currencies and coinages, and economic fluctuations. A few observations may indicate the pertinence of money to the physician's credibility.[41]

From the perspective of the patients and their families, paying for an examination was a major item, as we are able to infer from a supplicatory letter to the mayor and council of Arras in 1514. Martin le Roy wrote that, after his wife was found leprous by the town physician and surgeons,

> these sent your sergeant-at-arms, demanding money as their fee for having performed the inspection—and for the sergeant for having assembled them. Your humble petitioner is quite unable to pay, on account of the great losses and damage that he has incurred in the course of more than a year, exercising his office of university messenger between this city and Paris. In the latter city he was robbed by bandits from France, losing the sum of one hundred pounds in all, in silver, merchandise, and other valuables. In addition, since last Christmas he has lost two horses that were his livelihood.

Le Roy implored the council to cancel "the fees of the physician, surgeon, and sergeant": by doing so, they would "oblige me so much more to your service, which with my whole heart I wish to give, while praying to God for you." At least this part of Martin's sad story had a relatively happy ending, as, according to a marginal note, the council gave him 20 sous for settling the issue. This was almost certainly less than what he owed.[42]

Between 1508 and 1558, examination fees in Arras ranged from 12 to 40 sous for the doctor, from 6 to 24 sous for the surgeon, and from 6 to 16 sous for the sergeant. It may be noted that surgeons were paid anywhere from half as much as the doctors to the same amount. A median ratio is reliably documented in a 1475 memorandum from the city council of Evreux (Normandy), ordering the treasurer to pay, from tax revenues, "master Gieuffroy Lamy, priest and doctor of medicine, 15 sous *tournois,* and Philippot de Verretz, barber, 10 sous *tournois,* for having examined Jehan Fraynet, suffering from leprosy." The amount charged for a positive finding, such as this one, may sometimes have been twice that for a negative finding, but I have come across only one instance of this proration. Anyhow, Fraynet's expenses were covered by the town treasury; if he appealed and lost, he might have to pay, according to some regulations. In 1466, the council of Baux-de-Provence ordered Johannon de Bonafe, who appealed a nearly four-year-old positive verdict, "to have a qualified physician [*mege*] come, at the expense of the party that is proven wrong, so that, if Johannon is pronounced leprous by the physician, he will have to pay the entire expense and, if not, the village will pay everything." The relative heft of the cost is suggested by the fact that the council, "considering that the physician and the barber will want to be paid at that point, and in order to have the funds available," borrowed 6½ gold florins from seven villagers. Sad to say, two weeks later, Bonafe lost—both ways—when he was found leprous by a physician from Beaucaire and a surgeon from Tarascon, the "sentence" was announced publicly from the castle of the local lord, and he had to pay the bill.[43]

A dossier in the archives of Nevers allows a speculative estimate of the examiners' compensation. In 1545–1546, the municipal treasurer paid 35 sous *tournois* each "to master Jehan Lecler, doctor of medicine, and Antoine Turpin, master surgeon, for several visits to Nicolle Chardon, wife of Guillaume David, smelter, and Jehanne Brillot, wife of Louis Lefaulcheur, who were suspected of the disease of leprosy"; the same amount "to the master of Saint Lazarus [leprosarium], who supplied the linen, wood, and other necessary items for those visitations." The same cahier records a payment of 60 sous *tournois* to a doctor of theology for preaching the year's Lenten sermons. Even without certainty about the number of visitations and sermons, a comparison of the amounts suggests that the examiners' fees were moderate. It is likely, however, that rates were lower when set by municipal authorities than by the medical practitioners themselves.[44]

Examinations at the Cologne Faculty of Medicine were more expensive, as they were more prestigious; moreover, unlike in many other places, they were usually charged to the examinees. For at least a half century from 1491 on, total fees of at least 3 or 4 gold florins, in addition to at least as much in *albi*, or silver coins, were only the beginning; the examiners also expected food and drink. In exceptional cases of "alleged penury," the faculty granted partial or (rarely) complete exemptions and duly noted them—occasionally with a whiff of self-importance. After finding "the poor housewife Benigna de Heidelbergha" clean, and after collecting 8 marks, the Cologne doctors sighed, "[W]e did not receive anything for the [faculty] seal, because she had nothing to give." From the earliest extant records in 1491, the remunerations were euphemistically termed "customary gifts" (*presentie solite*). In 1519, however, the doctors found it necessary to guarantee these gifts, and they bypassed euphemisms in a sober and punctilious statute.

> By venerable custom, suspects to be examined for leprosy owe the following: First, to each doctor of the college, a gold mark—in gold of the right weight— and one mark to both beadles; with regard to the gifts, it is understood that each of those present will have his share, and that anyone who is absent will receive nothing.
>
> Item, one pound of sweetmeats and the same amount of confectionary cinnamon and jujube, with sugar.
>
> Item, one quart of superior wine for their lordships, and two quarts for the doctors' servants and the domestics of the house where the examination is held.
>
> Item, for the work and clean-up in the kitchen, six *albi.*
>
> Item, to the barber for the phlebotomy, four *albi.*
>
> Item, to each of the doctors' valets, one and one-half *albi.*
>
> Item, they will provide fruit according to the season, and wheat breads for one *albus.*
>
> Item, to the beadle who writes the testimonial letter, two *albi;* and to the dean, six *albi* for the seal.
>
> Item, the doctor in whose house the examination is held, will always have a double fee, that is, two gold marks.

The detail of these provisions, even if partly due to the context of faculty affairs, conveys the impression that, for the profession and public alike, diagnostic precision might become eclipsed by the formalities of examinations.[45]

The Physician as "His Honor"?

The ambivalence between decorum and diligence manifested itself quite clearly, and with a direct bearing on the examiner's trustworthiness, when it came to following up a comprehensive diagnosis with various tests (discussed in Chap. 7). A patient might insist on the performance of "experiments" as a criterion of credibility. In Ghent, when Christoffel Pychaut was declared leprous by the sisters of the Saint Lazarus hospital, he contested their competence because they limited themselves to a visual inspection whereas physicians performed tests on urine, drawn blood, and so on. On the other hand, the Cologne faculty neither mentioned such methods nor, for that matter, recorded the basic inspections of urine and blood which were a normal part of examinations. There are indications, indeed, that learned physicians tended to be reticent about routine diagnostic interventions and that some may have disdained the performance of tests.

Some members of faculties of medicine, who prided themselves on their power of reason and knowledge of the scientific tradition, said little about the thoroughness of their inspection. At Nîmes in 1327, the Montpellier masters considered the signs according to "the great philosophers and masters in medicine," without a word on tests; they concluded their affidavit with the assurance,

> This is our unanimous pronouncement, which we know for certain to be true, and which we masters are prepared to defend with the reasons and authorities of all the philosophers and sages in the science of medicine who have treated most fully of this matter.

In 1458, Heinrich Losser more than once drew a contrast between himself, the municipal physician of Frankfurt, and the dignified doctors of the University of Cologne, about whose judgments he had his doubts. When they cleared Hans Maderus of the suspicion of leprosy, Losser contradicted them by assuring that *he* had been able to recognize an incipient stage, by checking for a loss of sensation "by means of a needle." Losser's skeptical view of the Cologne academics did not mean that he, or other practitioners in the field, underestimated the importance of learning or, for that matter, the value of demonstrating privileged knowledge for securing their credibility. Not only municipal physicians but also surgeons frequently displayed their understanding, especially by explaining the causes of symptoms, when they drew up their affidavits. Nevertheless, the mental distance between Losser and Cologne suggested a ranking in the "honor" of medical judges. In addition, the distance ran parallel to an occasional division between the manual and intellectual intervention in the diagnosis of leprosy.[46]

Surgeons and barber-surgeons normally played an acknowledged—and often a prominent—role as examiners (see Plate 8 and interpretation in Chap. 7). It is all the more striking, then, that no surgeons participated in examinations by the medical faculties of Cologne and Tübingen, even if a barber drew the blood to be inspected. When surgeons did participate in judgments, they might draw varying degrees of condescension from physicians. Doctor Foreest, who had studied at the universities of Louvain, Bologna, Padua, and Paris, repeatedly pointed out that he left every manual act to "some surgeon," whose name he did not remember. "We inspected the entire body diligently, and ordered the surgeon to prick it with a needle. . . . We instructed the surgeon to rub the skin with his nails. . . . We commanded the surgeon to check behind the depressed tongue with a speculum." Thus, Foreest perfectly typified "the more fashionable physicians, spurning surgery," who, according to his illustrious teacher, Andreas Vesalius, "merely exercised a supervision" over manual procedures.[47]

Vesalius complained that surgery had been "handed over to laymen and people with no knowledge of the disciplines that go to serve the healing art." Physicians suspected, apparently more than the general public, that surgeons, with their lack of scientific knowledge and academic diplomas, might pose a threat to medicine and society. Credentials were easily forged by mountebanks who, for a price, would cater to the wishes of a community or authorities and would even dare to contradict a medical verdict. In 1435, the physician Pierre Contier found Louis Bermond clean, but the villagers of Ginasservis (Var) in Provence, determined to see Louis declared leprous, had him reexamined by two out-of-town barbers. These "so-called surgeons," in the words of Contier, pronounced Louis infected with leprosy. With his reputation at stake, not only did Contier maintain that his judgment had been fair and well founded, but he appealed to the Supreme Council, the highest court in Provence. The council ordered Louis to be examined once more, by Christian and Jewish physicians. These physicians agreed with the finding of their colleague. Thereupon, the brazen barbers were condemned to pay Pierre Contier a compensation in the considerable amount of 25 florins 10s. 4d.[48]

Dissension between physicians and surgeons surfaced in an experience that Felix Platter recalled, with characteristic perceptiveness, in his *Observationes.* The poignant story illuminates so many facets of the *iudicium leprosorum* that it deserves to be quoted with minimal abridgment.

There was a young married merchant from Colmar, whose mother we had found leprous in Basel in 1582, as was the case with his sister later. When pustules covered his face, he was declared leprous at the visitation in Colmar, separated from his wife, and sent out of the town. For three years, he lived secluded in a house outside town, in order not to be forced into the leprosarium. Meanwhile, the pustules and redness disappeared from his face, and no signs of disease remained. He went to the magistrate and arranged to be examined anew by the surgeons who had first declared him leprous. These, however, maintained their verdict, probably for fear that they might be sued for their earlier false judgment. There was no outward indication of disease, yet they maintained, among other things, that the blood was infected and teeming with worms.

If the surgeons really made this claim, their ignorance would have been obvious to anyone familiar with the medical discourse on leprosy. In any event, their intransigence—which may be contrasted with the physicians' tendency to hesitate—together with their pretense of authoritative knowledge, forced the young man back into exile for another two years.

During this time, not a single spot or anything else appeared when he bathed diligently and showed himself unclothed to his folks. On their advice, he traveled to Strasbourg for a medical examination, and he was declared free of leprosy. He brought the written testimony back to Colmar and showed it to the magistrate. But the surgeons, when questioned about this, were adamant. They declared that the man must have been carrying something that had the property to make the signs of leprosy less noticeable. Thereupon the authorities, bewildered by these contradicting opinions, requested the Basel magistrate to send me to Colmar for a resolution of the situation, just as they also requested the *lithotomus* from Fribourg.

In defense of the Colmar surgeons, one could argue that, even if they were mistaken, and stubborn in their error, they could not be faulted for suspecting that a patient might attempt to conceal his symptoms.

Be that as it may, the magistrate obviously found the surgeons credible enough to let their opinion carry the same weight as that of the Strasbourg physicians. The confusing draw called for extraordinary referees. The renowned Platter, together with the *lithotomus,* a surgeon (here, again, unnamed) who was probably famous for his special skills in cutting stones from the urethra and bladder, arrived in Colmar on September 12, 1589.

The authorities now let the young man be shaved, given new clothes, and brought into another house, so that he was not able to conceal or contrive anything that, as the surgeons asserted, could hide leprosy. Upon careful examination, we did not observe the slightest sign of leprosy. Therefore we delivered our judgment that the man was not, and had never been, leprous. We reported this to the attending deputies of the authorities. Thereupon the man was immediately sent back to his wife in town and, falling on his knees, he thanked God and us.[49]

Platter's reminiscence reflected the stereotypical—if not *always* inaccurate—view held by university-educated physicians that surgeons possessed limited knowledge, wisdom, and flexibility. At the same time, the narrated events dramatized the daunting difficulties and consequences of judging someone suspected of having leprosy. Examiners could be swayed by the presumption of heredity, the importance of signs in the face, the tendency of the disease to go into temporary remission, fear of losing credibility by changing a diagnosis, the reading of symptoms in the blood, and the suspicion of deceit by patients. Examinations, with exile as the most fateful consequence, also caused anxiety and agony for families, cries for justice in appeals, and social commotion. In other words, diagnosis was only a part of the *iudicium leprosorum*, in which medicine and justice were thoroughly intertwined.

The judicial stature that learned doctors acquired in the diagnosis of leprosy colored the language as well as the format of the procedure. Non-academic practitioners, too, adopted the courtroom vocabulary. When des Innocens set out the requirements for an examination, he sounded more like a judge than a diagnostician. His primary concern was

> not to incur the blame of having wrongfully condemned [*condamne*] a person of leprosy or, on the contrary, having absolved [*absous*] as healthy someone whom the Art will have convicted [*convaincu*] of leprosy. For here is the key of this whole discourse and the foundation of our entire enterprise.

Legal terminology permeated many medical records and guidelines. By means of "a proper inquiry" or "inquest" (*inquisitione*), examiners reached a verdict or "sentence" regarding whether the "suspect" would "deserve" (*mereretur*) to be expelled. The case was "to be pleaded" (*arguatur*) and supported by tests or "proofs" (*épreuves, Probe*) (see Chap. 7). The ultimate concern was less with the illness than with "justice" (*Gerechtigkeit*) or, more emphatically, "right and equity" (*recht und billicheit*). Numerous cases were, in the prefatory words of Guillaume des Innocens, "full of ambiguity and reasonable doubt." In such cases, the examiners—with the apparent exception of non-medical examiners—suspended judgment: they granted a "reprieve" (*Frist*) on condition of a return visit in six months to a year and, usually, a physician's supervision; the suspect was promised to receive, at that time, "a definitive, categorical sentence," although this promise was not always fulfilled. In cases with firm evidence (or, a cynic might interject, persuasive incentives), the suspect was either "condemned" (*damnatus, verurteilt*) or "absolved" and declared "innocent" (*unschuldig*). The last term is particularly significant because it harks back to the initiation of a case, when a potential patient was not only "suspected and reported" but also "accused" (*accusé, beklaget*), and even "inculpated" (*beschuldiget*), a label associated with guilt (*culpa*).[50]

The theme of culpability was maintained by Varanda when he turned to the signs "that our guilty person [*reus hic noster*] will display once the body is naked." This characterization apparently reached deeper than semantics, pointing to the presumption that the accused was guilty until proven innocent. Thus, when Ambroise Paré drew up a template in 1583, for a "Report of a Confirmed *Lepreux*," to be used by the Paris guild of surgeons, he piled up equivocal and unequivocal signs so indiscriminately that they favored a positive finding without leaving room for doubt or deliberation; in contrast, a template for a certificate of clean health was far more sketchy. Furthermore, the location of Paré's sample certificates underscored the judicial connection. They followed a template for a "Report on Girls, Whether or Not They Are Virgins," in the part of his works (bk. 28) which was intended to "instruct the young surgeon to make a proper report in justice": devoted to forensic medicine, the book concluded with "The Way to Embalm Dead Bodies."[51]

The concern with justice also increasingly colored the medical directives for conducting an examination. A voice of public authority—and even a moral tone—already resounded in an early outline of the examiner's options, drawn up by Guy de Chauliac in 1363. The scheme was adopted by later authors, who tended to raise the judicial and moral tone. Guy taught that, if suspected persons were found to have slight and equivocal signs of leprosy,

> they should be warned personally and secretly to maintain a good *regimen* and to obtain the counsel of physicians, lest they become leprous. If they have many equivocal and few unequivocal signs, they are popularly called *cassatus*, and they

should be warned sharply to maintain a good regimen under the supervision of physicians, and to stay in huts [*boreis*] or in their own dwellings without mingling much with people, because they are incurring leprosy. If they have many equivocal and unequivocal signs, they should be separated from the people and brought to the leprosarium, with kind and comforting words. If they are healthy, however, they should be absolved and sent to the authorities with a letter from the physicians.

The term *cassatus* calls for elucidation, since it might convey an erroneous impression about suspected patients who were given a reprieve. The word is readily interpreted as "broken" (*cassé*), as it was in the Middle English translation, "fordone or destroyed," in *The Cyrurgie of Guy de Chauliac.* This would imply, however, that the populace gave up on people whom physicians considered still curable. It is more plausible that the "popular" term *cassatus* in Guy's milieu meant something along the lines of "placed under house arrest," that is, confined to a hut or cabin (*une case*). Guy's reference to "boreis" inevitably evokes the *bories*, the famous age-old stone huts in Provence, the region where he spent much of his life. Moreover, this meaning is evident in the sixteenth-century French interpretation by Pierre Bocellin, as "reclusion" or solitary confinement.[52]

Bocellin, also a surgeon and Montpellier alumnus, simplified and sharpened Guy's outline. He intensified the examiner's first option, of admonishing the patient at the onset of the disease, into a "comminatory," or threatening, verdict. In the early seventeenth century another Montpellier author reinforced the impression that the synopsis pertained more to patient guidance than to medical deontology. Varanda—the "great physician of our time" whom Guillaume des Innocens endeavored to emulate in Toulouse (and from whom he borrowed liberally)—rearranged Guy's scheme, enhanced the moral tenor, and projected a more sanguine outlook. The examiner should grant "absolution to people who have been accused of being leprous, on account of a stupid and senseless suspicion, or envy, while they are totally free of this plague." The opposite outcome was the sequestration of confirmed patients who, "comforted by kind words, are ordered to withdraw from human intercourse and to keep themselves in special public hospitals, so that they do not infect others, and so that they may take care of themselves in the solitude of the place and the salubrity of the air." The genteel conclusion underscores Varanda's idealized view of the *iudicium leprosorum*, which seemed almost detached from the complex reality of the medical examination—all the more since he was writing when the official visibility of leprous patients had become insignificant.[53]

André Du Laurens, Varanda's contemporary, moralized most explicitly, while also advancing the strongest claim of medical competence in conjunction with judicial authority:

> Sequestration should not be enacted unless someone is first judged and condemned under the title of *lepra* by physicians and surgeons who, on account of the practice and experience of medicine, are the supreme judges in this tribunal, and are able to condemn to civil death. It is necessary, therefore, that they have extensive experience with this disease and perform the task with deliberation, lest they violate their conscience by separating a healthy person from the company of others without grave cause. Just as this is an act of impiety, it is inhuman cruelty to allow a leprous person to keep the company of healthy people.

The expression "judged and condemned under the title of *lepra*" (*Leprae nomine judicetur et condemnetur*) is more evocative of the judge's bench than of the doctor's clinic. At the same time, it invites us to inquire into the significance of this "title" or "name of *lepra*," and it whets our curiosity about what preceded the *iudicium leprosorum*. When, in the next chapter, we resume our exploration of the academic treatises and judicial records, we will find that the nomenclature of leprosy was more extensive and varied than for any other disease. A look at the labels for leprosy should set the medical judgment in the broad context of deep-seated, widespread, and timeless responses to the disease—and to its patients.[54]

CHAPTER THREE

The Many Labels of Leprosy

 We, Thomas de Saint-Pierre, Jean Le Lievre, and Robert de Saint-Germain, masters-regent in the faculty of medicine in Paris, have seen Jean de Bierville, master of arts from the diocese of Reims, who was suspected and reported by certain persons as having *the disease that is named and called lepra.*

We have inspected and palpated him carefully and methodically. Upon examining him thoroughly in accordance with the principles, signs, and conclusions of the Art of medicine, we find the aforesaid Jean de Bierville completely free of, untouched by, and not subject to *the said disease known by the name of lepra.* In fact, we were unable to find on him any true sign of *the disease called lepra.*

In view of this, the said Jean ought not to be disturbed, bothered or disquieted by anyone; nor should his company be feared on account of contagious disease. We are making this known, in accordance with the principles, signs and conclusions of the Art of medicine, to all whom it may or will concern; and we certify it with the present letter over our seals.

The three physicians issued this certificate on October 9, 1411, while meeting at the house of Master Thomas near Notre Dame. Their emphasis on "naming" in some way foreshadowed the reference of André Du Laurens, two centuries later, to a judgment "under the title of *lepra.*" The Paris masters, however, more clearly intended to stress the uniqueness of "the disease called *lepra,*" and they were addressing a suspicion and report rather than referring to a condemnation. Their action, while motivated by academic solidarity rather than mandated by public commission, combined echoes of the university milieu, the judicial climate, and social interaction.[1]

In this succinct and matter-of-fact letter, two sets of reiterations produced an air of lawyerlike forcefulness. One set, hammering away at "the name of *lepra*," underscored the burden on someone suspected of having "the said disease." Another repetition—equally significant, albeit less pronounced—drew attention to the authoritativeness of learned medicine. The physicians insisted that they had applied "the principles, signs, and conclusions of the Art of medicine" to the actual diagnostic examination as well as to their professional declaration. The "Art of medicine" was not an abstract phrase but the standard title of the university curriculum (*Ars medicine* or *Articella*) and, more comprehensively, of the body of learning which had become canonized, primarily, in the works of Hippocrates, Galen, and Avicenna. Some readers may find it surprising, especially for 1411, that these three practicing Paris masters of medicine were members of the clergy (one a priest).[2] In view of this fact, one may be even more surprised that the certificate is free of moralization—except, perhaps, for the admonition that one "ought not" to harass and shun Jean de Bierville. Indeed, the bare reference to the disease, without adjectives, stands in contrast with the value-laden terminology and the emotional modifiers we will encounter in some other documents and, occasionally, in the core of medical teaching.

The Power of Words
Rumors and Reputation

By the measured tone of their testimony, Thomas de Saint-Pierre and his colleagues might cause us to overlook the turmoil that had led to the examination. While it is safe to assume that Jean de Bierville was concerned with his health, it is important to recognize his real reason for visiting three physicians on October 9, 1411. He urgently wanted their help after being "suspected and reported by certain persons as having the disease." The practice of reporting suspected patients deserves special attention because it reveals a social setting that is largely invisible from an academic perch. Such "reports" pass almost unnoticed in medical treatises about leprosy, whereas they are amply documented—and often characterized as "accusations"—in health certificates, as we have seen in the preceding chapter. Their accusatory thrust was evident in a case that occurred in relative proximity, in space and time, to Jean de Bierville's problem in Paris. On March 18, 1416, the magistrate of Auxerre wrote to Master Johannes de Molino, licentiate in medicine,

> We have learned, from a report of the council proctors of the parish church of Chitry, that Thienotus Richo, dwelling in the said place, is suspected of leprosy, and bitterly denounced because he could pose a very great danger to the local inhabitants. . . .
>
> Therefore, with justified confidence in your experience, trustfulness, and diligence, at the behest of the proctors, and in keeping with our office, by this letter we call upon you to examine the above-named Thienotus, in accordance with what you have sworn before us on God's Holy Gospels.[3]

Thienotus turned out not to be leprous, after having been so "bitterly denounced." The magistrate's worried phrase serves as a reminder that being "suspected" and "reported" had implications that reached far beyond public health and epidemiology. Biblical echoes as well as modern testimonies may deepen our appreciation of these implications.

The Bible, in the Latin version that became known as the Vulgate, influenced certain medical responses to leprosy, as will become apparent in the following pages. The most consequential biblical influence emanated from chapters 13 and 14 of Leviticus, the book of laws and rituals for the priestly tribe of Levi. This influence was less visible in academic discussions than in official documents. Levite laws on ritual impurity prescribed that anyone with signs of *lepra* had to be "led" or "brought to the priest" (*adducetur ad sacerdotem*) (Vulgate Leviticus 13:2, 7, 9, and 19). Echoes of this command reverberated across Europe in certificates that sealed the fate of people after they had been reported to ecclesiastical or municipal authorities. The standing injunction, to "bring in" suspected patients, opened the door to ulterior motives. This factor undoubtedly affected the numbers of reported cases, even though motives such as animosity or vengefulness were rarely revealed until more recent times, when victims of such suspicions at last found their voice.

It is instructive to hear a modern victim reminisce, in 1978, about being reported to the authorities in Honolulu when he was twelve years old.

> We had no choice in going to see [the doctor]. I was reported, turned in to the Board of Health by my sister's ex-husband. It was a family matter, and he wanted to get even with my family. You see, after my sister was sent away to [the leprosy colony in] Kalaupapa, . . . my mother raised her grandchildren. But my sister's ex-

husband wanted the children. My mother wouldn't give them up. So, he turned
me in as a kind of revenge.

A hemisphere and more than five centuries away, malevolence played an equally
insidious, if less forcible, role in bringing a suspect to the examiners. On June
22, 1419, three members of the Faculty of Medicine in Cologne officially
recorded their finding after examining an "honorable man," whose name they
did not reveal. The man had "humbly subjected himself" to the examination
after being "defamed [*diffamatum*] by several of his rivals, by certain priests
and clergymen, and by some commoners and laypersons who were induced
by we do not know what spirit." After examining him "diligently and thoroughly," the doctors testified that the man was "very healthy" and that "he
should be taken by any sane person, and reputed [*reputandus*] in every way, as
being clean."[4]

Key terms in this testimonial illuminate the intertwined and nettlesome
corollaries of asserting that someone had leprosy. The meaning of *defamatus*
or *diffamatus* deteriorated, from having one's "fame" or reputation diminished
by an adverse report, to being denounced, and ultimately to seeing one's
name destroyed by slanderous charges. In a process that has occurred for
contagious diseases throughout history, civic vigilance could turn into unciv-
ilized persecution, and disease into guilt. This development was part of the
background when commissioners in Geneva examined a brother and sister
who were "suspected and inculpated [*suspecti et inculpati*], by their fellow parishioners and neighbors, of having leprosy." With the same connotation,
examiners in Constance agreed to exculpate ("excusamus") a woman whom
they had seen as "diffamatam" by her fellow citizens. In his grievance against
the physician Heinrich Losser, Hans Maderus complained to the Frankfurt
city fathers that he had been charged ("geschuldiget") with leprosy. He demanded compensation for his "grief and defamation" (*schaden und lümodt*),
and he drew attention to the social consequences by claiming that he had been
"defamed and avoided by people" (*belumut und von den luden gemeden*).[5]

The word *lümodt* (modern German *Leumund*), and the various derivations
that occur in the documents, have a plausible, albeit partial, match (and possibly a cognate) in the English word *blame*. Thus, arguing that Hans Maderus
had only himself to blame, Doctor Losser retorted that the plaintiff had "defamed" (*virlumodt*) himself by his importuning—although his misshapen face,
too, had "given him a bad reputation [*virlumodt*]." Reputation consisted of
several elements when it was at stake in charges of leprosy. According to the

Vulgate Leviticus (13:15), certain signs made it imperative that a person "shall be reputed [*reputabitur*] among the unclean" (we may note the divergent translation in the King James Bible, "pronounce him to be unclean"). It is difficult not to hear an echo of the biblical injunction in a verdict, issued by the *leprosi* outside Cologne, that a resident of Sonsbeck was "unclean and to be reputed [*reputandus*] leprous"—the judgment was overturned by the Cologne Faculty of Medicine in 1548. Judging putatively, that is, on the basis of what one tends to think (*putare*), gave free rein not only to prejudice, ulterior motives, and hearsay but also to groundless suspicion. The ominous potential became obvious when some terms were combined—for example, when the mayor of Nierstein sent a woman to Cologne for examination because she was "suspected and reputed to be unclean" (*verdacht und verlumdt, daz sie nit reyne sy*). In 1570, a resident of the French town of Ganges (Département de l'Hérault), about fifty kilometers north of Montpellier, voluntarily ran the risk of being found leprous when he requested "to be examined because, as he said, he did not want to remain suspected like this [*ainsy soubsonné*] any longer."[6]

Suspicions and reports were especially distressing when they flowed from anonymous sources. Conrad de Dannstet, physician of Duke Leopold of Austria, hinted at the layered opaqueness of hearsay in 1380 when he testified in favor of Peter Pirchfelder, a monk "whom several *suspicious* people have *allegedly reputed* to be infected with leprosy." The anonymity of the denouncers seemed to intensify the anxiety of Arnal Isern in Andorra when he sought to have his wife examined in La Seu d'Urgell because she had a facial lesion "and *some people* say that it is leprosy." The climate of allegations was captured most routinely in references to "rumors." The landgrave of Hessen sent his servant Iorg to the Frankfurt physicians because he was the subject of rumors ("in gerucht"). In the more common expression, someone was "rumored" (*beruchtigt*) to have the disease. Rumors gained strength by spreading. Examiners and notaries were aware of this, as we may gather from some of their stock phrases that were akin to the expression "hue and cry." The curate and parishioners of Saint-Croix in Lausanne reported Johannes Giliet to the magistrate because "the clamor and public rumor [*vox et fama publica*] ran, in the city of Lausanne and in the surrounding places," that he was infected. A man and two women came to be examined in Cologne "on account of the clamor of common talk" (*uss beruchtigung gemeiner Fame*). The consuls of Ganges sent Guilhaumes Vézye to their physician and surgeon because he was suspect "according to rumor, clamor, and common talk" (*suyvant le bruict voix et femme pub-*

licque). Official certificates, which could be displayed by examined individuals, were intended to counteract the spread of rumors. Occasionally, authorities might prescribe a more drastic remedy—for example, when the chapter of the Würzburg cathedral ordered a parish priest to "pronounce, publicly, from the pulpit of your church, that Margaretha Renner has been examined and is not in the least infected."[7]

Popular and Learned Names

The factors that played a role in suspecting, reporting, and judging a person of "leprosy" were as complex as the repercussions were far ranging. These factors included the power of words, which loomed in the names, definitions, and characterizations of the disease. A sixteenth-century French author highlighted this inherent power by his observation that people were terrified "just by hearing the name of the disease."[8] On a more intellectual level, medical language in action was illustrated in Jean de Bierville's bill of clean health. A salient feature in the Paris certificate was the physicians' repeated use of silent quotation marks in the expression "the disease that is named and called '*lepra.*'" Their quotation marks underscored a scientific category, while implying disregard for the equivalent but, most likely, different label that lay detractors would have applied to Jean (perhaps we may infer, therefore, that his detractors were university colleagues). Crossovers between learned and popular terminologies for leprosy were not uncommon in official documents. In treatises, however, they were less frequent than for some other notorious diseases, such as syphilis (*lues venerea* and *mal franzos*) or epilepsy (*epilepsia, morbus caduceus,* and *mal Saint-Jean*). In general, while learned terms originated in pathological phenomena and in scientific literature, vernacular names sprang from social manifestations, shared impressions, and pious traditions.

One popular designation, which was more common in sermons and hospital documents than in strictly medical sources, linked leprosy to Lazarus of the Gospels. The association sprang from the fusion of two New Testament figures. One, actually named Lazarus, was the brother of Martha and Mary and the friend of Jesus; he was raised from the dead by Jesus, after he had been buried for four days (John 11:1–44). The other was a fictitious person, featured as "Lazarus" in a famous parable (known for centuries as "Dives et pauper"); a beggar, "covered with ulcers," he was callously ignored by a rich man but carried into the bosom of Abraham by angels (Luke 16:19–31). The composite Lazarus became one of the most widely venerated saints, giving his name to many leprosaria—as did his even more beloved sister, Mary Magdalen (ubiquitous in France, as Sainte Madeleine). In a recent work, one historian has argued quite convincingly that the association with Lazarus gave greater dignity to poor patients. In any event, the saint was invoked for the comfort of suspects when they were about to be examined. Guy de Chauliac reminded the practitioner to begin the protocol by assuring the examinees that, "should they be found to be leprous, it would be a purgatory of their soul; and if the world hates them, God does not: on the contrary, He loved leprous Lazarus more than others."[9]

The derivation of Lazarus, *lazre,* which appeared as a common noun in French texts after 1150, mutated into *ladre* by the second half of the thirteenth century. Medical authors who still used the noun much later tended to evince a greater closeness to a lay milieu than to an academic setting. This was the case for the Toulouse surgeon Guillaume des Innocens, who, in his *Examen des éléphantiques ou lépreux* (1595), tended to employ the terms *ladre* and *ladrerie* when his discourse moved from theoretical to practical subjects. The synonymy of medical terms, which were rooted in academic and ecclesiastical usage, and popular appellations, which were derived from folk notions, remained worthy of notice as long as Latin alternated with vernacular languages in treatises and records. In the sixteenth century, for example, examiners in Lille referred to leprosy as "lepre ou ladrie" or, more explicitly, "leprosy, commonly called *ladrie*" (*lepre, vulgairement nommee ladrie*). By this time, however, *lazars* as a group were becoming causes of alarm in some areas, rather than objects of charitable piety; and they were of concern to government rather than to medicine. In 1560, Duke Wilhelm of Cleves referred to the generic leprous beggar as "the lazar" (*der Lazarus*) in an ordinance that was not related to public health but aimed at stemming an epidemic of begging. Around 1600, the Montpellier professor Jean de Varanda explained, with some academic aloofness, that leprosy "is popularly called the disease of Saint Lazarus, under whose protection these patients are thought to be placed, because we see them dwell in hospitals far removed from the edge of towns, and beg for alms."[10]

In the twelfth century, a climate of imaginative piety and charitable sentiments proved receptive to another term of endearment for people who were the poorest and most miserable among sufferers. The Latin noun *misellus,* a diminutive of *miser* which might be translated as "little wretch," gained currency, perhaps as an equivalent for the Arabic *miskîn.* It became the root of various vernacular names for leprosy as well as for its patients, and in medical texts as well as in colloquial usage. In an early thirteenth-century French translation of a Latin Salernitan treatise, the disease was referred to as "lepre," but the patient was referred to as "mesel" (plural "meseaus"). *Lepre* remained the choice of French translators of medical writings such as Bernard de Gordon's *Lilium medicine* (translated in 1377), but a century later Nicolas Panis preferred the term "mesellerie" when he translated Guy de Chauliac's *Chirurgia magna*. As for lay witnesses, Arnal Isern's neighbors in the Pyrenees were whispering that his wife suffered from "meselia." In the piedmont of the Erzgebirge (Ore Mountains), Hartmann von Aue's "poor Henry" was stricken by "miselsuht," literally "the disease of the *misellus.*"[11]

Different realities and sentiments gave rise to a uniquely German label that was coined and applied to leprosy later than *misellus* and its derivatives. The increasingly pronounced social fate of the patients accounted for the Middle High German neologism *ußsetzikeit* (modern German *Aussatz*). The literal meaning of the compound word was "the quality of being set outside," or "separateness": as the Marburg professor Heinrich Petraeus explained in 1615, "[I]n our German it is called *der Aussatz*, because those infected with this contagious disease are removed from the company of other people." In this sense, the label echoed the Mosaic injunction that the infected should be segregated ("separabitur") in the Vulgate. The vernacular term became more technical from the thirteenth century on, and it occurred in numerous official documents, including those drawn up by physicians. For example, in his formal rejection of the complaint of Hans Maderus, Heinrich Losser denied that he "had made an accusation of leprosy" (*geschuldiget han der ußsetztzekeit*). It would seem, however, that Losser adopted this word mainly when quoting someone else.[12]

When speaking for himself, Doctor Losser employed a vernacular name that monopolized the generic word for disease; this name, and its equivalents across linguistic borders, seemed to serve not as a euphemism but as hyperbolism, by singling out leprosy from among all the ailments. In his most solemn letter to the men of the Frankfurt council, in 1460, he declared that they had commissioned him "to examine people for *the* disease" (*dy luden czu besehen in der maladie*). In 1451, Archbishop Dietrich von Mainz requested the Cologne faculty to check whether a vassal of his was infected by "maletz" or "*the* disease" (*die maledij*). The infected became known simply as *malades* or *siechen*. The label for patients grew into *Melaten*. Melaten also became the name of Cologne's major leprosarium, which was established outside the city walls in the late twelfth century—on the site of today's famous cemetery. Many of the French hospices for leprous patients were known as *maladreries*. As if calling leprosy "the" disease were not forceful enough by itself, double names sometimes entailed odd redundancies. This was the case when the mayor of Annenberg sent a burgher to Frankfurt who was rumored to be suffering from "der malaczen sucht," or, in an admittedly facetious translation, "from the disease of malady." In a similar duplication in Flemish, Pierine Bruneel was admitted into the hospital of the Magdalene in Brugge upon being found "beziect ende melaetse," as if to say, "sick with malady." The pairing of names might indicate emphatic formality, linguistic ambiguity, or, occasionally, a transitional stage. Thus, Johann Huggelin wrote of "Aussatz oder Malatzey" in 1563, at a time when the latter name was becoming obsolete.[13]

In comparison with the proliferation of vernacular names, which bewilders linguists, the history of the Latin term seems simple and straightforward. This impression is superficial, however. The Paris masters in 1411 may have presented "the disease that is named and called '*lepra*'" as an immutable formula, but, ironically, if they had written their affidavit a century later or three centuries earlier, they might have hung virtual quotation marks around an entirely different, unrelated word. To be sure, the label *lepra* drew standing from its parentage, in an almost accidental meeting between medical and biblical traditions. It occurred in the Hippocratic writings, including the *Aphorisms*, which constituted a wellspring of Western medicine—but which were also concise and generic enough to leave ample room for interpretation. For example, listed among vernal diseases in *Aphorism* 3.20, the plural λεπραι covered a wide range of skin disorders that turned the skin rough, scaly, or flaky and which were comparable to today's psoriasis or eczema. While these ailments of the skin were "relatively benign," they became associated in the Aristotelian tradition with a repulsive lack of personal cleanliness. They were further characterized (in the pseudo-Aristotelian *Problemata*, for instance) as affecting the surface and therefore spreading by contact.[14]

The generic Hebrew term for skin diseases, *Zarã'at*, found a logical match in λεπρα when the Bible was translated into Greek, in the Septuagint. The Latin equivalent, *lepra*, was an equally natural choice for Saint Jerome when he produced the Vulgate. In the process, the translations and their wide dissemination stirred associations that far exceeded the province of bodily health. The venerable Mosaic laws, particularly in Leviticus, highlighted a disease that was not only contagious, repulsive, and related to uncleanness, like scabies, but whose diagnosis was equivalent to a judgment and whose chief remedies were ritual. The successive Greek and Latin translations of *tameh*, the

Hebrew word for "unclean," into ἀκάθαρτος and *immundus* fostered a shift
from the ritual sense to moral implications, which one may arguably discern
in terms such as *impure* or *dirty*. The Gospels, in addition, enhanced the image
of a dramatic affliction, whose only cure was by miracle (for example, the
"cleansing" of ten lepers in Luke 17:12–19). These associations colored perceptions of leprosy across the cultural spectrum because of the pervasive influence of the Bible. It is important, however, to realize that this influence
on medical learning and procedures was mainly indirect and latent, particularly in Latin writings. While examining hundreds of documents written by
physicians and surgeons, I have come across surprisingly few explicit references to Leviticus. In one of these, in fact, the biblical *"lepra Iudaeorum"* was
expressly set aside because "it was evidently not the result of a defect of the
spleen and liver, but a singular scourge of God; and it struck not only humans
and animals but also inanimate skins, rugs, and linen vestments" (a reference
to Leviticus 13:47–59). Mosaic and medical *lepra* were most likely to be linked
in vernacular contexts. The general tenor of these associations becomes evident in two representative instances, one in a popularizing treatise and the
other in an idiosyncratic certificate. Both of these instances, it should be
noted, lacked the charged biblical characterizations of leprosy as "a plague,"
"unclean" and "germinating," and "sin" or "trespass."[15]

   In his synoptic treatise *Practique sur la matiere de la contagieuse et infective maladie
de lepre,* the Savoyard surgeon Pierre Bocellin referred to scripture for a rhetorical expansion rather than logical support of his declaration that, "by definition," leprosy was contagious. Without explaining the premise of this definition, Bocellin elaborated, "[O]n account of the infectiousness, there is the
command in Mosaic law, that those who are afflicted by this disease must be
separated from the conversation of the healthy." A second explicit biblical
reference, with a quotation from Leviticus, occurs at the end of a certificate
that was issued in 1572 by municipal health commissioners of Ieper. The
physician Joris Aerlebout and the surgeon Esau Boudse wrote to the regents
of the leper hospital that they had examined young Waelraem Caemerlinck
at the request of his father, who was overcome by "the daily sight of his misery and wretchedness." The practitioners reported,

[H]aving seen the entire naked body and palpated it in several places, we have
found the kind of scaly and horrible sores that demonstrate the presence of true
lepreusie notwithstanding the good color of his face. In addition, a big sore of the

head which has lasted for many years, has come to such corruption that it will
not be cured in three or four years even if it is taken care of with all diligence.

Then the *physician* continued, switching from Flemish to Latin,

> And the very loathsome lesions and deep corrosions of this kind over the whole
> body seem not to be called *lepra* proper by the more learned men. Nevertheless,
> I see no apparent difficulty in the text of Sacred Scripture for one who reads and
> considers it correctly. Indeed, the text says, "when a man or woman has *lepra* arise
> in the head or beard, the priest shall see them; and if it is deeper than the rest of
> the flesh, and the hair is yellow and thinner than usual, he will mark them as
> unclean, because it is *lepra* of the head and beard." If such words cannot be con-
> tradicted because they are sacred, we, too, can regard this raised and deep lesion
> as *lepra.*

The physician appealed to Leviticus for practical reasons rather than medical principles. His real purpose in quoting the biblical lines was to support
his ad hoc (and avowedly unconventional) definition of leprosy, so that it
would be applicable to the case at hand and justify the boy's admission to the
leprosarium—as an act of mercy rather than exclusion. From the late 1500s
on, as we shall see, more examiners were prepared to identify different ailments as leprosy and, depending on the circumstances, to intensify or dilute
the power the label had acquired over the centuries. In fact, the label turned
out to be a case of mistaken identity. In order to untangle this confusion, it
is necessary to step back far enough for a full perspective.[16]

A Tangled Terminology

At the beginning of the Common Era, even as the term *lepra* began to spread
with the expanding orbit of Latin, an affliction far more serious than the
Hippocratic λεπραι gained notoriety. There is no need to reach for retrospective diagnosis in order to realize that descriptions of this newly noted
affliction corresponded, individually and cumulatively, far more closely to
the manifestation of Hansen's disease than to the skin conditions in biblical
and Hippocratic sources. Described symptoms ranged from discolored or
darkened embossments over the whole body to facial deformities and loss of
extremities. In the late Roman Republic, the naturalist Lucretius was vaguely

aware of the disease's existence in Egypt. Early in the first century CE, Aulus Cornelius Celsus reported, in his book *De medicina*, that it was a chronic ailment, "largely unknown in Italy and very common in certain lands." Pliny the Elder wrote of the condition as a new disease that was "native to Egypt" and "quickly died out in Italy." Citizens of the eastern provinces, particularly Aretaeus of Cappadocia, characterized it as causing horror and fear of infection. Rufus of Ephesus, as quoted three centuries later in the *Collectio medica* of Oribasius, wondered "how such a serious and common disease escaped the notice" of the ancients, even of "men capable of pondering everything in the tiniest detail." Celsus, Pliny, Aretaeus, and Rufus all used the Greek name of the disease, *elephantiasis*, which the Latin authors adopted or, occasionally, transformed into *elephantia*.[17]

In the second century, Galen of Pergamum reported that *elephas* was "widespread in Alexandria [Egypt] but very rare in Mysia [Asia Minor] and nonexistent in Germany and among the Scythians [north of the Black Sea]." However, it is far from certain that he was speaking *either* of what became known as leprosy *or* of what is still called "tropical elephantiasis," in which the Filaria worm causes the swelling of the extremities and the scrotum. In fact, the voluminous and influential works of Galen were the source of much confusion. They contained some thirty-one mentions of *elephantia*, with barely a passing description or definition. Even worse, these alternated almost indiscriminately with thirty-two mentions of *lepra*. The confusion deepened in the ambiguities and inconsistencies that marked many of his discussions, and it was compounded by the vicissitudes in the transmission of manuscripts and in the translations of the Greek originals. Moreover, some floating traditions gained stature by being ascribed to Galen. For example, according to the spurious Galenic treatise *Introductio seu Medicus*, "some of the older authors" had proposed a confusing division of *elephantiasis* into six skin ailments, one of which was *lepra*.[18]

The potential for confusion becomes apparent as soon as one considers two successive entries in Galen's glossary, known as *Definitiones medicae*.

CCXCV. *Lepra* is a change of the skin into an unnatural condition, accompanied by roughness, itching, and pain, and sometimes but not always by scales that are shed; occasionally it also wastes various parts of the body.
CCXCVI. *Elephas* is a condition that makes the skin thick and uneven; it causes a livid appearance of the skin and the whites of the eyes; the tips of the hands and feet are swollen and emit a livid and fetid sanies.

If overlapping definitions set the stage for a blurry nomenclature, matters were complicated by the fact that, in several of his works, Galen mentioned *lepra* and *elephas* together in the same sentence. In these instances, he suggested no distinguishing traits but, on the contrary, indicated identical features, including their essence as "melancholic diseases" of the skin, and their origin in bad diets.[19]

Notwithstanding the incipient confusion, medical authors continued for nearly another millennium to call the syndrome, which Aretaeus had described in detail, by the name that he had given it, *elephantiasis.* The meaning of *lepra,* as bequeathed by the Vulgate and cultivated in miracle stories and moralizing imagery, did not affect Latin nosology until the eleventh century. Meanwhile, the Aretaean sense of *elephantiasis,* though with its pathos muted, prevailed among disease categories. Even when biblical exegeses and illustrations presented *leprosi* as cripples, and therefore as suffering from more than skin afflictions, early medieval medical compilations kept applying the label *elephantiosi* to patients fitting the description by Aretaeus. This is the case for texts in two now celebrated manuscripts, the eighth-century "Lorscher Arzneibuch" (in a Bamberg codex) and the mid-twelfth-century "Treatises on Medicine" (at the United States National Library of Medicine). At the same time, medical authors maintained the term *lepra* for superficial ailments, akin to "stains." In Salerno, Gariopontus still treated *elephantia,* in the Aretaean sense, as a major disease and worthy of a chapter in his compendium that became known as *Passionarius.*[20]

It was in the Salernitan orbit, however, that the first signs of transition emerged, together with the first advance of Arabic influences. Constantine the African is credited with the translation of a treatise on leprosy which was probably written in Cairo by Ibn al-Jazzār (fl. 1000). While this translation acquired the Latin title *Liber de Elephancia,* the term *lepra* was substituted in two of its manuscript copies. In the *Viaticum,* which was Constantine the African's translation of Ibn al-Jazzār's *Zād al-musāfir,* the major disease was called *lepra;* it was given a "fourfold" division, with *elephancia* as one of the forms. The Salernitan master Bartholomeus (fl. 1120) mentioned "*elephantiasis,* which the common people call *lepra.*" Within decades, Johannes Platearius and Gilles de Corbeil adopted *lepra* as the principal disease with four subspecies, among which *elephantiasis* still ranked as the gravest and bore the Aretaean symptoms. One salient marker of the changing usage between the eleventh century and the twelfth was the adaptation of a formula that had been copied for at least three centuries. Gariopontus declared, "We recognize *elephantiosos* from spots, resembling impetigo, which appear in the beginning." A century later Johannes de Sancto Paulo, also at Salerno, revised this into "We recognize *leprosos* from spots, resembling impetigo, which appear in the beginning."[21]

The rise of the term *lepra* in medical usage was no doubt boosted by the cultural environment of biblical authority and Vulgate vocabulary. The label's primacy was sealed, however, by the Latin translations of Arabic encyclopedias, above all, the *Canon of Medicine* of Avicenna and the *Liber nonus ad Almansorem*, the ninth book of a compendium by Rhazes, which covered diseases in a head-to-foot order. These encyclopedias presented *elephantiasis* primarily as an affliction of the legs, and quite secondarily as a stage or a manifestation— one of several—of leprosy. They covered a profuse variety of skin diseases, many of which found no ready counterpart in Latin or even in the European experience. One label, for example, *al baras*, slipped into Latin untranslated, and it drifted back and forth between chapters on leprosy and others on various skin diseases.[22] The Arabic writings presented a rich vocabulary that bewildered the translators. At the same time, they largely reduced the identity of *elephantiasis* (*al-judam*) to the syndrome that has retained the name until now. As a result, the default term *lepra* offered manageable simplicity in the expanding nomenclature of Western medicine. In spite of its semantic spuriousness, it remained the dominant term in Latin and spoken languages, in learned treatises and official documents, until modern times. It even withstood an eloquent resurgence of the historically more accurate name, *elephantiasis.*

The return to *elephantiasis* was arguably typical of Renaissance humanism, with a renewed interest in language, a pronounced criticism of immediate predecessors, and an express return to classical sources. In 1545 Paulus Iuliarius of Verona lamented,

> [P]hysicians have written in so many ways about *lepra* that we can hardly recognize what it is. The moderns have mistakenly followed the Arabs, who greatly misuse names and have created enormous confusion in the identification of these kinds of diseases. Now they confuse everything because they believe that, when the ancients described *elephantiasis*, they were speaking of *lepra*. They do not understand that the disease *elephantiasis* (which, according to Pliny, attacked no one in Italy before the time of Pompey the Great) is one disease, and *lepra* another.[23]

As an indication that this was a time of transition, we may note that Iuliarius still used the label *lepra* in his title, *De lepra et eius curatione*, even though he identified the disease as the classical *elephantiasis.*

One year later, in 1546 (sixteen years after he wrote his famous treatise on syphilis), Girolamo Fracastoro raised an additional terminological issue. He scolded recent writers who, "with irrational obstinacy, have recorded their opinion that *elephantia* and the French Sickness are one and the same disease." He claimed that they were "led astray" because they insisted

> that the ancients wrote on *lepra* and *elephantia* as two distinct diseases. Hence they thought that the ancients meant by the word *lepra* the disease which we now commonly call leprosy, and they did not know what *elephantia* could be.

When the Greeks wrote about *lepra*, Fracastoro explained, they

> meant another and far milder affection which they discuss along with psora, i.e. scabies. This affection may indeed lead to *elephantia*, which, strictly speaking, is that ailment which not only the common people but also recent medical writers, whether Latin or Arab, call *lepra*.

In 1547, at some distance from Fracastoro's critical thinking, an ambiguous certificate by a Flemish *medicus* exemplified the continuing confusion. Jacobus vander Burch wrote to the regents of the leprosarium of Ieper that, after careful examination, he had found an orphaned young man to be

> leprous in accordance with the description and interpretation of *lepra* by Galen and all the other ancient and most learned doctors, even though he is not *elefantiacus*. *Elefancia* is what Avicenna and some Arabs call *lepra*, and something we do not have in this region, but it is also that disease *elefancia* of which we have found certain signs in Pierkin de Hooge.

Compounding the potential for confusion in documents from Flanders, the adjectives *lepreus* and *leproos*, equivalents for "leprous," could be stretched to include individuals with a contagious disease who were to be separated, as in this reference to syphilis, "lepreus ex morbo gallico." The humanist Jodocus Lommius managed to avoid the ambiguities, both of a dual terminology and of indeterminate usage, by employing exclusively the term *elephantia* when he described the symptoms of leprosy in 1559. A generation later, Pieter van Foreest (1522–1597), who had studied medicine in Louvain, Padua, and Paris, summed up the received but still open-ended terminology. After devoting a chapter to "*elephantia* and *lepra* of the Arabs," giving the impression that they were separate entities, he recorded a consultation "for a very noble eight-year-old girl," on the subject of "*lepra* of the Greeks which, for some expert authors, is a very bad kind of impetigo."[24]

Fully unraveling the linguistic and historical entanglements required more time. André Du Laurens (1558–1609), chancellor of the University of Montpellier and physician to Henri IV, observed that, while *elephantiasis* was "the more regular appellation among the Greeks," it remained an "ambiguous" term. In addition, the Greek noun with which the Arabs replaced it, namely, *lepra*, referred to "a particular affliction of the skin for Hippocrates, Galen, and Paulus Aegineta," and it corresponded to "the *albara nigra* of Avicenna, and the *impetigo* of Celsus." Further confusion might result from Galen's teaching "that *lepra* can degenerate into *elephantiasis* and, elsewhere, that *elephantiasis* can abate and turn into *lepra*." Even after concluding that, "to untie the knot with one word, the *lepra* of the Arabs is the *elephantiasis* of the Greeks," Du Laurens continued to use the now supposedly discredited term. His vice-chancellor at Montpellier (and, later, dean), Jean de Varanda, was more resolute in adopting the Greek replacement. In a straightforward summary, he eliminated all ambiguity:

> *Elephantiasis* signifies that disease that the Arabs, Latins, and the common people of nearly every nation, have called *lepra* or leprous condition: quite different from the *lepra* of the Greeks, which is an infection of the skin only, and which can either precede or follow *elephantia*, as Galen teaches.

When he described the disease, Varanda adhered closely to "the very ancient author, Aretaeus," who had been forgotten from the fifth to the sixteenth century. One and a half centuries after Varanda, the Montpellier alumnus François Raymond, writing in French, methodically followed and elaborated the terminology and description of Aretaeus, "a great painter of nature," whom he read in the original Greek.[25]

Neither the erudite return to Aretaeus nor terminological correctness was able to dislodge *lepra* from its entrenched position, or even to introduce *elephantiasis* as an alternative in official usage. Well into the eighteenth century, scholarly physicians attempted to maintain a comprehensive nomenclature. In 1733, professors of medicine at the University of Louvain in Brabant (now Belgium) certified that Nicolaus van Goordaert was "afflicted by that kind of *lepra* which the Greeks call *Elephantiasis* or, in vernacular parlance, the Disease of Saint Lazarus."[26] These attempts, however, appeared increasingly archaic and oblivious to the consent that medicine had reached on the identity of the disease. We may be inclined to consider the eventual monopoly of the noun *lepra* as merely a matter of words, but we should not forget the exceptionally charged meaning of this particular label. When it migrated from superficial ailments to a ravaging disease, the contagiousness and repulsiveness of the Hebrew *Zarā'at* merged with the destructiveness and totality of the Latin *elephantia*. Moreover, as the biblical legacy permeated a substratum of the medical tradition, notions of impurity and judgment tainted some discussions of causes and consequences. This baggage revealed itself, in a rudimentary fashion, in various attempts to interpret the word by rationalization, etymology, or analogy; it was aggravated by the attribution of certain adjectives to the noun; and, in varying degrees, it warped supposedly objective definitions.

Interpretations and Characterizations

Much of premodern medical learning relied on the belief that words were keys to the knowledge of reality and, inversely, that there was a factual reason for every name. The physician Alberto de Bologna (Albertus de Zanchariis), for example, whom Boccaccio deemed "brilliant," professed this belief when he commented on Avicenna's chapters on leprosy. Master Alberto listed nine criteria by which diseases were named and from which one could learn something about their location, cause, appearance, and, presumably, essence. There is some irony in the fact that he drew these criteria from Galen and Avicenna. Galen warned his readers "not to become ensnared in incorrect names but to endeavor to know the thing from its nature." Avicenna's chapters on leprosy evinced little interest in words—with two exceptions that will be noted shortly. Some of the criteria for assigning names to diseases were obvious, such as location for the naming of pleurisy, but others required rationalization. Two favorite tools for reconstructing the reality behind words were etymology and analogy, and metaphor was a major element in both. The reconstructions interwove logic, stereotype, poetic imagination, and reported observation, with the ambivalent effect of both illuminating and dimming the experiential knowledge of the disease.[27]

Isidore of Seville (ca. 570–636), an etymologist whose influence spanned nearly a millennium, assured that "everything can be more clearly comprehended if you know the origin of a word." It was quite common to seek the origin simply in an association between the way the word sounded and some characteristic of the named object. The predominantly fanciful nature of these associations does not preclude their relevance, for they reflected preconceived notions and, at least occasionally, actual observations; moreover, even semantic dabblings could shape further perceptions of the disease. Pierre Bocellin did not wish to dwell much "on the multitude and contradictions of the etymologies," and he preferred "to give the readers complete freedom to follow this one or that one as they saw fit." His convenient summary of "l'ethimologie des anciens interpretes" showed that, according to one interpretation, *lepra* was named "from a bad lesion" (*lesio prava*); according to another, "from a stony lesion" (*lesio petrosa*) because the affected tissues become hard as stones; according to yet another, more dire interpretation, "from the wolf [*a lupo*], because it devours the limbs like a wolf"; or, through a strange late thirteenth-century dictionary entry with biblical references, from the Greek word "*leporia*, which, in our common language, we may interpret as calamity or danger." Bocellin also mentioned the most intriguing speculation, by Guy de Chauliac, who had suggested that the name was derived from the (presumably harelike) "tip of the nose [*a lepore nasi*], because this is where the signs appear sooner and with greater certainty." It so happens that the "destruction of tissues within and around the nose" has always been one of the most distinctive marks of the *facies leprosa*.[28]

Three manifestations, linked to leprosy since Aretaeus, fueled particularly vivid interpretations, in which imagination supplemented observation and tradition. When focusing on *elephantiasis*, Aretaeus had described the elephant in great zoological detail in order to demonstrate how much it differed from all other animals in being so huge, dark, misshapen, and rough skinned—and even in kneeling backward. It is worth noting that medical authors did not develop the prejudicial potential of these differences, even when they adopted and elaborated the comparison between the form of the disease and the appearance of the animal, which Aretaeus had "observed so elegantly" according to Du Laurens. Most of them concentrated on the similarly rough, cracked, and knobby skin. Some, however, embroidered metaphorically or, in the words of Gilles de Corbeil, "by means of trope," on the immensity, force, and destructiveness that *elephantiasis* had in common with the elephant.[29]

Authors took greater liberties with another manifestation that, as Aretaeus reported, was called *leo* or *leontiasis* because, in an early stage of leprosy, "extreme wrinkles of the forehead" made the patient resemble a lion. Rufus of Ephesus, a contemporary of Aretaeus, offered a less accurate and more suggestive explanation that the resemblance lay in thickened lips, collapsed cheeks, and bad odor. Relatively early amplifications presented the stage as "named after the lion because this animal is warmer than the others"; or they depicted patients as "reddish-yellow, like lions." While most of the analogies referred to the forehead, Aretaeus opened the door to metaphor by observing that the patient's "brow contracts so furiously that it covers the eyes, as it does in angry persons or lions." Freely expanding the interpretation of "*leonina*," Avicenna asserted that it "makes the face terrifying" and "holds on to its prey with the rapacity of the lion" and, quite intriguingly, that it "occurs most often to lions." Gilles de Corbeil surmised that the manifestation was named "metaphorically, for the lion's ferocity." Other medical writers proposed that, just as "the lion is a fierce and very fast animal," this form "kills swiftly"; not only did these authors allow poetic license to negate empirical evidence, but they also contradicted Aretaeus, who had noted that the lengthy course of the disease mirrored the longevity of the elephant.[30]

A third manifestation, on which authors speculated most wildly, was the one mentioned most concisely by Aretaeus. The appellation originated in observed facial features, but it metamorphosed into feverish fantasy, with far-reaching implications. Aristotle had taught that, when nature is unable to distribute nourishment proportionately through the body, limbs may "develop irregularly," sometimes to the extent "that not one of them resembles what it was before"; this was similar to what happened in "the disease known as satyriasis, in which the face appears like that of some other creature—a satyr—owing to a quantity of unconcocted humour being diverted into parts of the face." Aristotle was trying to explain facial distortions, including a thickened forehead and protruding cheekbones. With less physiological or descriptive precision, Aretaeus stated that *elephantiasis* was "also called *satyriasis* on account of the red cheeks and the insatiable and shameless desire for sexual intercourse." Imaginations naturally accelerated on the slope of this premise, in spite of early clarifications that this condition was "different from the disease of the genitals called by the same name." In addition, meanings became confused by the inconsistencies of Galen and the misunderstandings of his translators.[31]

From *De accidenti et morbo*, the Latin version of one of Galen's most widely cited works, readers learned that leprous patients, with the dilation of their nostrils and the thinning of their ears, "become like the animal that is called *sathon*, that is, the elephant"! In Galen's treatise *De tumoribus*, translated from the Greek by Niccolò da Reggio, one could read that, at the onset of *elephantia*, swollen temples made the patient look like a satyr, "but sometimes there are swellings around other parts, such as the unnatural extensions of the pudenda which some call *satyriásmus,* but others *priapismus.*" Guy de Chauliac, with his usual knack for summarizing, drew together several linguistic strands, but he also intensified the allusions. Citing Galen, he stated that "patients with *elephantia* (that is, with leprosy) generally become like satyrs. The satyr, or *saton* in the Arabic translation, is an animal of frightful appearance" or, in another sense, "with a horribly fixed stare." Also referring to Galen—but drawing on an intermediary source that I have not yet been able to identify—Velasco de Tharanta (fl. 1418) introduced yet another identification for "the animal that is called *saton,* that is, the badger (*taxus* [presumably for *taxoninus*])." Huggelin added a few imaginative touches by explaining that "the patient's face resembles that of satyrs (animals in India, with a human appearance)" and that patients "have a propensity for the work of nature and love, like satyrs, because they have much sperm." Varanda, more expansively, drew attention to "the distortion of the face, which matches depictions of satyrs, and the ready and insatiable lust for intercourse, which is due to the air and sharp fluid that, usually, abound in their bodies and induce almost permanent arousal." There was no longer any mention of the facial features at the origin of *satyriasis* as a manifestation of leprosy when Du Laurens noted—also with some expansion—that some chose the appellation because the patient had "the male member constantly distended and erect, with incredible itching and desire for venereal congress. They seem to have borrowed the name from the satyrs, which the old poets and sculptors always depicted with an erection."[32]

Medical authors, whether classical, Scholastic, or humanist, gave voice to imaginations that circulated far beyond the literate minority and which included obstinate myths about the excessive libido of people with leprosy. These myths pervaded the vernacular literature and reflected deep-seated fears. They may have influenced medical rationalizations, affected dietary injunctions against sex by leprous patients, and inspired restrictions on sexual activity or punishment for sexual misconduct. Inversely, the speculations by learned practitioners probably reinforced widespread stereotypes. It is difficult, however, to map or measure the interaction between popular notions and the imaginative semantics of authors. This difficulty is even greater with regard to the general medical characterizations of the disease and those who had it, which fell below the level of nosological definitions and diagnostic descriptions, and which Jean de Varanda called "the common qualities" (*vulgares qualitates*). The modifiers employed by physicians seemed hardly different

from the simple predicates in nonmedical sources. Nevertheless, the characterizations merit attention here because, even when they merely reflected
"common" perceptions of leprosy, they added up to a composite image that
was more overwhelming and—with greater consequence—more prejudicial
than for any other disease. Their cumulative impact justifies an attempt to
assess their place in premodern medicine.[33]

Attributes

The qualities attributed to *lepra* in a medical framework fall into three major
categories with somewhat distinguishable though overlapping slants. First,
in the realm of public measures, a range of predicates revolved around a concept of "unclean" which, in varying proportions, combined sanitary, ritual,
and moral concerns. Leviticus provided an authoritative source for the concept and concerns, even if (as we have seen) it was rarely cited in a medical
setting. Second, in the area of popular responses colored by intuition,
impression, and emotion, adjectives ran a gamut of repulsive qualities. Several
of these adjectives, such as *repugnant* and *ugly*, also enlivened the storytelling
of Hartmann von Aue and other poets.[34] In a third category, favored by academic writers, rhetoric and natural philosophy supplied justifications for calling leprosy "bad" in the fullest sense. This category remained largely speculative, except for one particularly threatening aspect of badness that, as we
will see, became a persistent predicate for practitioners as well as scientists
until modern times. In general, medicine seems to have played an ambivalent
role in the accumulation and perpetuation of negative images of leprosy, by
sanctioning and intensifying certain attributions while often mitigating their
prejudicial potential.

   The stigmatization of the *leprosus* as unclean ("immundus"), which was
conspicuous in the Vulgate Leviticus, found echoes in the language of examination reports, though without the ritualistic overtones. The label was absent from treatises, with one notable exception. In her medical discussion of
lepra, in a work titled *Causae et curae*, Hildegard of Bingen (1098–1179) juxtaposed patients "who are unclean [*immundi*] from wrath" with others whose
lust is uncontrolled (*de libidine incontinentes*). As for the records of medical examinations, German documents indicated a particular fondness for the contrast between clean and unclean ("reyn" and "unreyn"); for the fluidity between clean and pure ("reyn" and "sauber"); and—as late as 1701—for the
emphatic effect of pairing uncleanness with impurity ("unreinigkeit und

unsauberkeit"). This apparent propensity surfaced even when the certificates were composed in Latin, for example, by the Cologne physicians who judged the chancellor's nephew to be "totally unclean and leprous" (*immundus omnino et leprosus*) in 1531. To the surprise of the modern reader, these documents mentioned no links between uncleanness and lack of body hygiene. Medical treatises were equally silent about a possible relationship between bodily cleanliness and leprosy. Even Galen connected the two only obliquely, in his observation that "unwashed and filthy" (*illoti et sordidi*) people were predisposed to pruritus, which he presented as a concomitant of *lepra, elephas,* and scabies.[35]

One connection was too inherent to be ignored. The characterization of a disease as unclean automatically raised the specter of a spreading uncleanness or, in more properly medical terms, the complex notion of infection. The layers of this notion found precedents in ancient medicine. They will require closer scrutiny in a later chapter, when we examine the explanations or causes of leprosy. Suffice it here to introduce the layers as they appeared in general characterizations. A single passage in Leviticus 13, arguably the most memorable, combined the potent elements. The terminology of the Vulgate, which was authoritative for most of the period and area examined here, is largely lost in the King James Version or modern English translations. Therefore, I offer the following translation from Latin (with the key words also quoted in Latin) for the sake of faithfulness rather than elegance.

> Whoever shall be blemished [*maculatus*] by leprosy, and separated according to the judgment of the priest, shall have torn clothing, the head bare, and the mouth covered with a cloth. He shall shout that he is polluted and filthy [*contaminatum ac sordidum*], and as long as he is leprous and unclean [*inmundus*] he shall live alone, outside the camp.

Whatever influence this passage may have exerted, it adumbrated the medical categorization of the *leprosus*, which was more outspoken in public documents than in academic treatises. The stigmatization of the patient—as well as the visual recognition of the disease—hinged on the interpretation of *macula* and *maculatus*. Macules raised the first public suspicions, even though their indeterminate nature diminished their diagnostic value. *Maculatus* was a fluid term, with meanings that ranged from "spotted" and "blotched" to "stained" and "soiled" (the last two may be brought into relief by recalling their opposite, "immaculate"). Escalating interpretations would cause a person with blemishes or spots, first, to be suspected and referred to authoritative judgment; then, to be deemed stained, tainted, and contagious; and, ultimately, to be excluded as dangerously unclean.[36]

We have contemplated, earlier in this chapter, the language and ramifications of suspecting someone of having leprosy. A first step toward confirming the suspicion was to recognize the disease, as the Salernitan master Gariopontus taught in the eleventh century, "by the stains [*ex maculis*] that appear in the beginning." Therefore, a positively sick person was designated as blemished, *maculatus, befleckt,* or *entaché.* Inversely, the very absence of "the stain" might be sufficient for clearing a suspect. Thus, "finding no blemish [*nullam maculam*] of the disease" on Margaret Flach, the master and collegium of the poor lepers outside the walls of Constance certified in 1397, in this case without involvement by medical practitioners, "[W]e securely clear her of all and every blemish of leprosy." Usually, however, a subsequent and more crucial stage was determining whether, beyond superficial appearances, the suspect's body was in fact spoiled, tainted, or "infected" by the disease. For emphasis, officials might declare a patient "spoiled *and* infected" (*corruptum et infectum*), a phrase that occurs four times in one document, concerning Johannes Giliet in Lausanne in 1396. Equivalent terms were *attainct* and *infecté* in French, and *bedorven, gheinfecteert,* and *besmet* in Flemish. In German, *befleckt* covered the sense of infected as well as blemished; at the same time, *behaftet,* which originally referred to someone "burdened" with leprosy, also took on the meaning of "infected." Indeed, "infected by leprosy" became the default designation. From at least the thirteenth century, this suggestive designation matched and overtook more neutral expressions, such as "suffering from leprosy." It also became more closely associated with contagion, as we will see in the following chapter.[37]

For now, it is sufficient to point out that many allusions to contagiousness were indefinite. This was the case when Parisian master Jean de Bierville was called "untouched" by the disease, in the quotation that opened this chapter. A similarly indeterminate sense was conveyed by the inquiry of Arnal Isern, in La Seu d'Urgell, whether his wife was "touched by the disease of leprosy" (*tocada de malautia de maselina*). Between *infectus* and *tactus* there was a third, somewhat more specific predicate: *contaminatus.* This term had a precedent in the Vulgate version of Leviticus 13. There, however, it carried not only the implication of ritual pollution but also the added meaning of a solemn verdict, as in verse 11, "contaminabit itaque eum sacerdos," translated in the King James Version as "and the priest shall pronounce him unclean." In contrast, medical documents treated infection, in its various guises, as physical fact. Thus, even if one insists on hearing undertones, it would be difficult to argue that Flemish physicians were alluding to ritual pollution in 1559, either when they phrased the question whether Boudewijn de Cuupere was "infected by the stain [*gheinfecteert van de smitte*] of *lepra*" or when they decided that he was "contaminated and stained [*ghecontamineert ende besmet*] by the disease of true *lepra*." Similarly, the famous Nuremberg physicians Hermann and Hartmann Schedel relied on mundane perception when they summarized the entire notion of infection in one brief document, in 1483. The uncle and nephew jointly wrote a deposition regarding Elizabeth de Streyberg, who wished to know "whether she was infected by the disease of *lepra*." The doctors decided that, as the woman showed signs of an early stage and was "disfigured and blemished" (*defoedata ac maculata*), and as "*lepra* by itself is a contagious disease according to the authority of all the physicians, she should be duly avoided."[38]

A Vocabulary of Impressions and Emotions

The aspect of disfigurement leads us into a second major category of qualities, which were attributed to leprosy on the basis of impressions. These attributions were the most salient in medicine as well as in literature, and in academic discourse as well as in public records. They concentrated on the offensive and fearsome ravages caused by lepromatous leprosy. In the Latin translations both of Aretaeus and Avicenna, the disease was called "feda," an adjective whose meaning ranged from "ugly" and "foul" to "hideous" and "abominable." Medieval authors expanded the register with "vilis" and "turpis," odious qualities that have their equivalent in modern expressions such as "gross" and "ghastly." Diagnoses underscored the *defedacio*, or "spoiling," of the face, skin, or extremities; and they linked manifestations such as "a foul ulcer" or "a foul disposition" to the repulsiveness of *lepra*. The close association between the disfigurement and breakdown, or "corruption," of the body accounted, as early as the eighth century, for the harsh remark that leprous patients "become putrid and foul" (*putrescunt et foediscunt*). These medical characterizations were summed up by the poet Petrarch (1300–1374), whose famous animosity toward physicians was matched by an intense, albeit selective, interest in health. He mused that, "if leprosy invades a putrid [*putre*] and wretched body," the patient's "unbearable pain is made heavier by the loathsomeness, nausea, and shame [*foeditas, fastigium, et pudor*] of the disease." Four centuries later another nonmedical author evoked similar impressions.

Adam Smith, in *The Wealth of Nations* (1776), drew a parallel between the "muti-lation, deformity, and wretchedness" of cowardice and leprosy, and he char-acterized the latter as a "loathsome and offensive disease."[39]

For centuries, people with leprosy were singled out for a particular offen-siveness, namely, to the sense of smell, in a world that was probably not very sweet scented. The tradition adhered to the comment by Aretaeus that they "have foul breath" and "fetid ulcers." In actual circumstances, fetor occasion-ally triggered more response than the disease itself. In January 1557, after four examinations in as many months, two physicians of the town of Ieper concluded that a poor girl, Janneken Brassers, "had fallen into such misery, wretchedness, lameness and corruption of *laderie* that no one would take charge or care of her, on account of the great stench, because it might bring someone infection or corruption." Therefore, the physicians requested the town's "Regents of the Common Poor" (*regieders ende gouverneurs vanden Ghe-meenen Armen*) "to provide a place where she could be separated from conver-sation with the public and could receive good treatment to protect her body and to feed her properly." It would appear, however, that from the sixteenth century on, authors alluded less frequently to fetor while showing greater concern with assaults on the sense of sight. While finding Latin precedents in the Aretaean description of *elephantiasis* as "ugly to behold" (*visu foedus*) and in Galen's phrase "fetid and horrible to see," they placed increasing empha-sis on the visual aspect. In 1540, Pierre Bocellin repeatedly stated, for empha-sis, that leprosy destroyed "the body's beauty." Five years later, Paulus Iulia-rius believed that it was uncomfortable to look at patients ("aspectu sunt ingrato"). As late as 1767, François Raymond largely adhered to the descrip-tion of *elephantiasis* by Aretaeus, but he made it more graphic by elaborating that the body is "disfigured by hideous [*hideux*] tumors" and "dreadful [*affreux*] ulcers."[40]

The ability of leprosy to strike people with dread was a theme that re-curred most regularly, as a grim refrain. For Aretaeus, the disease was "terri-fying in every way, like that monster, the elephant." At the other end of our chronological scope, in a 1778 frontispiece that is reproduced in Plate 4, *lepra* was portrayed as "more horrible than death." Ironically, in the work in which this plate appears, Dutch surgeons concentrated on leprosy in Surinam and the Americas, as an exotic affliction. The etching underscores the visual aspect of leprosy's fearsomeness, as if, for people in Utrecht and Leiden in 1778, the "horrific appearance" was a greater cause of fear than the threat of conta-gion. In a more specific description, the face looked "terrifying," like that of

G. G. SCHILLINGII

DE

LEPRA

COMMENTATIONES.

RECENSUIT

J. D. HAHN.

HORRIDIOR   MORTE.

A. Delfos

LUGDUNI BATAVORUM,
APUD SAM. ET JOAN. LUCHTMANS,

TRAJECTI AD RHENUM,
APUD ABR. VAN PADDENBURG,

MDCCLXXVIII.

PLATE 4. "Horridior morte": frontispiece of G. G. Schilling, *De lepra commentationes,* ed. J. D. Hahn (Leiden: Sam[uel] and Joan[nes] Luchtmans, 1778). At the National Library of Medicine, Bethesda, Maryland.

the lion, or "horrific, deformed in the same manner as in statues of satyrs." Of course, the disease itself, too, was "full of horror," causing fright by its "savage" nature, especially in earlier centuries. Bocellin considered *lepra* most "contagious and horrible," although he added that both characteristics also applied to scabies and morphea (a scurfy eruption). Bocellin's ambiguous stance on the exceptional abhorrence of leprosy typifies once more, however narrowly, the reciprocal relationships between science and folklore, and between medical language and popular stereotypes, particularly when it came to the role of emotion. While common perceptions evidently colored scientific categories and labels, their intensity faded in the process of rationalization. Inversely, while characterizations by physicians no doubt reinforced popular emotions, medical authors and practitioners often restrained impulsive public responses and suspicions. Velasco de Tharanta, for example, an eminent summarizer of then current thought, mitigated the uniqueness of leprosy's threat by pointing out that it was but one of many ghastly and contagious diseases. Medical examiners, for their part, expressly dismissed various conditions that, for the impressionable layperson, raised suspicions of leprosy. They regularly underscored a "clean" bill of health with clauses such as "notwithstanding the disfigurement" or "notwithstanding a horrible appearance."[41]

Framing the differences between partial or deceptive impressions and the genuine disease was primarily an academic task. These differences were at the center of the attempts to define and explain *lepra,* and they shaped the criteria for diagnosis and therapeutics, as we will see in the following chapters. The challenge to differentiate also led authors, if not practitioners, to speculate about the distinctive gravity of leprosy. The tone was set by the proposition, opening Avicenna's seminal chapter, that "[l]eprosy is a bad disease" (*[l]epra est infirmitas mala*). No modifier was more fluid than *malus,* with meanings that ranged from merely "not good" to "evil" and "foul." This was also an elastic attribute, for it accepted as nearly synonymous adjectives that were more evocative, including "malignant," "burdensome," "vicious," and "disgusting." Several masters anticipated the question, with Alberto de Bologna, of why the label should be reserved for leprosy, "since *every* illness is bad." They attributed an intrinsic malignancy to the essence, causes, course, spread, and effects of the disease.[42]

Lepra was called "bad" because it combined the three principal types of disease, namely, a bad "complexion," or faulty combination of fundamental qualities; a bad "composition," or destruction of the form of the body and

of the function of the organs; and a "dissolution of integrity" (*solutio conti-
nuitatis*), or loss of body parts. Worst of all, it was essentially cold and dry
(as the definitions will show in the next chapter) and thus diametrically op-
posed to the warm and moist "complexion" of life. Causes, with the elusive
malignancy of poison, included heredity, unclean catamenia, and noxious
diets. Advancing stealthily, the disease "violently eats [*rodit*] the skin" and,
deep inside, decomposes the "noble internal organs, especially the liver and
spleen." It spread by "contagion," a frightening notion whether it was understood as internal "dispensation" through the body or as propagation through
the community. Social consequences made leprosy worse than relatively
comparable afflictions because it turned patients into "exiles from human
intercourse." Once the disease was manifest, moreover, either the inevitable
outcome was death, or it was rooted too deeply to be cured without the
benefit of medicine, unlike most diseases, which could be cured by nature.[43]

In sum, as Du Laurens reported, "some of the ancients called leprosy a
'Herculanean' disease, because it is more harmful and more ferocious than
almost all other diseases, stubborn, resistant to medications, and not to be
vanquished by any but herculean strength." He observed further that "it was
the opinion of nearly all the ancients that this evil is a divine punishment"
and that "ancient laws counted *leprosi* among the accursed [*maledictorum*]." Jean
Fernel adapted the observation to his own era, stating that *lepra* "yields to
no remedies, and the common people are wont to beseech the heavens for a
cure." It is significant that Du Laurens and Fernel distanced themselves from
the ancients and the populace. In doing so while, at the same time, echoing
the popular hyperboles, these physicians underscored the coexistence of similarities and differences between commonplace and medical responses to leprosy. Although a thirteenth-century dictionary called the disease "accursed"
(*execrabilis*), medical authors did not characterize it as inflicted by God, and
they seldom employed the biblical term for a divine scourge, "plague" (*plaga*),
which was so readily available in Leviticus. It is a striking paradox that, as late
as 1734, Württemberg health commissioners still condemned a woman as
having "the loathsome, hereditary, and disastrous disease," after their learned
colleagues had been searching for centuries to define and explain leprosy in
rational terms, as we will see in the following chapter.[44]

Chapter Four

Definitions and Explanations

 The lord bailiff has summoned the physician, master Deuslosal, and the master surgeons, Deuslogar and Vitalis, Jews and residents of the city of Urgell. He has ordered them to perform an examination on the woman Ramona, properly and lawfully, and in accordance with the art and science of medicine. After doing this, they have presented the following unanimous report, in the presence of the lord bailiff and the undersigned witnesses.

Said medical practitioners declare under oath that they have all three made blood to be drawn from the said Ramona, seen and inspected her entire person and her urine, collected her blood, and palpated all her members. In all these parts they have found no sign of *lepra* according to what is written in the fourth book of Avicenna. Therefore, the ailment that she has in the nose is a kind of cancer, which is particular to one organ. *Lepra*, on the contrary, is universal, because it proceeds to the hands, feet, nails and skin. Moreover, it is a certain change or corruption in the blood, which at present is not to be found in Ramona according to our art and science.

This encouraging turn of events, in La Seu d'Urgell on December 14, 1372, was confirmed at the request of Ramona's husband, Arnal Isern of the Valley of Andorra, in a letter drawn up by a notary—with the fitting name of Petrus de Ministrelis. Arnal owed the exoneration of his wife not only to the methodical investigation by the medical practitioners ("metges phísics e surgeans"), which he had sought, but also to their precise interpretation of the defini-tion of leprosy. The physician and the two surgeons avowedly anchored their "art and science" in the sovereign authority of Avicenna.[1]

A CONCEPTUAL PATHOLOGY

Foundations: Avicenna's Formula, Galen's Teaching, and the Physiology of Humors

The reader might infer from the 1372 certificate that the practitioners con-sulted Avicenna for making their diagnosis, but this impression is inaccurate, for two reasons. In the first place, the *Canon of Medicine* offered only a sum-mary chapter on the signs of leprosy, which was rarely cited, and which was far shorter than the *Canon's* other two chapters on the subject—proportion-ately, one of the shortest in all the medical literature on *lepra*.[2] Second, the examiners' verdict did not spell out the signs but, rather, underscored the essence of what distinguished Ramona's condition from leprosy. At issue was the *definition* of her disease, rather than the symptoms in themselves. The au-thoritative source was Avicenna's first chapter, whose title, in the Latin trans-lation by Gerard of Cremona, was didactically formulated as a question, "What Is *Lepra* and Its Cause?" Master Deuslosal and his colleagues evidently drew their references, directly or indirectly, from Avicenna's chapter, which opened with the declaration,

> *Lepra* is a bad disease caused by the spreading of black bile through the whole body, which corrupts the complexion of the members as well as their shape and appearance. . . . It is like a cancer common to the whole body. . . . Sometimes, the black bile is expelled to one member, and it leads to a hardening, sclerosis, or cancer, according to its dispositions. . . . The deepest efficient cause is a bad com-plexion of the liver, when it tends to become too warm and dry so that it burns the blood into black bile.[3]

These lines provided, for at least four centuries, the structure and substance for identifications that constituted the "pathology" of leprosy.

Unlike its modern counterpart, the early pathology of leprosy was con-ceptual rather than empirical. As a result, it was also more open to rational-izing speculation, beholden to authoritative tradition, and, above all, suscep-tible to ideological and cultural trends. The variety of identifications irked

Paulus Iuliarius, physician of Verona, so much that in 1545 he admitted going around muttering,

> [I]t would be better for human welfare if, with few exceptions aside from Hippocrates and Galen, all the published books on medicine were burned in the fire. So much confusion has arisen from this that no medicine seems to be left when the discord of practitioners has come to promote ignorance. Physicians have written in so many ways about *lepra,* that we are hardly able to recognize what it is or how to treat it.[4]

The following pages may cause the reader to agree with Iuliarius, but they will also, I hope, give an idea of the challenges presented by the identity of the disease, the intensity of the responses, and the far-reaching practical implications of seemingly abstract concepts.

In order to situate the premodern pathology of *lepra* in a conceptual framework, and to appreciate its constants and variations, it is useful to recall some fundamentals of natural philosophy and premodern physiology. In the basic view of nature, as taught by Aristotle, everything was composed of four elements (fire, air, water, and earth). Thereby, everything was subject to "corruption," or, literally, breakdown. In addition, living compounds depended on a balanced coexistence of warmth and moisture: while too much heat resulted in death by withering and "burning up," too little warmth allowed moisture to lead to decay by "putrefaction," or rotting. Life itself was a process similar to "coction," or cooking. Galen built his physiology on these foundations, assigning a more pronounced role to the interaction of "innate" or "natural" heat and basic or "radical" moisture. He further interwove the Aristotelian principles with an elaborate scheme of the four humors that were part of the Hippocratic tradition. In this scheme, each humor possessed a different combination of primary or elemental qualities, reddish bile or choler being dry and warm, blood moist and warm, phlegm moist and cold, and black bile or melancholy dry and cold. Figure 4.1 gives a general outline, greatly simplified, of the processes and failures in the maintenance of human life as they were envisioned by Galen.

In Galenic physiology, human life was maintained by an adequate nourishing of the body, which required the complete conversion of food and drink into the substance and quality of every part. This conversion, or digestion, began after the initial dissolution of the food in the mouth, and it proceeded in distinct stages. Each stage was also regulated by a specific "faculty" or "power" (*virtus*)—Galen's most addictive and unproductive term for a host of poorly understood agencies. After drawing in the food with an "attractive power," the stomach turned it into chyle, which was easier to absorb. Then, the liver, "attracting" most of the chyle (after surplus was sent to the intestines for excretion), processed it into blood. This was the most vital of the humors, to some extent composed of the others, but essentially different from them—in too many ways to pursue here. A healthy liver also strained out most of two secondary and potentially troublesome humors. Reddish bile, rising to the surface, was attracted by the gall bladder, while black bile sank to the spleen. The gall bladder and spleen turned the respective biles into something akin to digestive agents or catalysts. Phlegm, while lined up with the other humors in Figure 4.1 for the sake of graphic clarity, had little, if anything, to do with the liver; there was no comprehensive or consistent account of its origin and processing. Be that as it may, the distribution of phlegm as well as of the purified blood and the refined biles through the body was governed by a "disseminative faculty." Once the humors were in place, the "unitive faculty" caused them to be bound to every part of the body. Ultimately,

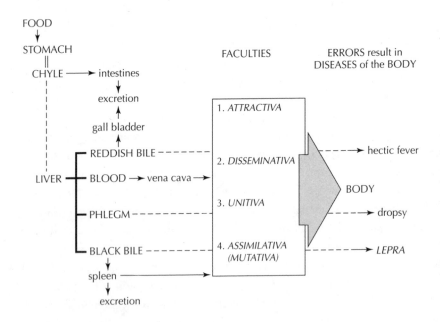

Figure 4.1. Digestion according to Galen: from food to body

thanks to the "assimilative faculty," they were absorbed and changed into the nature of the organs and tissues.[5]

This construction of physiology set the stage for errors in the "digestive" process, by which baffling diseases were explained. If the purified humors were not distributed, the body overheated for want of basic moisture, to be gradually consumed by a permanent and wasting fever, which was called "hectic" and often associated with the disease later identified as tuberculosis. If the nutrients were distributed but not "united" or attached to the parts, they floated around idly in a body swollen with edema or with one or another form of dropsy; the problem consisted primarily in an accumulation of phlegm. In the third—and ontologically most perverse—scenario, if the humors were delivered and attached to the parts but failed to be absorbed and transformed into the substance of the tissues, they caused neither the consumption nor drowning of the body but its steady disintegration, in the "corruption" and "putrefaction" of *lepra*. It is easy to see how, with this scenario, medical theories reinforced associations of the disease with decay, rotting, stench, and slow death. The details of the failure at the transformative stage remained vague, but they left no doubt that the chief villain was black bile that had been overcooked or burned ("adust") and left in the blood. This "unnatural" or spoiled black bile now threatened the body, as dead ashes, a toxic substance (*venenum*), and, worst of all, the cold and dry opposite of life itself, which is warm and moist.

Divergent Doctrines and the Diversity of Definitions

The hypothetical scheme of health and disease—and of life and death—left ample room for diverging interpretations, which we will survey shortly. At this point, however, a brief look at one major divergence, which became part of the authoritative tradition, may prepare us for the thrust of the definitions and explanations in general, and for their changing relationship with authoritative sources in particular. When Galen placed the root cause of leprosy in an error of the assimilative faculty, he did little to correlate this account with his other attributions: he had also attributed leprosy to "a bad complexion" (about which more shortly), to the burning of black bile, and, most tantalizingly, to a malfunction of the liver. Raising this last suggestion by Galen to the level of a theory, Avicenna proposed that organic failure caused the liver to overheat and to overcook or burn the blood and black bile. As could be expected, this apparent disagreement between two authoritative explanations challenged the "conciliatory" ingenuity of Scholastic commentators. Bernard de Gordon reconciled both viewpoints, without naming their proponents and with a slight nod to Galen, in a one-line clarification, "[A]n error of the second digestive faculty in the liver can be the remote cause of *lepra*, but the immediate cause is an error of the immutative faculty in the flesh."[6]

We may be more surprised to hear echoes of the doctrinal discord far beyond strictly academic circles. Guy de Chauliac, an alumnus but not a teaching member of a medical faculty, opened the chapter on *lepra* in his *Chirurgia magna* with Galen's definition (see Table 4.1), but he added, with a nod to Avicenna and in contrast with Bernard (on whom he drew heavily),

> I understand the failure of the assimilative faculty [to be the cause] in the immediate sense, because the indirect cause may be a failure of the digestive and blood-producing faculty in the liver. This is why Avicenna calls the failure of the liver's power the prior [*antiquiorem*] efficient cause.

At a greater distance from academe—and probably depending on Chauliac but with diminished precision, Hans von Gersdorff also noted the discrepancy when he compiled his surgical manual, *Feldtbuch der Wundtartzney* (1517). First, he deferred to "the words of Galen" in defining *lepra* as a disruption in the process of assimilation ("ein zerstörung aller der gleych"). Then, however, Gersdorff elaborated, "[H]ere I understand the failure of assimilative power [*die irrung der gleychenden krafft*] as a means, and by these means the cooking of the blood in the liver may be the cause." He concluded, "[T]herefore, Avicenna calls it an error of the power of the liver [*die irrung der krafft der leberen*]."[7]

In addition to surfacing in Latin and vernacular theoretical paragraphs, the difference in emphasis between Galen and Avicenna left traces in practical areas, not only in prescriptions for treatment but even in records of actual situations. Thus, in a sixteenth-century case, two inspection teams explicitly based their judgments on academic traditions, yet they gave different accounts, one attributing the disease to a deficient power and the other to a malfunctioning organ. First, two physicians and two surgeons in Ieper relied on the Galenic view when they took the feebleness, or "powerlessness and total misery" (*onvermuegentheit ende alle miserie*), of a patient as basis for deeming him leprous. They testified, "[W]hereas the Greeks and the principal authors of medicine have classified this kind of corruption of the blood and putrefaction of all the members under leprosy, therefore we reasonably judge that he is leprous." The practitioners' pleas to have this man admitted to the leper hospital were evidently not accepted by the regents, for the following month he was examined in Lille by a physician and a surgeon. These examiners confirmed their colleagues' judgment, but their opinion adhered more closely to the teaching of Avicenna. They found the patient truly leprous because they had "taken careful note of the extreme warmth of the liver [*prins garde a lextreme challeur de foye*] and the rampant inflammation of the blood, which was followed by utter consumption of all the muscles."[8]

While they may have espoused different interpretations, the medical examiners in Ieper and Lille agreed not only in judging the patient leprous but also in defining the disease itself as a complete breakdown of the blood and consumption of the limbs. Practitioners depended on a clear and authoritative definition of leprosy for the credibility of their diagnosis—and of their professional pronouncements in general. "The principles, signs and conclusions of the Art of Medicine," in which the Parisian masters anchored their exoneration of Jean de Bierville in 1411, began with definitions.[9] Indeed, it is good to keep in mind that much more than scholastic acuity or dialectical exercise was at stake when authors labored on a precise identification, and when teachers argued about explanations. Even when they were too prone to drift off into hairsplitting speculation, learned practitioners were endeavoring to transcend volatile and powerful emotional responses with rational inquiry. They were struggling to grasp the nature of an omnivorous and multiform disease. Furthermore, their widely varied efforts to capture the essence of leprosy demonstrated unexpected degrees of individuality. When we collate a score of formulations, coined between the second and seventeenth century, the broad diversity becomes apparent immediately. Although the collation in Table 4.1 is selective, and although it telescopes a long stretch of one and a half millennia, it contradicts general portrayals of the entire era as monolithic. The definitions are presented in their original Latin or vernacular wording in order to give a faithful idea of the shared vocabulary and concepts, as well as of the variations in format and content.

The collation, a simple list of twenty-two definitions with their respective authors (in capitals, the four who were quoted most frequently) and dates, calls for at least two caveats. First, by quoting only a few initial words, the table cannot do justice to the bewildering diversity of the definitions—let alone the explanations. Suffice it to compare briefly two complete formulations. One of the most carefully structured definitions was drawn up by

Bernard de Gordon, who in 1305 presented *lepra* as "a systemic [*consimilis*] disease that breaks down the appearance, shape, and composition of the members and ultimately dissolves the integrity; it results from melancholic matter spread through the whole body."[10] Bernard's deliberate structure becomes more evident when contrasted with the formulas of some other authors, such as Paulus Iuliarius, who—admittedly, two centuries later and in a less academic environment—mixed pathology, etiology, and semiology when he defined leprosy as

> a melancholic defect that originates in melancholic blood, for when black bile becomes more abundant than blood, those who suffer from the disease emit a strong smell and have an unpleasant appearance; some of them also develop ulcers; it is said to be a universal cancer of the entire body.[11]

It is readily apparent that both of these definitions borrowed heavily from the same preeminent source. However, their differences, and the diversity of the definitions in general, should make us wary of facile generalizations about medieval medicine, and they call for careful collation. When a dozen or so formulations are collated, even a rough parsing of the result may allow us to recognize the variety as well as the continuity in Scholastic rationalizations of *lepra*.

As a second caveat, it should be pointed out that the selection in Table 4.1 is only partially representative of the literature on leprosy, for one reason because several authors dispensed with a formal definition, preferring to concentrate immediately on diagnosis and therapeutics. This practical preoccupation marked texts that are scattered across many centuries. An early medieval guide, extant in the eighth-century "Lorscher Arzneibuch," adopted by Gariopontus, and adapted by the later Salernitan master Johannes de Sancto Paulo, opened with the declaration, "This is how we recognize a leprous patient [*elephantiosum*]." Hildegard of Bingen, though mixing medical ideas and advice with moralization, did not define the disease in her warnings about links between *lepra* and gluttony, drunkenness, and other vices. A few authors seemed somewhat ambivalent about the importance of defining the disease, for they headed straight into prescriptive statements but then, almost as an afterthought, returned to the definition. In 1563, Johann Huggelin dedicated his brief tract to the head of a hospital in Basel, as expert advice for the judgment of suspected patients. Then, however, he paused, "[B]efore I tell and pursue the signs, I wish first to show what *Aussatz* is, how many forms and kinds of *Malatzey* there are, and also from what cause this disease results most commonly." With very few exceptions, when authors did not formally define *lepra*, they based their advice on one or another of the definitions collected in Table 4.1.[12]

When one proceeds from an initial glance to a word-by-word comparison, each examination of Table 4.1 reveals new variations among the formulas. Their diversity is all the more remarkable because they occur in didactic texts, which tended to be derivative in character. The variability begins at the very opening of the formulation, with the majority of authors identifying leprosy as an entity, in a straightforward predication, while a few referred to it, rather, as an event. In the first category, which shared the elementary proposition "Lepra est," the predicates covered the gamut of common terms for disease, from the prevailing *morbus* to alternatives such as *egritudo* and *passio*. This terminology was mirrored, not surprisingly, in archival documents, occasionally with as much redundancy as variation. In Pistoia in 1288, a notary certified that Master Laurentius, *medicus*, was asked by Johannes de Stagiana to see whether his daughter Pasquese was suffering from "the disease of the infirmity" (*infirmitatis morbo*) of leprosy; after inspecting the signs of her "illness" (*egritudinis*), the physician declared that the girl was indeed infected with "the disease" (*morbo*).[13] In vernacular usage, the predicates had fewer derivatives than equivalents, with the latter ranging from *maladie* to *kranckheit*. It is more important to note, returning to Table 4.1, that authors did not copy their word for "disease" from the Latin version of their two sovereign authorities: they adopted neither the term *affectus* from Galen (which referred more often to a state of the mind than of the body) nor the utterly common noun *infirmitas* from Avicenna.

Even when a definition was expressly attributed to a canonical source, it might be altered subtly, but sufficiently to suggest variances that extended beyond mere terminology. Theodoric Borgognoni (d. 1298) stated, "*Lepra*, as Avicenna says, is a bad complexion resulting from a bad disposition of bile in the entire body." It is immediately apparent that, notwithstanding his claim that he was quoting Avicenna verbatim, Theodoric introduced two variants: he substituted "mala complexio" for "infirmitas," and "ex mala dispositione cholere" for "ex spersione cholere nigre." If one argues that the first substitution may have been inadvertent, due to the nearby mention of "complexion" in the *Canon*'s defining paragraph (quoted earlier in this chapter), it is worth noting that Theodoric compounded the change by making complexion the predicate of *lepra*—that is, he defined *lepra* as a bad complexion rather

Table 4.1. Formal Definitions

Lepram mutative virtutis errorem nimium esse dicimus.	—GALEN (ca. 129–200), *De accidenti et morbo*
Lepra est cutis mutatio in habitum qui praeter naturam sit cum asperitate . . . Elephas est affectus qui cutem crassam atque inaequabilem reddit . . .	—GALEN, *Definitiones medicae*
*Scias quod Lepra fit ex superfluitatibus colere nigre . . .	—Serapion (Yuhanna ibn Sarabyun, ca. 873)
*Lepra est nascens passio de colera nigra incensa et putrefacta apparens in corporis superficie et nascens de quatuor humoribus, sed tamen incensis et corruptis, et in coleram nigram mutatis . . .	—Constantinus Africanus (ca. 1080)
*Lepra est membrorum corruptio ex humoribus putrefactioni habilibus effecta . . .	—Johannes Platearius (ca. 1100)
Templum naturae perimens contagio lepra pluribus ex causis oritur . . .	—Gilles de Corbeil (ca. 1200)
*Lepra ex corruptis humoribus consurgere in corpore consuevit . . .	—Roger Frugard of Parma (ca. 1200)
*Lepra est infirmitas mala proveniens ex spersione cholere nigre in corpore toto quare corrumpitur complexio membrorum et forma et figura eorum: et fortasse corrumpitur in fine ipsorum continuitas . . .	—AVICENNA, *Canon* (Latin ca. 1180)
*Lepra, sicut dicit Avicenna, est mala complexio perveniens ex mala dispositione cholere in toto corpore . . .	—Theodoric Borgognoni (ca. 1280)
*Lepra est morbus consimilis corrumpens figuram et formam et compositionem membrorum et finaliter solvens continuitatem, proveniens de materia melancholica spersa ad totum corpus . . .	—BERNARD DE GORDON, *Lilium* (1305)
*Lepra est morbus turpis ex materia melancholica aut reducta ad melancholiam corrupta corruptione incorrigibili . . .	—Henri de Mondeville (1308)
Omnis lepra [fit] a colera nigra vel a sanguine adusto vel ab alio humore adusto . . . , ex spersione huius humoris supra membra simplicia . . . , ex incinerata materia vel grossa vel viscosa . . .	—Jordanus de Turre (ca. 1320)

Continued

Table 4.1. Formal Definitions, *continued*

*Lepra est morbus universalis corporis totius proveniens ex dominio melancolie non naturalis.	— Niccolò Bertucci (ca. 1330)
Lepra est vilis membrorum corrupcio et humorum, inicium habens a vasis . . .	—Anon. *De lepra* (ca. 1350)
*Lepra est error maximus virtutis assimilative, qua forma corrumpitur in toto . . .	—GUY DE CHAULIAC (1363)
*Lepra est morbus corrumpens complexionem et formam et figuram . . . turpis contagiosus ex melancolia corrupta spersa per totum corpus . . .	—Velasco de Tharanta (1418)
*Est lepra egritudo mala proveniens ex spersione cholere nigre . . .	—Bartholomeus de Montagnana (ca. 1430)
*Lepra das ist die Maltzey oder uszetzigkeit ist ein zerstörung aller der gleych des menschen, und do durch allen gleychen ire Krafft genommen würt und auch davon zerstöret . . .	—Hans von Gersdorff (1517)
[La lepre] est selon Avicenne une maladie corrompue, corrumpant la complexion, forme, et figure des membres; Galenus la dict estre maladie tres grande, faisant errer la vertu assimilative en la chair.	—Pierre Bocellin (1540)
*Grave imprimis et horroris plenum vicium elephantiae est: ut primum id invasit, summa pars corporis crebris maculis inspergitur, crebrosque tumores, maxime iuxta summam frontem . . .	—Jodocus Lommius (ca. 1550)
Elephantiasis est intemperies iecoris sicca calidaque, cum certo putredinis aut venenositatis gradu iuncta, per quam humor atrabiliarius copiosior omnes corporis partes corrumpit . . .	—Jean de Varanda (ca. 1610)
Tertium genus [scabiei] a squamis Leprae nomen acquisivit . . . Quartum genus scabiei summum et foedissimum, ab Arabibus Lepra, a recentioribus autem Elephantiasis nuncupatum, cancer universalis vocatur . . .	—Samuel Hafenreffer (1660)

* Denotes the opening line of treatise or chapter.

than as a disease caused by a bad complexion. It is possible, I believe, that this reflected the impact, in the 1270s and 1280s, of recent Latin translations of nearly three dozen of Galen's works. This "new Galen," according to the historian who bequeathed both the finding and the phrase, "provoked a real intellectual upheaval in the approach to medical subjects." Two of the newly translated treatises, which played a particular role in the attempts to define leprosy, reinforced the central role of complexion in medical theory and its applications.[14]

The Galenic Construct of *Complexio*

The concept of *complexio* was available to the West since the heyday of Salerno. Nourished by natural philosophy, the notion envisioned the "makeup" of things or, more dynamically, the "mix" and interplay of the primary qualities in a great number of entities, from medicinal herbs to planets. In one of three sermons "for *leprosi* and the rejected" which the Franciscan friar Guibert de Tournai wrote around 1260, we detect some resistance of ascetic theologians to the burgeoning physicality of medical thought. Guibert warned against the diabolical temptation to shun austerity, fasting, and "everything contrary to the flesh" on the pretext that "this upsets the complexions and causes diseases." As if alluding to scholastic debates in the young faculties of medicine, Guibert scoffed, "[T]he devil has become a physician carrying on a disputation about complexions." For the sake of perspective, it is useful to consider that Guibert had taught in the volatile atmosphere of the University of Paris, where clerical responses to Aristotelian natural philosophy were complex and contradictory.[15]

In his seminal work *De complexionibus,* Galen drew together basic strands of the Aristotelian and Hippocratic traditions. He maintained that a balanced or "tempered" mix (κρασις, *temperamentum*) of heat, cold, dryness, and wetness constituted health. In the human body and in each of its parts, the mix was affected by age, gender, and other factors, which accounted for many "complexional," or, as we would say now, "constitutional," differences. Therefore, when Theodoric defined *lepra* as "a bad complexion" rather than "a bad disease," he transferred the onus, perhaps unintentionally, from an agent to the body's constitution. Furthermore, he implied that the entire body was affected equally. This position, while held by Avicenna, was just then raising questions, in the wake of another Galenic work with the title *De malicia complexionis diverse.* In this treatise, Galen argued that, in the case of a lapse of the complexion (*malitia* or *malicia complexionis*), the imbalance (*discrasia, intemperies*) was most often uneven (*diversa, inequalis*), neither equally displacing the natural complexion in all parts of the body nor impairing every normal function. If the imbalance proceeded to becoming uniform (*equalis*), then destruction threatened the entire body.[16]

In this framework, and very early in his treatise, Galen assigned *lepra*, together with gangrene and cancer, to the category of "uneven imbalance in any one of the members" (*malicia complexionis diverse in unoquoque membrorum*). By appearing here to make it a disease of one or other part of the body, the treatise introduced a challenge to the established notion that it was a disease of the entire body. This notion was sanctioned by the authority of Avicenna, and it was firmly buttressed by his catchy phrase "a cancer of the whole body." The phrase stimulated two separate questions, one about the relationship between cancer and leprosy, and the other about whether leprosy was to be treated as a disease of the entire body, of the flesh, or of the skin. While the contrast between a localized and a generalized disease ignored the modern understanding of metastasis, the linkage to cancer called forth fears that are not nearly as recent as we might think. The second corollary, namely, the differentiation between a systemic and a superficial ailment, often determined whether leprosy was allocated to one or another part of medicine, as we will see when we examine taxonomies (Chap. 6). The concept of an all-body disease was standard in medical encyclopedias. However, the allocation to ailments of the skin (*cutis*)—in what we now call dermatology—was not uncommon: it was graphically underscored in the title page of Samuel Hafenreffer's *Nosodochium*, which is reproduced in Plate 5 (together with an affidavit of an examination for leprosy, at which he presided). The differentiation between systemic and superficial also affected the fate of patients, as Doctor Losser indicated: "[T]o be separated are those who have signs of permanent *uβseczczekeit*, not only of the skin but of the flesh—when there is no hope." Reserving the two important corollaries to Avicenna's phrase for detailed analysis in later chapters, we will concentrate here on the humoral interpretation of the disease.[17]

THE COURSE AND CONSEQUENCES
OF THE HUMORAL INTERPRETATION

It is possible to sample the responses of universities to the new challenge of Galen's complexional doctrine, to survey the changing definitions of leprosy,

PLATE 5. (*Top left*) Frontispiece of Samuel Hafenreffer, *Nosodochium in quo cutis, eique adhaerentium partium, affectus omnes...fidelissime traduntur...Renovatum & plurimis in locis auctum,* Ulm, Germany: Balthasar Kühnen, 1660; at the National Library of Medicine, Bethesda, Maryland. (*Bottom right*) Certificate of health for Maria Jacoba Maunesin, witnessed by Samuel Hafenreffer in 1656; Universitätsarchiv Tübingen, Signatur 20/10, No. 119a.

and even to gauge some practical implications of the conceptual structures by following the discourse at one faculty of medicine. The discourse could be reconstructed as it developed in Bologna, from the time of Theodoric and his contemporary Taddeo Alderotti (d. 1295), through the first decades of the fourteenth century, in the careers of Taddeo's students, most notable among them Alberto de Bologna, Bartolomeo da Varignana, Dino del Garbo, Guglielmo da Brescia, and Niccolò Bertucci. Each of these masters responded not only to the relation between Galen and Avicenna but also to the implications for *lepra,* and their responses await methodical comparison—a project all by itself. Meanwhile, the discourse is more readily traceable at the University of Montpellier, both because the groundwork has been done and because the school retained a clear identity over many generations. The impact of the "new Galen" on Montpellier, in the decades around 1300, has been the subject of methodical study. Moreover, it is possible to map currents over three centuries through the views on leprosy of ten Montpellier-related authors, five of whom wrote as professors, three as surgeons, and two as practitioners.[18]

Late Thirteenth to Early Fifteenth Century

An early response—possibly the first—to Galen's work *De malicia complexionis diverse* is preserved in a dozen formal questions (*dubitationes*) on the book. According to the colophon in a manuscript that contains their summaries, the original author of the questions and their corresponding theses (*positiones*) was Bernard de Hangarra. The colophon calls this Bernard a "former chancellor" of the University of Montpellier, and it is most likely that he lectured on Galen's treatise around 1290. The second question is, "Whether *lepra* is a bad complexion of the entire body." In reply, the tersest and most general of the theses states, "The Chancellor said that it is not, because the order of nature is to give the finer nourishment to the principal organs, so that the extremities are left with the bad, from which they are infected first."[19]

In the 1290s, the stellar Montpellier master Arnau de Vilanova wrote a full-fledged commentary on *De malicia complexionis diverse.* He noted the seeming contradiction between Galen's classification and "what is commonly said of *lepra,* namely, that it is a universal cancer of the entire body, as Avicenna clearly states in *Canon* IV." Adding chronological perspective, and at least implicitly distancing himself from Bernard de Hangarra, Arnau continued,

[T]he answer used to be that it is particular to one kind of organ, such as the flesh only, as Galen teaches in the sixth book on *Disease and Symptom* [*De morbo et accidenti*], where we read the true definition of leprosy, "an imbalanced complexion in the flesh, causing the assimilative faculty to fail," or "a failure of the assimilative faculty in the flesh, caused by a cold and dry imbalance." If *lepra* is to be the name of a disease, it should be defined in the former manner; if the name of a symptom, in the latter manner.

The introduction of the answer as outdated ("used to be"), and the concluding rejection because it "does not suit the proposition," reveal a progressive outlook on teaching.[20]

Arnau's own thesis merits extensive quotation, not only for comparison with subsequent explanations but also because it clearly shows that for him— if not readily for the modern reader impatient with Scholastic abstraction— a definition was intrinsically linked to the physician's assessment of disease *and* patient.

It should be said that the physician sometimes judges *lepra* to be partial, sometimes universal. Since the physician makes a judgment about the disposition of the body to be treated [*sanabilis*] on the basis of what sense perception presents, he then deems *lepra* to be a partial imbalance when it manifests itself only in one part of the body: this is the case, most often, in the beginning of its manifestation. . . . However, when it manifests itself in the entire body, [the physician] deems it to be universal, when it is confirmed. Therefore, it should be said that Galen is speaking of *lepra* in the beginning of its manifestation . . . ; on the other hand, when it is called universal, this is when it is confirmed.[21]

Over the decade that followed, Montpellier masters apparently worked further on forging a definition as well as an agreement between Avicenna and the "new" Galen. In his *Lilium medicine*, Bernard de Gordon gave the impression that the discourse had covered quite some distance by 1305. He devoted not one but two questions, and three times as many words as Arnau, to the issue. First, he raised the standing *dubitatio* "whether *lepra* is a disease of the whole." Like his more famous colleague, he reconciled Galen and Avicenna by declaring, "[I]t can be called a disease of the whole and of a part, each in a manner of speaking." Bernard also agreed with Arnau in linking his answer to sense perception. On the one hand, he argued, leprosy "presents itself in the integral parts that constitute the whole," and which range from the skin to the liver; on the other hand, "in some [patients] it presents itself first in the nose but not in the eyes, and in others first in the eyes but not in the nose." These general agreements with Arnau, however, were offset by clear-cut differences in tone and position.

Bernard admitted that leprosy could be understood either as an imbalanced complexion in itself or as a consequence, but he added dismissively, "[L]et it be called a disease or a symptom, I do not care." On the more substantial question, whether it was a partial or a total disease, he was not satisfied with Arnau's simple solution of matching the dichotomy with the incipient and advanced stages. Instead, Bernard based his answer on the essence of *lepra*, and he left little ambiguity about the totality of the disease. He reasoned that the pernicious matter was dispersed throughout the body and became manifest in every part. Beginning inside, the disease emerged to the surface and then returned inside to cause death. It destroyed all the faculties: the psychic (*animalis*) by causing delusions, terrifying dreams, and loss of sensation; the life-sustaining (*spiritualis*) by interfering with the voice and respiration; and the nutritional (*naturalis*) by complete disfigurement. *Lepra* could be considered a partial disease only to the extent that it affected "the flesh and not the heart or bones": the natural red flesh, warm and moist, failed "to assimilate" the noxious nutriment; instead, it produced "cold, phlegmatic, granulous and disconnected [*separatam*] flesh." We may recall here, from the beginning of this chapter, that the Andorran medical practitioners solidly based their judgment, in accordance with "what is written in the fourth book of Avicenna," on the distinction between Ramona's localized or "particular" cancer and leprosy, which, "on the contrary, is universal, because it proceeds to the hands, feet, nails and skin.[22]

Bernard buttressed his position on the all-inclusive reach of *lepra*—and underscored his distance from Arnau—by the methodical pursuit of another *dubitatio*, which would have fit more logically in the latter's commentary than in a medical encyclopedia. Paradoxically, while examining this part of Galenic doctrine more closely than Arnau, he also displayed a slightly greater degree of independence from Galen—yet he did not compensate with an obviously deepened deference to Avicenna. Raising the question "whether the complexional balance is uneven or equal in *lepra*," he argued firmly that it was uneven (*diversa*). An excessively cold and dry complexion was so inherently contrary to the warm and moist "principles of life" that, if it uniformly affected the entire body, it would cause swift death—and, as anyone could observe, dying of leprosy was anything but swift. In addition, the burned and corrupt melancholic matter was so harmful that, if evenly distributed through the body, it would cause immediate suffocation and loss of limbs.[23] These arguments dovetailed with the carefully constructed tripartite definition in the first section of Bernard's chapter. He defined *lepra*, first, as a "consimilar disease" (*morbus consimilis*) because it affected the flesh, which was a homogeneous, or "consimilar," part of the body according to standard doctrine. Second, it was an "organic disease" (*morbus officialis*) because it destroyed "the shape, form, and composition of the organs." Third, as a "total disease" (*morbus communis*), it caused the progressive loss of limbs, which would be total if death did not intervene. Almost as an afterthought, Bernard extended his definition to the root cause, which was "melancholic matter dispersed through the entire body."[24]

The coda of Bernard's definition points to a shifting choice of heuristic tools, from qualities and powers to materials and processes. This transition, which was neither linear nor uniform, followed the general direction of fourteenth- and fifteenth-century medical teaching, as the "new" Galen became absorbed in the curriculum and eclipsed by comprehensive and structured works such as Avicenna's *Canon*. At the University of Montpellier, definitions of *lepra* exemplified this shift in a variety of ways. More and more authors transferred the "malignity" (*malicia*) from the imbalanced complexion to the offending matter. Fewer paid attention to the failure of a faculty, and more to the failing organ. In addition, ideas of a bad disposition or a morbid condition yielded the stage to dynamic processes that, though envisioned centuries earlier, had played secondary roles. These processes ranged from the relatively abstract "change" (*mutatio*) to the descriptive "spreading" or "dispersion" (*spersio*), and from the natural-philosophical "breakdown" (*corruptio*) to the more earthy "rotting" (*putrefactio*). Even in the public forum, and for many centuries to come, academic physicians were likely to demonstrate how important they considered the new core of the definition, from condition to process. Thus, the Cologne Faculty of Medicine protested in 1574 that the leprosarium matrons were not qualified to decide the fate of a suspected patient. The matrons failed to understand that leprosy consisted not so much— and here the doctors switched to magisterial Latin—"in a change of complexion" (*in mutata complexione*) but far more in "the destruction of form and appearance" (*in corruptione figure et forme*).[25]

A year or so after Bernard de Gordon published his *Lilium*, and possibly with his encouragement, Henri de Mondeville began an effort that would stretch over many years, to write a textbook on surgery, the *Chirurgia*. When the time came to devote a chapter to *lepra*, he professed,

> [T]his subject is difficult, and it pertains far more to medicine than to surgery. Therefore, I will tread lightly through the matter, except to the extent that people sometimes come to us asking about the signs and judgment of *leprosi*, and so that we may treat palliatively where it is feasible.

Consistent with this standpoint, his definition was straightforward, even categorical, as if to head off doubt or debate.

> *Lepra* is a shameful disease resulting from matter that is melancholic or turned into melancholy, and destroyed by irremediable destruction. It is to the whole body like a cancer is to the cancerous member. Hence, just as a cancer cannot be cured without the total destruction of the affected member, neither is *lepra* cured without the destruction or excision of the entire body—but this is impossible.[26]

The minimalism of Henri's formulation makes some features all the more salient. He makes the totality of the disease more radical by moving from pathology, where Avicenna had introduced the analogy with cancer (both diseases being melancholic, corrosive, and chronic), to therapeutics, where Henri turns it into an argument *ad absurdum* based on the notion of destruction. By inference, he views *lepra* in essence not as a complexional problem but as a process of destruction or breakdown. His brief definition contains four references to destruction, including the emphatic alliteration "corrupta corruptione incorrigibili." A further significant feature of his definition is that matter (*materia*), rather than unbalanced qualities or errant faculties, holds the key to understanding.

Another author, somewhat younger than Henri de Mondeville and with a more fully documented (and more checkered) role in the Faculty of Medicine at Montpellier, was Jordanus de Turre (fl. 1313–1335). He, too, was preoccupied with practical matters when he wrote his "notes on Lepra," not only at the request of "our beloved colleagues, bachelors and students" but also "mindful of the benefit of the commonwealth." He focused on signs and treatment, paid no heed to dialectics, and favored tangible items. The elements of a definition were interwoven with his "general rules" for managing the disease, and each element flowed from the principle that "all *lepra* is caused by corrupted matter." *Materia* took the place that *causa* normally occupied, in the declaration "there is a double matter in all *lepra*, namely antecedent and conjoint." Jordanus specified this causative matter as "burned humors," and further as "matter that is incinerated, congested, and firmly stuck in the simple organs." The Galenic standbys, *malicia* and *virtus*, did make an appearance, but each only once. Morever, both allusions converged on the involvement of a defective liver: "in all *lepra*, there is default [*malicia*] and weakness of the liver first, of the spleen second"; and "in all *lepra*, the natural faculty [*virtus naturalis*] is injured first, both in serving and in being served, so that one should always treat the liver." In the context of other indicators, we may read the convergence on the liver as a reflection of Avicenna's ascendancy.[27]

A generation after Jordanus de Turre, Guy de Chauliac (who had studied at Bologna under Niccolò Bertucci) was still willing to reconcile—or, to be more accurate, to juxtapose—Avicenna's focus on hepatic failure with Galen's blaming of the "assimilative faculty." Still, as we have seen earlier in this chapter, Guy favored the teachings of the *Canon medicine* over the glosses on *De malicia complexionis diverse*. Adherence to Avicenna ("sequendo Avicennam") was his reason for judging that Bernard de Gordon had "treated the matter very well."[28] He blurred or simplified the philosophical distinctions of the venerable Montpellier master, even though he borrowed most of his definition from the *Lilium medicine*. Also, when he mentioned, in passing, "a bad cold and dry complexion," Guy did not proceed to the underlying principle of *malicia*. In sum, Guy de Chauliac reinforces our impression of certain trends, which emerged over three or four generations of authors in the Montpellier orbit. It would appear that, at least for the discourse on *lepra*, the deference to Avicenna expanded, the pursuit of complexional explanations slowed, and the interest in material agents and processes intensified during the fourteenth century.

These trends are conveniently telescoped in a medical compendium, the *Practica sive Philonium*, compiled by Velasco de Tharanta, native of Portugal and physician to the counts of Foix. Citations firmly establish his connection with Montpellier, particularly in the person of Jean de Tournemire, the author's acknowledged teacher. By the range, selection, and interpretation of his sources, Velasco gives us a glimpse into the core of medical doctrine at the time; by his idiosyncrasies, he reveals the softness at the edges. In his extensive chapter on *lepra*, explanations rely more on material causes than on qualities and faculties. The chapter reflects an "Arabized" medical literature, as it mentions Avicenna by name about twenty times more than Galen, in addition to authors such as Haly Abbas, Mesue, Rhazes, and Serapion. On quite a few occasions, it appears (even if we allow for errors by copyists) that Velasco cared more about the comprehensiveness than the precision of his compilation. It would lead us too far here to disentangle the knots in his compilation, even within the boundaries of our subject. One brief excerpt will suffice to illuminate a few salient points. After presenting a classification of "melancholic diseases," which deviated substantially from other taxonomies (see Chap. 5), Velasco asserted,

> [I]f the superfluous black bile is sent to the entire body, if the digestive faculty fails there and is thus unable to convert it into the substance of the part, and if [the black bile] does not putrefy, it causes *lepra*. In these statements Gordonius is in agreement with Avicenna.[29]

Actually, for the corresponding part of his definition, Bernard de Gordon did not cite Avicenna but came to his conclusion that it "was clear enough from Galen" after drawing explicitly on two of Galen's most widely quoted treatises. In a further imprecision, Velasco substituted an excess of "superfluous" black bile for the "corrupt and unfit melancholic matter" that, according to Bernard, failed to be properly assimilated into tissue. This substitution was at odds with Velasco's (somewhat ambiguous) declaration, later in the chapter, that "black bile is the material cause of all *lepra*, in conjunction with its cold and dry complexion." It may also be noted that, by his choice of categories and terms, he diluted the Galenic notions on which Bernard had built. In additional simplifications, Velasco settled for the generic "digestive faculty" instead of the specific "disseminative faculty," and for the more colloquial verb "convertere" instead of the technical "disseminare."[30]

Sixteenth and Seventeenth Centuries

Conceptual and linguistic "dilution" in definitions of leprosy was a feature, no less liberating than impoverishing, which paralleled developments in the vernacularization of medicine. Among the authors who moved closer to non-academic readers by writing in the vernacular, some, paradoxically, also returned to the set core of the medical curriculum. This "conservatism" characterized the first full treatise on leprosy in French, *Practique sur la matiere de la contagieuse et infective maladie de lepre*, which was compiled by Pierre Bocellin in 1539. Resuming the Montpellier discourse after a century-long hiatus, Bocellin was similar in several ways to his earlier fellow alumnus, Velasco de Tharanta—whom he mentioned repeatedly, in addition to Arnau de Vilanova, Bernard de Gordon, and Guy de Chauliac. Bocellin, too, collected didactic sources while he was fully involved in practice and at a physical distance from the academic milieu. His leading authority was still Avicenna, whose definition took precedence over Galen's. At the same time, however, Bocellin's explanation suggested a resurgence of Galenic categories. In his implicit solution of the dilemma, whether the root cause of the disease lay in a faulty faculty or a defective organ, he defined *lepre* as

> a failure of the blood-making faculty [*la vertu sanguificative*] with regard to the liver, and of the assimilative faculty with regard to the skin and the flesh, eventually destroying [*corrumpant*] the union of all the parts.

Except for an occasional effort to integrate authoritative positions, as in this quotation, Bocellin mostly assembled teachings without venturing an opinion, but with the objective of passing them on to a public far beyond academic circles.[31]

We return to the academic milieu and, ironically, to the semblance of a déjà vu for concluding our survey of the discourse on leprosy at the University of Montpellier. During Bocellin's lifetime, "medical humanism" was gathering momentum, with proponents seeking a return to textual precision and "original" sources. With new translations from the Greek and with celebrated publications of collected works, the movement caused an "enhancement of Galen's reputation, accompanied by a good deal of denigration of the medieval Arabo-Latin versions of Greek medicine and of Avicenna." By the end of the sixteenth century, medical humanism contributed to what we might call "a second coming of the New Galen" to the Faculty of Medicine at Montpellier. Across the board, and on the subject of leprosy in particular, responses were far more complex and less conciliatory than they had been to the "first coming." Ambiguity marks the words of our two early seventeenth-century Montpellier authors, André Du Laurens and Jean de Varanda, and their attitudes toward their sources and predecessors.[32]

Du Laurens presented his treatise *De lepra* as a commentary on the chapter of Guy de Chauliac, to whom, however, he paid only passing attention—and backhanded compliments. He prefaced his essay with a historical synopsis, which differed markedly in outlook from Guy's brief excursion into history, and in which he mixed classicist civic-mindedness with almost baroque pathos.

> Among all the diseases, which attack and afflict the human body, none is more loathsome or more to be feared and deplored, than *Lepra*. Some call it "civic death," because one who is infected by it becomes separated from all human society and intercourse, or is held for dead, and hence is barely deemed worthy of the name "human."

From this philosophical opening, duly complemented by the Aristotelian dictum that man is "animal Politicum," Du Laurens switched to an extrapolation of Leviticus, in which he seemed to take liberties with the thrust of the Mosaic legislation. "In the opinion of nearly all the ancients, this evil was a divine punishment, since the ancient laws counted *leprosi* among the number of the doomed and ordered them to be sent away, far from all gatherings of people." His sweeping generalization, together with the subsequent part of his historical survey, makes us wonder whether he intended to underscore the distance between primitive times and his own age, or to recommend biblical opinions and measures as exemplary.[33]

After reporting that the disease had come to Rome rather late according to Pliny, Du Laurens returned to earlier antiquity, and he sounded increasingly censorious. His tone was more moralizing than that of any of his medieval predecessors when he suggested that Hippocrates would never have seen a leprous patient,

> either because of the excellent constitution and balance of the air and region in which he lived, or because of the temperate way of life which was customary to the people of that era. In our time, however, this evil is all too widespread [*nimis frequens*] across Europe, because of the intemperate use of things, even licit ones, because of the neglect of an orderly way of life, and to some extent because syphilis [*lues Venerea*] has not been cured thoroughly enough.

The chapter that followed was surprisingly free of moral undertones. It dealt with nomenclature and the ambiguity of each term ("vocabulum ambiguum").[34]

When Du Laurens finally came to the question "What is *Lepra*," he displayed some ambivalence of his own toward the supposedly principal source of his definition. His characterization of "our Guy, following the Arabs," seemed to relegate his author to the other side. Yet he followed and even out-paced Guy by reintroducing and largely resolving all the Galenist dilemmas, of disease versus symptom, faculty versus organ failure, total versus partial disease, and equal versus uneven imbalance. Du Laurens devoted an entire paragraph to *lepra* as symptom, which he defined as "a failure of the assimilative faculty, resulting in the destruction of the form of the parts." This failure involved "the production of blood" (*sanguificatio*), in the second digestion, or "cooking," which took place in the liver. It made sense, therefore, that Guy "our author, agreeing with Avicenna, establishes a failure of the liver as the intermediate cause of *Lepra*." On his way to defining the disease itself, Du Laurens made a strange detour, introducing the Parisian Galenist Jean Fernel (1497–1558), "a physician of our time who is not without renown." He quoted and briefly explicated Fernel's quite commonplace definition, "a hidden, malignant, and contagious disease of the entire substance, which makes the skin resemble the hide of an elephant." Then, however, he bluntly rejected this formulation ("Improbatur") with the claim that this rejection was not just a personal quirk but an academic, albeit recent, position: "to tell the truth, this definition is not accepted by our men [*a nostris*]" because by now all ambiguity "should be banished from our school."[35]

Building on Avicenna's phrase, Du Laurens anchored his own definition in the formula "a universal tumor, caused by a burned and melancholic humor." On this foundation, he constructed a threefold identity of *lepra* as an "imbalance" (*intemperatura*) of a hot liver which burns the blood and causes the parts of the body to be cold and dry; a "malformation" (*mala conformatio*) resulting in growths and disfigurement; and a "loss of integrity" (*solutio continui*) causing ulcers, cracks, and the like. This definition inevitably steered Du Laurens to another controversial issue that had first been raised, three centuries earlier, by the advent of *De malicia complexionis diverse*, namely, the question of whether leprosy was spread evenly or unevenly. By this time, Renaissance translators had replaced the medieval phrase for an uneven complexional imbalance, *malicia complexionis diverse*, with a more elegant term, *intemperies inaequalis*. After revealing a transitional stage by referring to the evenly present "*intemperatura*, as described by Galen," Du Laurens opened up Guy de Chauliac's dense formulation of the issue, and he embroidered it with his own version of Galen's categories.

Our author calls *lepra* an even and uneven imbalance, since different things are involved. If the bodies of leprous people are considered as a whole, you will find them unevenly imbalanced, where some parts are excessively warm, like the liver, others excessively cold, like the solid parts, which are entirely cooled and dried out. If, on the other hand, [the bodies] are considered exclusively in their solid parts, we shall find that in *lepra* there is an even imbalance, because they are evenly cooled and dried out.[36]

This may have been the last attempt to settle, or even to raise, the dilemma.

Jean de Varanda was Du Laurens's junior by only a few years, but he established an amazing distance from his Montpellier colleague—and predecessors—with his *Tractatus de Elephantiasi seu Lepra.* He began his chapter "What Is Elephantiasis" with a bold critique of previous teaching.

Many writers of medicine, when they have examined the essence of *Elephantiasis,* have proposed various definitions, by which they seem to me to have inadequately expressed its proper nature and disposition [*affectum*], kind, and causes.

Citing the standard formulations verbatim, Varanda set them aside one by one; it did not matter whether they had been derived from Galen and Avicenna, or propounded by prominent Montpellier masters for three centuries. Guy de Chauliac and André Du Laurens were unmistakably included (see Table 4.1) in a first group of authors who fell short. "Some define [leprosy] by the most remote symptom, when they say that it is a very great error of the assimilative faculty [*errorem maximum virtutis assimilative*]." A vast second group, which ranged from Avicenna to Henri de Mondeville and, inferentially, Du Laurens, inadequately defined leprosy "by the resemblance with another disease, when they call it a cancer of the entire body."[37]

Jean de Varanda disapproved of yet another group, which was particularly attentive to semiology and had gained stature since the mid-sixteenth century with authors such as Jodocus Lommius: "Some, giving more consideration to externals and the syndromes that follow diseases, have presented a description of the fearsome face, tuberous skin, loss of feeling in the extremities, and similar accompanying or ensuing dispositions." One category of authors stood at least a step higher in Varanda's esteem: "The greater experts have said that it is a consimilar disease, destroying the shape, form, and composition of the members, and ultimately causing the dissolution of continuity, due to melancholic matter that is spread through the entire body." Bernard de Gordon headed this category, as may be gathered from collating the quotation with Bernard's definition in Table 4.1 ("morbus consimilis corrumpens figuram et formam . . . spersa ad totum corpus"). These experts, however, in Varanda's estimation "still did not reach the true essence of the subject, because they did not say what kind of imbalance it was, in what part it is primarily seated, and from where such great abundance of black bile may flow into the whole body."[38]

Varanda's sweeping challenge to entrenched categories, from *accidens* to assimilative faculty, was matched by his magisterial self-confidence. "Now, let us see if we can establish a better definition." He would found it on the double consideration that, first, it was impossible for a principal organ not to be affected when the function of that organ was impaired throughout the body; second, "beyond the obvious [*vulgares*] qualities and the imbalance in the parts, there are certain more latent factors of contagion and malignity, or agents that destroy the substance." This second pillar was buttressed by the authority of Galen (while Avicenna is notably absent from the treatise), who taught that "poisonous humors are sometimes produced in us" and that these destroy substances. With these premises in place, Varanda formulated his own definition:

> *Elephantiasis* is a dry and warm imbalance of the liver, joined with a certain degree of decay or poisonousness, by which a more abundant black bile is generated and gradually destroys all the parts of the body.[39]

The explanation concentrated on two points, with far-reaching implications that will become progressively clearer in the following chapters.

First, the notion of *intemperies* led Varanda to insist that one could not call *elephantiasis* "simply a cancer of the entire body" because this would mean that the imbalance was evenly present. He conceded that it might be called a cancer "in some way, because it comes from the same matter as the particular cancer, and it invades—though unevenly—not only the internal but also the external parts, above all, the flesh, skin, and muscles." In his second clarification, he confirmed his allocation of the imbalance specifically to the liver, which stood in contrast with earlier circumlocutions, particularly with the elaboration by André Du Laurens of Guy de Chauliac's statement. Varanda anticipated, and answered, challenges to his focus on the liver. His rebuttal of one particular objection is of special interest. One would ask, Why would the liver necessarily be involved, since *elephantiasis* was contracted "not only by heredity but also by contagion or contact; how could it be that those corpuscles, which flow from infected bodies, reach the liver, deep in the body, rather than the skin and other parts?" In reply, "we say that those corpuscles, the seed bearers of contagion [*contagionis seminaria*], can stick to any part in their way, but not change any part in such a way to produce the same disposition, unless they have an affinity with it [*nisi cum illa habent Analogiam*]." One example of such affinity was the diuretic effect of blistering flies or beetles (*cantharides*), which, though attached to the skin, excited the bladder by "a specific antipathy" (*antipathiam peculiarem*). Inversely, the "contagious running" of ophthalmia inflamed the eyes but not any other part of the face.

> Hence, what would prevent a particle and the spirit of blood, in the skin or other external parts, to be infected by contact with this contagious outflow, and then to carry this contagious stain [*labes*] through the veins as a channel, to the liver itself? These external parts are like a medium, but the liver is the terminal to which they are carried.

Against another objection, Varanda argued that, by remedying the imbalance in the liver, it was possible to cure patients, at least those who did "not yet have confirmed *elephantem*."[40]

With this argument, Varanda completed his scan of three issues that, from the fifteenth to the seventeenth century, steadily gained prominence in the attempts to define leprosy. The trend to characterize the disease as contagious, hereditary, and incurable, which calls for further exploration, was increasingly pronounced in academic and public medicine. Far beyond the definitions, these predicates carried the heaviest onus in widespread perceptions, and they persisted into recent times. In treatises and, even more frequently, in official declarations, these characteristics became not merely secondary or contingent attributes but essential properties. They also often overlapped conceptually. Above all, ideas of reproduction and sex provided various links between contagion and heredity, as we will see in the following chapter. Heredity, in turn, raised the issue of incurability, at least indirectly, by the association with determinism. Niccolò Bertucci stated the link in more general terms: "[T]his disease is rarely acquired but most commonly present through generation, so that it is difficult to cure."[41] We will examine the attributions of contagiousness and heredity in the following chapter. The issue of incurability, on the other hand, was associated with "confirmed" leprosy, one of the categories that will be discussed in Chapter 6; inversely, the issue of curability was at the center of prognostication and treatment, which will be the subject of Chapter 8.

The End of an Era: From Conceptual to Empirical Pathology

Jean de Varanda was one of the last medical authors to attach great importance to an elaborate conceptual pathology of leprosy. His outlook was overtaken by a growing emphasis on the presentation of the disease rather than on an understanding based on natural philosophy. The shift in emphasis had begun outside the Montpellier Faculty of Medicine, at the time when that medical faculty apparently fostered a resurgence of Galenism. The most vocal critic of Galen, Paracelsus (1493–1541), had challenged the humoral framework of definitions. Subsequent authors gradually replaced this framework with a more empirical approach to the pathology of leprosy. In particular, Jodocus Lommius initiated the primacy of signs (semiology), and Ambroise Paré and Felix Platter promoted the use of description. These tendencies converged and culminated, to some extent, in the 1765 thesis of Isaac Uddman, who rejected the "extreme inconstancy of the authors" and of received tradition. Deeming "personal observations" (*propriae observationes*) paramount, Uddman presented a new type of "definition." Rather than one tight-knit formula, it was a descriptive mosaic of the features and developments he had witnessed in patients in Finland and Sweden. With intriguing synchronicity, Doctor François Raymond took the same approach in Provence, supplementing the description of *elephantiasis* by Aretaeus with "the examples observed in the different regions of the world" and "the cases that I have had occasion to see."[42]

The preference for descriptions over definitions was naturally accompanied by an increasing elasticity in the identification of *lepra* or *elephantiasis*. This elasticity became most strikingly evident in the seventeenth and eighteenth centuries, in treatises but above all in official documents. In a 1636 thesis, Johann Ruland claimed that he could "conveniently" call syphilis "venereal *lepra*," long after any potential for confusing the two diseases had evaporated. In official documents, the evidence of loosening boundaries—and of various ramifications—was widespread and diverse, so that a couple of examples will have to suffice. In Waldsee in 1697, a physician and a barber-surgeon examined

Anna Weberin, age sixteen, whom we found afflicted with a leprous scab [*ainer leprosen raudten*] on the surface of her body so that, even though she is clean in the mouth and throat, she should be separated from the healthy. Item, we have examined Ursula Weberin, age forty, whose blood has become corrupt as a result of the stoppage of her menses, so that we found her afflicted with a scurvish running scab [*ainer scherbockhischen flissigen raudt*], and she should be likewise separated from the healthy.

In 1761, the municipal physicians of Antwerp expressly closed the door on the leprosy of yesteryear as well as on its "international" incidence—and they abandoned a centuries-old definition—when they spoke of a "*lèpre* of these times" or "*lepra* of this country" (*hujus patriae*); they were referring to a newly appearing skin disease that may have been akin to a complication of psoriasis. In 1777, the sworn physician of Wurzach reported that "Franziska Grabherrin was afflicted with a very painful, malignant, and loathsome skin disease, which we could well call *a kind of Lepra.*" Indeed, in the last hundred years of the period under consideration, examiners tended to base their judgment on attributes of leprosy rather than on the defined identity of the disease. They relied on two criteria in particular, namely, the attributes of contagiousness and heredity, which became fused into the essence of leprosy under the pressure of social concerns. Suggesting the thrust of this fusion, four practitioners in Ravensburg in 1688 found "that Regina, about sixty years old, has in her body a real predisposition to an *aussatz* or hereditary disease, on account of which she should justly be separated and avoided." The projection of leprosy as contagious and hereditary, often with little regard for a precise definition of the disease, was loaded with implications and consequences. These constitute the subject of our next chapter.[43]

CHAPTER FIVE

"Une maladie contagieuse et héréditaire"

Be it known that, by the order of the Lord High Steward of Wolffegg, we the undersigned sworn examiners, Johan Georg Brix, Ph.D., M.D., physician of the city of Wangen in the Holy Roman Empire, and Ignatius Brigel and Antonius Berle, approved barber-surgeons, have examined Judita Hilpertzhauserin, née Bitschin, of Somhof, age fifty-nine. . . .

We have found her *functiones naturales, vitales,* and *animales* to be impaired, and all her external parts horribly disfigured.

Therefore, we conclude that the named Judita is burdened and infected with the foul and hereditary disease of *aussatz,* and that she should be completely separated from healthy people, so that these may not become likewise contaminated.

Dated November 20, 1734, this certificate, like many others, attests to a firm medical certainty that leprosy was both contagious and hereditary. We will see, in the following pages, that this certainty grew in a slow and uneven process, influenced by social trends. The certificate also bears witness to the public fear of the disease's presumed two-pronged threat to society. This fear, indeed, outlived medical knowledge until recent times, in official measures as well as in popular perceptions. Although premodern sources routinely paired contagiousness and heredity, we will tackle each characteristic separately for the sake of clarity.[1]

Contagion: A Coalescing Certainty

The changing understanding of "infection" and "contagion" has a long and complicated history. Fortunately, premier historians of medicine have unraveled most of the complexities in recent decades, for a broad scope of communicable diseases, though with plague at the center. Rather than attempting to summarize their elucidations, I will let them guide my exploration of academic treatises and notarial deeds, while keeping a tight focus on the growth of the conviction that leprosy was infectious and contagious. Around 1600, as we saw toward the end of the previous chapter, the Montpellier professor Jean de Varanda voiced this conviction in greater detail than his predecessors. Public documents baldly matched this precision. In 1570, for example, the commissioners in Ganges, not far from Montpellier, were quite direct about the case of Guilhaume Vézye. When the doctor and the surgeon examined Guilhaume, they questioned him about "his contacts" (*sa conversation*) and "whether he had at any time kept company with people infected by the same disease." Thorough examination led the commissioners to conclude, "[W]e declare him completely leprous, and admonish him in the name of God to withdraw from the contact and company of healthy people, and to make his dwelling in a remote place in the fields, in order not to harm the commonwealth."[2]

Neither this nor any other official document, however, spelled out the technical detail with which Varanda imagined the person-to-person communication of leprosy. Combining various long-standing hypotheses, he visualized the transmission as caused by "corpuscles, the seed bearers of contagion" (*contagionis seminaria*), a receptive "affinity" (*analogia*) in the exposed body, and the conveyance of the infection, or "stain" (*labes*), through a conducting medium. No matter how much he insisted on the plausibility of indirect conveyance, he found it difficult to elucidate the idea. His difficulty became obvious in his mismatched comparison with the way in which "the power of the torpedo fish happens to flow into the fisherman's hand through some intermediary part." The analogy, inaccurately borrowed from Galen, omitted the crucial detail of the (metal) trident as the "intermediary part"; moreover, Varanda's inclusion of an external transmitter would have vitiated his own point about conveyance *inside* the body.

When we consider the obstacles to understanding how leprosy was communicated, we can hardly be surprised that learned explanations, even as they indirectly validated the belief in contagion, did not readily permeate from academe into society at large. This was the case for the notions of *seminaria* and *analogia* (alternating with the antonym *antipathia*), which Varanda and others applied to the transmission of leprosy. In both terms, readers may discern echoes of Galenic tenets or, at least, of ideas entertained by sixteenth-century Galenist authors. These authors included Girolamo Fracastoro, who used the phrase "seminaria contagionum" in his treatise *De contagione et contagiosis morbis*, published together with his essay *De sympathia et antipathia rerum* (1546). Around the same time, Jean Fernel cited ophthalmia as an example of a flow that was contagious according to the consensus of the authors. Varanda's own contemporary Santorio Santorio (Sanctorius, 1561–1636) listed scores of effects produced by "sympathy" in a commentary on Galen.[3]

Early Attempts to Understand Transmission

One medieval author who had shown some familiarity with the idea of contagion by sympathy was Arnau de Vilanova. In his commentary on *De malicia complexionis diverse*, Arnau seconded Galen's speculation that disease could make one person's complexion hostile to another's, "so that in contagious diseases, one individual, by an affinity [*per similitudinem*], injures and even corrupts another with his or her humidity: this is the case for the sperm of a leprous patient and the saliva of an epileptic person or of someone with rabies."[4] Except for scattered instances such as this, however, a vast distance separated Galen's inchoate suggestions from the detailed speculations of his commentators and, even more, from the early seventeenth-century certainty about the spread of leprosy. Differentiations, between external and internal agents, between infection and contagion, and between metaphorical and physical senses of "contact," developed very slowly. The period under consideration was a transitional stage, between the vague sense that leprosy was somehow communicable and the widespread modern conviction that it can be transmitted by mere touch. During the transition, suspicions centered on the air as a medium, while maintaining a certain ambiguity about the tangibility and intangibility of this medium—in other words, about transmission as a material process or a consonance.

The ambiguity about communicability may be illustrated by the contrast between two nearly contemporary viewpoints. While Arnau de Vilanova believed in contagion by affinity, the Catalan physician Martí de Soler held, as Michael McVaugh calls it, a "materialist view of transmission." When Martí cleared a reported *leprosus*, he assured that the man was no threat to the com-

munity because "*lepra* was not present in his body at the moment, and therefore it could not pass from him to someone else." The position of Velasco
de Tharanta, nearly a century later, was still ambiguous, although he generally insisted on the contagiousness of leprosy, as we will see. He answered
the question of "how *lepra* can be caused by exterior touch, even when the
liver is not yet injured," with this rather oblique statement:

> I say that the internal humors can be infected and corrupted by inhaled air, lead-
> ing not only to a putrefying fever, as we have seen in the chapter on pestilence,
> but many other diseases, as is evident in pulmonary consumption [*ptisi*] and *lepra*,
> in which those infected humors come to the liver and infect it; the infected liver
> produces much black bile and then sends it through the entire body.[5]

Touching a leprous person was hardly dramatized in medical sources as
the ultimate horror, notwithstanding the facile and widespread assumptions
in popular stereotype and pseudohistory. Physicians or health commissioners did not commonly link clothing to the spread of leprosy, in sharp contrast with the measures that officials prescribed against plague. I have encountered only one explicit connection, and it was made relatively late, in the early
seventeenth century. Varanda instructed examiners to question a suspected
patient "about things that are able to induce contagion, namely whether he
has used the clothes of *leprosi* for a long time." It seems also puzzling that
medical authors were silent on manual contact, and particularly that they did
not mention gloves even though some municipal ordinances required them.
The municipal accounts of Brugge show that, from 1305 to 1580, the magistrate provided gloves to people suspected of having leprosy—before they
were examined. It is safe to infer, from the contrast between medical reticence
and public measures, that, at least for a time, physicians showed less concern
with touch than officials did.

One exception to the silence of medical authors is probably no mere coincidence. The Flemish surgeon Jan Yperman (ca. 1260–ca. 1330) made the
sweeping—and somewhat cryptic—assertion that "the hands of a *laserschen*
person would make healthy people *lasers*, and for fear of this [patients] are
removed from the community: this is ordained in the Old and New Testament." It was unusual, to say the least, for a surgeon to associate manual touch
with fear of contagion. Moreover, Yperman's association is amply balanced
by the evidence of palpation and other contact, which we will introduce in
the chapters on the diagnosis and the treatment of patients. For graphic and,

PLATE 6. "Contagious Breath," priest hearing confession from a leprous patient (detail): the central scene in the *Lepraschau*, Nuremberg, 1493 (see Plate 2).

arguably, convincing disproof of the supposedly universal and absolute fear of touching or being touched by a *leprosus*, we need only look at the woodcut titled "Inspection of Leprous Patients" (Plate 8 in Chap. 7, where I will comment on the plate more fully). In this depiction, first printed in Strasbourg in 1517, the surgeon is placing both hands on the patient's head. At the same time, the surgeon's facial expression may be explained by "the stench of breath," which the caption lists as one of the symptoms, and which would have been associated with contagion by air rather than by touch.[6]

The vignette in Plate 6 powerfully evokes both the dread of smelling the patient and the fear of being infected by his fetid breath. This scene is a detail from the center of the woodcut, printed in 1493, which we have examined in Chapter 2 (see Plate 2). At first glance, the person shielding his nose and mouth appears to be a physician, for he is wearing the kind of bonnet that identifies physicians in general iconography—and even in another scene from the same woodcut. Furthermore, some instructions made it clear that "the physician should always protect himself from the breath of the examinee, lest he become infected." It does not take long, however, to realize that the seated person is, rather, a priest. The print, in which he is the central figure, was distributed as a prayer sheet at one of the crowded examinations that were staged annually in Nuremberg. The patient, with visible facial lesions, is kneeling in a position of supplication, with his hat and staff lying next to him and a clapper attached to his belt. It is most likely that he is either beginning to make his confession or asking for absolution and consolation. Thus, in addition to visualizing a dual notion, namely, that leprosy was transmitted through the air and that fetor was one of its most repulsive effects, the scene dramatizes a general popular rather than a specifically medical response. It also opens a vast dimension beyond medicine. The detail, and the other inspirational scenes together with the captions of the woodcut, illuminate associations—rarely expressed but undoubtedly adopted by medical practitioners—between suffering and redemption, the flesh and the soul, and disease and sin.[7]

In view of currently keen associations between contagion and sex, and of past theories about confusion between leprosy and syphilis, it is important to emphasize at this point that sexual transmission was not integrated into the premodern definition of leprosy but, rather, included among sundry causes—which will be the subject of our next chapter. It is equally important to realize that, for a long time, contagiousness was but one of the secondary characteristics of the disease in the medical discourse—in contrast with the ecclesiastical decrees—and that it became a primary hallmark through a very uneven process. Furthermore, nonmedical circumstances were most instrumental in transforming contagiousness from an attribute in the definition to the principal stated reason for isolating patients. To be sure, as early as the first century CE, Aretaeus had exclaimed, "[W]ho would not flee them? [W]ho, be it a son, father, or own brother, would not turn away when there is fear that the disease may be communicated? This is why many have taken their dearest ones into deserts and mountains." However, the context proves that this was an expression of horror rather than an affirmation of communicability. After all, the same was true in the twelfth century, when Hartmann von Aue's poor Henry "became so ugly that no one wanted to look at him." In any event, the remark of Aretaeus left no trail in medical discussions.[8]

Harbingers of Change

For well over a millennium, there was little evident change in the vagueness of opinions on the transmission of diseases in general, and of leprosy in particular. The explanation of leprosy as essentially a complexional and chronic disease did not encourage curiosity about its communicability, or about the ramifications of infection. When classical Latin authors associated "touch," in the words *contagio* and *contagium*, with the spreading of disease, they set the stage for a metaphorical understanding of the term. In a narrower framework of *physica*, from the first to the early thirteenth century, "contagion" referred to an indefinite external agent or, in an even more restricted sense, to the internal spread of corruption. External agency drew greater attention in the course of the thirteenth century, in a changing social framework. For example, the emergence of towns fostered consciousness of "the other," and canon lawyers labored to classify human conditions and relationships.

Signs of change, though inchoate and uneven, surfaced in the medical discourse. Salernitan writings, together with newly available translations from Arabic, drew attention to the role of infected air and of the proximity of patients in causing leprosy. Johannes Platearius listed several causes "by which the spirits are infected or the humors of the members are corrupted, so that impurity will follow, and corruption of the body, and thence *lepra*"; these causes included "corrupted and infected air" and "the presence [*accessio*] of *leprosi*." The Latin version of Avicenna's *Canon* confirmed the role of "the corruption of the air," and it explained that "the proximity of *leprosi*" was a cause "because the disease is invasive." It is possible—but far from evident—that speculations on the causative role of inhalation (*inspiratio*) and exhalation (*respiratio*) in the "New Galen" further enhanced thoughts about surrounding air and nearby patients. In his treatise *De differentiis febrium*, Galen taught that "vaporous and smoky superfluities, constantly pouring forth," were drawn in and that surrounding air could be "contaminated by putrid evaporation." These explanations, however, pertained to pestilential fever, and early Scholastic exegesis seems not to have dwelled on their implications for the spread of *lepra*.[9]

Around 1300, a medical work linked, apparently for the first time, infection by air with the expulsion of patients, thereby "irreversibly crossing a line," according to François-Olivier Touati, currently the leading authority on medieval leprosy. The medical work, titled *Breviarium practice*, probably originated in Italy and, I believe, in a nonacademic urban milieu. The author of this "breviary of practice" claimed that

> *leprosi* infect the air, and this infected air infects in turn when it is drawn in and enters the bodies of people who converse with them. Because the air infects those people, *leprosi* must be separated from the conversation of healthy people; therefore, let them be forced to live in remote places.

In 1305, Bernard de Gordon moved a step closer toward the idea of epidemic communicability when, enumerating several causes, he stated that the introduction of *lepra* "is possibly due to bad, corrupt, and pestilential air." Bernard seems to have been the first to add the modifier "pestilential" to the phrase "corrupt air" of his sources. The addition is worth noting, all the more as it stands in contrast with the omission of the phrase itself by Theodoric, only a few years earlier. Bernard also modified "the proximity of *leprosi*," from Avicenna's enumeration of causes, into "too much chatting" (*nimia confabulacio*) with them.[10]

Henri de Mondeville, while adopting Bernard's references to air and prolonged conversation, was characteristically more blunt about cases of advanced *lepra*.

> Neither the physician nor the surgeon should get involved in these, unless he is persuaded by urgent prayers or a very high fee, and only after a formal prognostication, because it is a most vile and contagious disease, and because the *leprosus*

takes great pleasure in chatting with physicians and getting close to them. Also, if people find out, they despise physicians who deal with *leprosi* and, very often, they hold them for rotten and accursed.

For all his bluntness, Henri still—like Bernard shortly before—qualified dangerous contact as "prolonged" conversation. In addition, Henri did not connect the dots between the presumed dangers of contagion and the need to sequester patients. The incompleteness of these connections was also evident in the intolerant setting of 1321, when the Jews were accused of plotting to spread leprosy, not by infecting people but by poisoning wells. Around the same time, Jordanus de Turre seemed to ignore the danger of spreading infection when he recommended that patients "be moved frequently from place to place." In the 1330s, Niccolò Bertucci remained ambiguous: he envisioned an internal process rather than external transmission when stating, "[P]hysicians call it a bad disease because the healthy body is stained by its contagion"; and he retained a degree of reserve when listing "constant habitation" among patients, rather than casual contact, as one of several common but indirect causes of *lepra*.[11]

Bertucci's onetime student, Guy de Chauliac, was the first textbook author to make a straight connection, in 1363, by labeling the disease "contagious *and* infectious," and to include air and contact, jointly, among the primary causes. It stands to reason that the Black Death (1347–1350) was a watershed for medical positions, especially with regard to the spread of disease, by imposing "a new interpretive framework," as Touati argues. On the other hand, it is striking that Guy was still silent on the need for sequestration, while officials who recorded medical interventions—and even some authors of vernacular medicine—began to link this need with contagiousness *before* 1347. This linkage apparently emerged in the closing decades of the thirteenth century. It was still absent, in midcentury, from the oldest extant record of an examination by physicians (rather than by nonmedical referees, as was common practice). According to this record, a notarized report of 1250, four physicians examined a man named Pierzivallus in Siena, "at the command of the podesta and court." They "judged and sentenced him as being infected" (*iudicaverunt et sententiaverunt eum pro infecto*), but they were silent on the notion of contagion as well as on the consequence of expulsion. Rather, poor Pierzivallus received a "subvention and alm" of 10 shillings for when he "went to live with the residents of Saint Lazarus." This detail does not preclude the possibility that some people in Siena wished to remove Pierzivallus from their midst;

nevertheless, it supports the most current and most informed thesis that, until the late thirteenth century, leprosaria served more as charitable shelters for the incurable than as isolation wards for the contagious.[12]

A rising concern with contagion, as well as a gradual convergence of official and medical views, may be adumbrated by the sequence of some documents. In 1287, the statutes of the town of Chianciano ordained that someone suspected of having leprosy "should be seen by a knowledgeable physician, and if he is judged to be leprous, he should be entirely expelled from the company of the healthy within fifteen days." The first factual evidence in which infection and segregation appeared together, in a forum that was both medical and official—though still without contagion as an explicit nexus—was a medical certificate issued in Pistoia in 1288. It was dictated by Master Laurentius, physician, who had examined Pasquese de Stagiana at the request of her father. The physician, "invoking God's name, officially pronounced that the said Pasquese was infected with *lepra*, suffered from the disease, and had to be separated from healthy people."

A further stage in the pre-1347 escalation of expressed concern with contagion was reflected in a medical decision that was recorded between 1334 and 1337 in Bavaria. It was the first certificate to originate outside an Italian urban setting, and the first of numerous German clearances (*Gutachten*) that were issued until the early nineteenth century. In the document, a high church official informed a rural dean that "bearer has been found, in an examination and declaration by physicians in our presence, to be healthy, clean, and free of the infection of *lepra*." Since the subject appeared "somewhat disposed" to incurring the disease, he was ordered to return in a year. Meanwhile, the dean was enjoined to proclaim this result from the pulpit and to admit the man "to the communion of other healthy people." One practitioner who, also before 1347, alluded to the nexus between infection and sequestration was Thomas Scellinck (d. ca. 1350) of Tienen, in Brabant (east of Brussels). In his Middle Dutch surgical manual, *Boec van Surgien* (1343), Scellinck drew a sweeping inference from Galen's teaching that the complexion of leprous patients is "evil and corrupted." He inferred,

[H]ence, their thoughts and thinking are evil and poisonous. This is why they should be separated from healthy people, and not be allowed near the healthy or to walk among them. And Galen says that their breath corrupts the air, and from this air healthy people may become leprous [*laserich*].[13]

Official procedures and some vernacular treatises convey the overall impression that they firmly linked infection and sequestration decades before academic authors propounded the connection. If this impression is accurate, how can we explain the lag? A plausible answer is that practitioners, such as the surgeon Scellinck, were close to popular fears, and that public officials took the initiative in spelling out the details and dangers of contagion; university men, on the other hand (and, perhaps, sooner in their role as practitioners than as writers) would have followed the trend by expressing increasingly categorical views. This hypothesis, and even the premise, calls for two qualifications, one with regard to official measures and the other with regard to medical positions. There continued to be public records in which contagion and separation were not connected, mentioned, or even implied. This was the case in Vienna in 1380 when Conradus de Dannstet, physician to Duke Leopold of Austria, certified that Petrus Pirchfelder was "entirely free of the infection of *lepra*," in contradiction with the suspicion of many. In fact, Doctor Conradus showed little fear of contracting the disease, for he examined Petrus "in a room of my own house" and in the presence of two other practitioners.[14] We should further qualify our hypothesis by the realization that doctors often warded off spontaneous fears, even if academic authors eventually joined not only the preoccupation with contagion but also the social momentum toward segregation.

Several documents bear witness to the restraining influence exerted by doctors. In 1357, for example, three examiners of the city of Cologne issued a levelheaded verdict.

> We have diligently examined Mr. Johannes named Junghen, priest of Bunna and pastor of the church of Kruppach, who was charged with the blemish of *lepra*. We have found, observing the signs of the Art of Medicine, that he is clean of the said disease. Therefore, he may safely circulate among the people, and he does not have to be shunned or separated on account of the aforesaid ailment.

In 1411, the Parisian professors resisted prejudicial exclusion when they assured that Jean de Bierville was free of *lepra* and that "his company should not be feared on account of contagious disease." In Nuremberg in the 1470s, Hermann Schedel calmly reiterated his earlier finding that, in spite of a badly ulcerated foot, a man was not infected, even while admitting, "I understand that Conrad has spent some time in the leprosarium with others who were infected, and I suspect that he has been quite adversely affected in the humors of his body by the prolonged stay with *leprosi.*" In Cologne in 1521, in a similar case that would naturally and logically have triggered fear of contagion, three members of the Faculty of Medicine evinced an air of almost blasé calm in their decision. They declared that an examined monk did not have to be sequestered, "notwithstanding redness of the face, various macules, and tuberosities in several parts of his body, particularly near the left hand, which he admitted having acquired in the service of a superior who had died of *lepra.*"[15]

Rising Public Alarm

Even if physicians calmed popular fears of contagion and restrained over-hasty sequestration, official concerns became increasingly epidemiological from the late fourteenth century on. Authorities, and then their health commissioners, kept raising the alarm level, introducing preventive measures, and widening the scope of their interventions. Some authorities, however, were and remained more worried about threats to order than about risks to health. In 1371, Charles V of France issued an injunction against the "great number" of leprous beggars who were milling around in the crowded streets of Paris. It appears that the injunction was neither heeded by the beggars nor enforced by the supposedly endangered burghers, and even that the promiscuity spread beyond Paris. In 1413 Charles VI warned, more forcefully than his predecessor—but still without allusion to contagion—that

> many *meseaux* men and women, infected with the disease of *lepre,* are constantly, day in day out, going and coming through the towns, seeking their livelihoods and alms, eating and drinking in the streets, crossroads, and other public places where the greatest masses of people pass. This goes so far, that they quite often bother people, and hinder them in passing through and in going to their businesses.[16]

The shift to a more epidemiological concern may be illustrated by the juxtaposition of two documents that were issued in the same place but separated by twenty years. In 1375, the residents of Lausanne's leprosarium filed a petition against Johannes Dalphin, who, they claimed, "has, by mistake, been unduly judged leprous and, as a result, separated from the company of healthy people" and ordered to spend the rest of his life in the leprosarium. The residents complained that, while staying with them, he was "consuming their goods and alms," and they "urgently requested" the court to order a medical examination because the *leprosi* should not be burdened with "the additional affliction of this consumption." The court official, concerned, "above all, that someone should not be infected with the contagion of *lepra* and that the goods of the *leprosi* should not be consumed improperly," commissioned an examination by the physician Jacques de Cors, who judged Dalphin not to be infected "at the moment."[17]

Two decades later, the curate and parishioners of Lausanne's Holy Cross parish reported Johannes Giliet to the court as being suspected of having leprosy.

> Public rumor and clamor run in the city of Lausanne and the surrounding area, that Johannes Giliet is corrupted and infected by the disease of *lepra*, so that, if it should happen that Johannes conversed with the healthy, the contagion of *lepra* could cause those healthy people conversing with him to fall prey to the same disease. This would turn into a great danger and burden for the common good.

The parishioners wanted the court official

> to declare: that Johannes Giliet is leprous, and infected and corrupted with the disease of *lepra*; that he must be separated from the company of the healthy; and that we must compel him, with the remedy of law, to observe the decision.

As he was requested further "to provide for appropriate remedy," the official summoned a physician and a barber, "sworn to us for the investigation of such or other diseases," who judged Giliet "to be leprous and corrupted and infected by the disease of *lepra*."[18] A comparison of the two documents, even without word-by-word parsing and with allowance for their different occasions and outcomes, readily reveals the change in focus and tone. Especially the appeals to "the common good" and "remedy of law" suggest a higher level of nervousness in late fourteenth-century Lausanne.

In the fifteenth century, contagion loomed ever larger as a public threat, in official measures and, in varying degrees, in popular fears. In 1416 the magistrate of Auxerre commissioned the examination of Thienotus Richo because "he could pose a very great risk to the local inhabitants, unless we provide a remedy, especially since the said disease of *lepra* is definitely contagious [*contagiosus existat*]." When calls for examination and separation originated in the community, they often gained in intensity when broadcast by authorities. Official warnings about the person-to-person spread of leprosy reflected not only rising popular fears but also the expansion of public health mechanisms. It is also likely that some authorities alarmed the populace for political purposes. Moreover, any attempt to gauge the extent of fear should be guided by two caveats. First, concerns about contagion by proximity were neither constant nor universal, and the spread of fear was neither a linear development nor a universal phenomenon. Second, it is important to keep in mind that calls for removing a suspected source of contagion, even when they appealed to fear, often had some ulterior objective. One such objective, as we saw in some *iudicia* in Chapter 2, was to secure charitable shelter for suffering indigents. Another ulterior motive for appealing to fear also shines through—in an almost comically different way—in a suit that the dean and chapter of Troyes cathedral filed in 1516,

> against two leprous priests from the burgh of Saint Siria, who are not residing in their own cottages outside the burgh but among the people in the very burgh. What is worse, they keep concubines with them, who openly bring the ecclesiastical vestments to the church of Saint Siria for the masses to be celebrated by those *leprosi*, to the *scandal and peril* of all the people of the said burgh.

The chapter members sternly requested the Troyes court official to take action, "or else, lords such as the temporal lords of said place will apply a remedy in accordance with law."[19]

A Newly Crystallized Axiom, "The Disease *Is* Contagious"

The threat by the Troyes cathedral chapter, and the inferred readiness of the temporal lords to enforce order, typify the tightening of social structures over the course of two centuries. From the decades around 1400, this tightening had an increasingly visible impact on medical positions. Treatises presented a hardened identification of leprosy *as* contagious; certificates, paradoxically, yielded to equivocation. Authors within and outside the university came to treat transmission by proximity as axiomatic, though usually still emphasizing conveyance by air rather than touch. They adopted contagiousness as an a priori framework for experience, even when they appeared to be observant. Note, for example, the nexus between "we see" and "because" in a collection of dialectical yet perceptive *questiones*.

We see that healthy persons become infected by leprous ones, because the disease is contagious. Certain diseases are contagious, and they can be known from the following verses: "Acute fever, phthisis, scabies, falling sickness, sacred fire, anthrax, *lipa*, *lepra*, and *phrenesis* tell us the contagious ones." Even a healthy person who is not predisposed can be infected on account of continued presence among the infected because they corrupt the air and consequently the blood of the healthy person.

While the list of "contagious" diseases was not new, it was framed with a suggestion of personal observation, especially when the unknown author reiterated the qualification that the transmission required prolonged proximity ("continuam moram cum ipsis leprosis").[20]

Medical Authors

In 1418, Velasco de Tharanta underwrote a definition of *lepra* as contagious. While he was not "original" but derived the ingredients from various sources, he welded them into the most comprehensive formula to date: "I say that it is a shameful, contagious disease," more contagious than scabies and similar diseases. Specifically,

primary causes are the corruption of the air, such as there is in pestilential air; also, dealings and conversation with *leprosi*, because the inhaled and exhaled air, corrupted and infected by them, corrupts the surrounding air and thus infects someone who draws it repeatedly: because this is a disease that invades others, as Avicenna says.

Avicenna, we should note, was far more ambiguous. In the Latin version of the *Canon*, proximity was among the "contributing" (*iuvative*) rather than primary (efficient and material) causes, and the disease was called "invasive" without "others" as passive object. Velasco, however, was insistent in his precision, and he specified further that the causes included "cohabitation, conversation, and frequent eating with *leprosi*, because it is a contagious disease that infects others [*contagiosus et aliorum infectivus*]." In a more elaborate and, arguably, more ominous clarification, he ranked *lepra* with ophthalmia and pulmonary tuberculosis when treating it as a prototypically contagious disease.[21]

Sixteenth-century authors moved farther away from the teachings of Avicenna and Galen, at least on this issue, by integrating communicability ever more fully into the essence of leprosy. Pierre Bocellin, who vernacularized much of what Velasco had written, presented this integration as established doctrine. In a surprising but revealing twist, however, he anchored the doctrine not in medical tradition but in religious law.

> It is said, *by definition*, that *Lepre* is contagious, because its infectiousness accounts for the command, even in Mosaic law, that the afflicted should be separated from the conversation of the healthy.

For further rhetorical effect, Bocellin then switched to Latin to cite the rationale, "lest a few cattle infect the entire herd." Ironically, the sentence "ne paucae pecudes omne pecus inficiant" was not a biblical adage but a commonplace proverb, with possible Roman origins in Juvenal and Virgil, which evoked age-old pastoral experiences with communicable diseases. Ominously, however, the sentence also echoed expressions in church documents, one a condemnation of unruly *leprosi* in Melun in 1201, and another an excommunication by the archdiocese of Sens in 1312. Thus, Bocellin's use of the sentence suggested the convergence of medical notions about contagion with popular calls for preventive separation as well as with ecclesiastical measures of exclusion.[22]

Ambroise Paré, like Bocellin, often reflected vernacular notions; as surgeon to kings, however, he was much better placed to reinforce those notions. In his *Briefve description de la lepre ou ladrerie*, he spelled out, in almost effusive detail, the aspects and implications of communicability. He attributed the higher incidence of *lèpre* in certain regions, in part, to frequent congress with patients,

> for the sweat and vapors that issue from their bodies are poisonous. And so is their breath, and drinking from the cups and other vessels from which they have drunk: for their mouth emits a bloody saliva from the gums and teeth, which is poisonous in nature, in the same manner as the drool of a rabid dog. For this reason the magistrates enjoin them to drink only from their own barrel, and it is my wish that all the *ladres* would do this, so that they might not have occasion to infect anyone this way.

In the strongest suggestion of an epidemic disease so far, Paré asserted that leprosy was "a contagious disease, almost like the plague [*quasi comme la peste*], and totally incurable, like the plague, too, often is."[23]

In a chapter that was devoted entirely to "separating the *ladres* from the conversation and company of the healthy," Paré exemplified both an author's conformity with the cultural context and his ability, in turn, to influence his milieu. He weighed in with his personal advice, and he reinforced the connection between fears of contagion and official measures, by applying the emphasis of reiteration and descriptive evocation.

> In view of the danger inherent in conversing with such people, the authorities [*les magistrats*] must see to it that they are separated and sent out of the company of the healthy, all the more because this evil is contagious almost like the plague, and the surrounding air, which we inhale and draw into our bodies, may be infected by their breath and by the exhalation of the excretions that issue from their ulcers. The healthy person attracts this air when conversing with them. . . . For this reason, it is good and necessary to have them separated.

In one of the few instances in which authors mixed medical and biblical justifications, Paré added that the sequestration "does not run counter to Sacred Scripture, for it is written [Numbers 5:2] that the Lord ordered the *lépreux* to be separated from the host of the children of Israel, and the same is commanded in Leviticus." Paré noted two differences between the treatment in Mosaic law and in his own society. Instead of having to wear torn clothes and to pronounce themselves unclean, the condemned "today are given clappers and a cup, so that they will be recognized by the people." In a second chronological differentiation, Paré managed to move farther into religious territory while, at the same time, insinuating that the Old Testament fell short in humanitarianism.

> However, I do advise that, when we will want to separate them, we should do this as gently and amiably as possible, keeping in mind that they resemble us, for, should it please God, we would be touched by a similar disease, nay, even worse. And we should admonish them that, however much they may be separated from the world, they are nevertheless loved by God while patiently bearing their cross. In truth, Jesus Christ, while on this earth, was willing to communicate and converse with *lépreux*, giving them health.[24]

Another author who, probably more than any other, interwove medicine and religion while linking contagiousness with sequestration was Guillaume des Innocens. In 1595, while making it clear that the biblical *lepra* was a different disease and the result of "the will of God, justly angered at the sinner," the Toulouse surgeon nevertheless saw continuity between the Old Testament and his own times.

> Emulating the Hebrew people, all our Christians as well as everybody else, infidel and pagan, in a well governed society [*en leur République bien policée*], today conduct themselves in a similar extraordinary way toward known or alleged *Eléphantiques* or *lépreux.*

For des Innocens, more explicitly than for Paré, the purpose of giving a clapper to leprous patients was, "so that, by its noise, the neighbors, bystanders, or others within earshot, be warned to make way and to stay far from the path, air, or breath of those poor folks, giving them space and, from time to time, some alms." While he echoed Paré in explaining that the patients' oral ulcers infected the air, des Innocens added greater emphasis by pointing out,

> [I]f Hippocrates and Galen have said that the breath of consumptives is very contagious for the bystanders who will inhale it, how much more should we fear the breath and fetid, infectious exhalations of *ladres,* in whom the evil is not only seated in the lungs but has also ravaged, by its poison, the vital, natural, and psychic parts.

Des Innocens took a significant step further than his predecessors when he identified the initiators of protective measures. Instead of "les magistrats," he credited—though somewhat tentatively—academic medicine with the first response to the danger of infection. "This is why the school of physicians (in my opinion) has invented such procedures for safeguarding the public interest, by preventing the conversation of the *ladres* with the healthy and clean people."[25]

Medical Examiners

Whether the initiative of protecting people from infection lay with "the school of physicians" or with the magistrates, it should not eclipse the role of the medical examiners. It is easy to imagine that these, while generally directed by stern notions of contagion, were often caught between academic restraint and public pressure, and between theoretical premise and practical situation. Heinrich Losser applied medical doctrine when he observed, in

Frankfurt in 1456, that confirmed patients "must be isolated" because their "lungs and breath are infected or poisonous, and this may cause harm." In Regensburg in 1483, Hermann and Hartmann Schedel summarized the entire notion of infection when they clarified a decision. On January 3, the uncle and nephew jointly wrote a deposition regarding Elizabeth de Streyberg, who wished to know "whether she was infected." They found her "disfigured and blemished by some preliminary signs that tend toward *lepra*." Two days later, "since doubt might arise—among both genders—about this judgment," the doctors felt constrained to clarify.

> By the tenor of our judgment the lady is in the beginning stage of *lepra*, and since *lepra* by itself is a contagious disease according to the authority of all the physicians, it is right that Lady Elizabeth should be avoided.

However, the two-step intervention by the Schedels raises several questions about the particular situation. Did these sworn physicians of Nuremberg make an exception by handling a request in the imperial city of Regensburg? Did their allusion to possible doubts "among both genders" carry any significance? And why did they deem it necessary to appeal to universal medical authority? These questions make it tempting to suspect that this case concealed some intrigue and that it unfolded in a smaller social circle rather than in the community at large.[26]

The anecdote from 1483 Regensburg also illustrates the paradox that, as notions of contagiousness became more firmly set, their application to individual cases became more open to doubt. Indeed, with increasing frequency—and perhaps in conscious resistance to public pressures—medical examiners declared a suspected patient neither positive nor negative. They issued inconclusive reports instead and advocated temporary remedies. This trend is exemplified by more than one-quarter of the three hundred or so inspections that I have tallied from the fifteenth through the eighteenth century. The earliest instance of this precautionary ambiguity which I have found was the Auxerre case of March 1416, which, as we saw, triggered fear of "very great risk." The practitioners claimed that Thienotus was "not completely leprous *but, to be on the safe side*, he should be given a deadline in September and be warned meanwhile not to go into crowds or large gatherings." Both the length of the reprieve and the restrictions on circulation might be left vague. In Frankfurt in 1456, Doctor Losser and his colleagues gave Rupprecht Snyder an undetermined "reprieve [*frist*] with the understanding that, meanwhile, he will restrain himself, keep to himself, and not be too sociable." In Lille in 1559, Doctor Valerand de Courouble and Master Surgeon Gilles Beudon found it necessary that Boudewyn Cupre "be separated for some time from communication and conversation with the healthy." Most commonly, however, the term for reexamination was set at six months or one year, and the separation was delineated with precision.[27]

Of the scores of provisional reports, none is more illuminating than one that was part of a rather intense episode in Marseille in 1464. The community of Auriol requested Jacobus Buadelli, licentiate in medicine, and Johannes Grivilhoni, master surgeon, to examine Bartholomeus Giraudus. Finding no unequivocal signs, the practitioners claimed,

> [W]e know and declare that Bartholomeus should not be separated from the healthy, with the proviso that we order him to be absent from the congregation of people, namely, in church, the town square, the market, and other places, for one year only, counted from the present day. In addition, we also order that Bartholomeus should abstain from certain foods, which we have identified for him, for the same duration of one year; that he should have himself purged by expert physicians, so that he does not fall into that grave disease; and that at the end of that year he should be visited and tested by experts in medicine, to see whether he is then entirely cured or touched by the said disease.

Right after this deposition, the bailiff and council members "protested solemnly" that they would hold the physician responsible, in his person and goods, if it turned out that the suspect was really infected and should have been separated immediately. The same day, Master Jacobus, "after listening more conscientiously to diverse arguments against Bartholomeus, who had reportedly disobeyed earlier directives," ordered that, "if and when he was found not to observe the decision, he would immediately have to leave the place for a year."[28]

Master Jacobus Buadelli revised his judgment only minimally and conditionally, in a diplomatic concession to public pressure. Many examiners, on the other hand, issued their reports in obvious deference to the magistrates, with resulting overtones of alarm. In Geneva in 1489, when three practitioners stated their commission, they evidently recited the formula that the authorities had employed. They were ordered to examine two suspected patients, Petrus Martini and his sister Anthonia, "so that we might avoid the ill that might befall many [*malum multorum*] if no measures are taken." Reflecting similar public anxiety, two physicians of Ieper in 1557 found Janneken Brassers so greatly ill that no one wanted to take care of her "because it would bring infection and corruption to everyone [*een ighelyck*]." It is worth mentioning that, with some ambiguity, the Ieper physicians blamed the abandonment as well as the contagiousness of Janneken on "the great stench of her bodily affliction." Three years later, they decided that Jacob Mahieu should be separated on account of "the great threat of contamination which emanated from him" (*het grote inconvenient ende besmettijnge van dien uutspruutende*). In 1570, the Ieper commissioners judged that Walram Camerlinck was so infectious that his continued presence in the community would present a "grave danger" (*groot dangier*), and that Pieter vanden Briele was "very dangerous" (*seer dangereux*). The awareness of a threat eventually produced an official (if temporary) German neologism for the danger of infection, *Ansteckungsgefahr*. In the often intriguing way of etymologies, the verb *anstecken*, also used for lighting a candle, suggested an interpretation of infection which combined the agencies of air, touch, and a medium (fire in this case).[29]

COMPLETING AND TRANSCENDING THE NOTION OF CONTAGION

In certificates and treatises of the sixteenth century, the notion of contagion culminated in a simple juxtaposition of conversation and contact. This was a natural outcome, as medical authors, of Latin and vernacular writings, increasingly took it for granted that leprosy was intrinsically and dangerously communicable. Girolamo Fracastoro simply classified *elephantiasis* among the contagious diseases in his treatise devoted to this category. André Du Laurens, citing Jean Fernel, presented a formal definition of *lepra*, as

> malignant and contagious, because this evil is not always generated or born with us, but it is more often communicated, as the putrefied humor, which is the cause of the evil, acquires a certain poisonous property [*venenosam quandam proprietatem*] and this, in turn, is the cause of the contagion and propagation.

Du Laurens further specified that the causes included "*contact and* conversation with leprous persons, and even mere inhalation." About a decade later, Petraeus also juxtaposed "contact and conversation." He further underscored the equation between the essence and the communicability of the disease, "[W]e define it to be a loathsome cachexia, very contagious indeed"; in addition, he echoed the idea that communication was by means of a venomous agency or, more precise, "a poisoned quality" (*venenata qualitate*).[30]

Speculation on the action of some "poison" was neither original nor exempt from a qualitative outlook, yet it indicated that medical authors were searching for material agents in the transmission of leprosy. This search intensified in the seventeenth century, climaxing as late as 1765 in the intelligent and too little known dissertation of Isaac Uddman.[31] His thesis cannot receive justice in the limited space of this book, but three salient aspects can provide valuable perspective. Firmly rooted in tradition, Uddman applied to leprosy the "hypothetical reconstructions" of disease transmission which had been attempted over the ages, while, also like many before him, "exploiting a range of analogies drawn from all aspects of life."[32] With compelling logic, however, he abandoned the use of metaphor as well as the reliance on intangible forces in the spread of leprosy, and he concentrated exclusively on the agency of physical elements. This step became a leap when he argued that these material agents were alive and constructed concrete premises from observed data.

Uddman was hardly the first to suggest the transmitting role of "minute animals" or "little beasties," as Varro had called them almost eighteen centuries earlier. However, the Finnish doctor developed the suggestion more methodically than previous authors. He began with a mundane agricultural observation.

> Even the simplest farmer knows that the wild thistle, a common companion to barley, is so contagious [*contagiosam*] that grain, otherwise unspoiled but coated during threshing with one speck of the kind of black dust, yields nothing but wild thistle when it is sown in the earth.

When he moved from vegetable analogy to animalcules, Uddman proceeded in the direction of infinity, to agents of steadily diminishing size ("subtilitatis infinitae"). In addition, he largely replaced the speculative arguments, which had been commonplace since antiquity, with the logical device of inference. He inferred, for example, that the effectiveness of certain remedies, the most notable of which was the use of mercury against lice in children, was based on their lethal effects and thereby proved the presence of living agents, even if these were invisible. He further argued that visibility itself changed with time. Thus, scabies became understood as caused by mites, which could be seen with the naked (*non armatus*) eye but which the ancients had failed to recognize. Now, moreover, "armed" with the microscope, the eye was able to see previously undetectable tiny "eel-shaped worms," living or even revived after dormancy. "Hence, might there not be still smaller animalcules? Who would deny it? Or who will determine where this marvelous scale of nature, gradually and to an almost infinite degree advancing to the largest and the smallest, may begin or end?"[33]

These provocative questions might make us overlook a second feature of Uddman's thesis, one that is actually more directly pertinent to our interest in the degree of premodern associations of contagiousness with leprosy. If his systematic discussion of transmitting agents led us to believe that Uddman was preoccupied with contagion, quite the opposite is true. He was trying to move away from a preoccupation with human transmission and to see the spread of leprosy in the broader framework of infection. In this attempt, he was more in tune with one tendency in Scandinavia, namely, to deny the communicability of leprosy, which Peter Richards has called "the Atlantic view" because it apparently marked attitudes in Norway and Iceland. By the same token, Uddman was at odds with "the Baltic view" of his own milieu, since the belief in contagiousness prevailed in Finland and Sweden. Richards proposes, perhaps in somewhat sweeping terms, that this belief "arose directly out of the late medieval fear of infection."[34]

Uddman distanced himself from this fear in his cultural environment, as may be seen in two subtly impatient expressions. After pursuing several case reports that entailed potentially productive questions about the spread of some noninfectious diseases, he interrupted himself: "[B]ut the common folk will urge us to return to contagious diseases." In a second remark, he revealed his dissent with popular beliefs while also shedding light on attitudes in the past—and demonstrating that he was not living in an ivory tower. He reported that his attention had first been drawn to leprosy because it was "a disease familiar enough to my countrymen in Ostrobothnia but not quite known to the other provinces" of his native Finland.

The force and incidence of this disease is such, especially in the northern part of the province, that in 1631 it was necessary to build a hospital in Kronoby parish for those who suffered from this illness, so that they might be kept away from daily commerce with people in order not to infect others, for it is the popular belief [*vulgo creditur*] that this disease, just like some other scabies, cannot but be contagious [*non posse non esse contagiosum*].

In addition to venting his exasperation at the *vulgus* by his choice of words, Uddman disapprovingly but unmistakably revealed that the fear of contagion and the wish to isolate patients, rather than the charitable intention of sheltering them, had led to the building of the Kronoby leprosarium in 1631.[35]

Fear of Transmission: Widening the Scope; Heredity

In their effort to stave off the danger of contagion, officials and examiners often cast a net that extended far beyond leprosy. For example, practitioners in Lille advised the temporary sequestration of Boudewyn Cupre in 1559 because he might infect others, not exactly with leprosy but "with the same disease as the one vexing him, which is caused by salty phlegms and the burning of blood, which is called *impetigo* in Latin and commonly *morphea*, and which is the road to true *ladrie*." This kind of preventive stance dovetailed, methodologically and chronologically, with an increasing elasticity in the identification of leprosy, which we noted at the end of the preceding chapter. The medical team of Ravensburg demonstrated a particular looseness with its criteria for sequestering someone. In 1688 the members of this team ordered sixty-year-old Regina to be "separated and avoided" because she showed "a positive inclination to a leprosy or hereditary disease." In 1693, they determined that forty-four-year-old Anna should be separated because she had "a strong and acute leprosy or hereditary disease." The phrase "strong and acute leprosy" was an oxymoron, even then, because the chronic nature of the disease was well known. In addition, the indefinite article in "*a* leprosy" (*eine Aussatz*) underscored the open-endedness of the reason for exclusion. Most significant, both of these verdicts were equally—and more ominously—open-ended in their use of the phrase "leprosy *or* hereditary disease," which suggested that the two terms were interchangeable.[36]

Heredity, like contagiousness, became an established feature and a defining characteristic of leprosy. At Wangen in 1734, as we saw in the beginning of this chapter, "the foul *and hereditary* disease" of leprosy was the reason for sequestering Judita. The characteristic was applied with increasing rigidity from the fourteenth century on. Earlier authors, at least since Constantine the African and Avicenna—and with the notable exception of Galen—associated generational transmission with the disease. However, they listed it as one of a wide range of factors. The emphasis on a hereditary origin of leprosy evolved in a contorted process rather than along a straight line, and a clear picture of the changes requires a somewhat schematic analysis of the medical discourse.

Heredity was entangled with contagion and other extrinsic issues. In addition, it sat across a fundamental etiological divide, namely, the distinction between becoming leprous "in the womb" and afterward, or between "before and after birth."[37]

The picture became more complicated by a further differentiation, at least implicit, within the onset in utero. One set of uterine causes ranged from conception during menstruation (!) to the intercourse of a leprous man with a pregnant woman. These causes, consisting mainly of incidental and preventable events, dovetailed with various other extrinsic factors, which will be the subject of the next chapter. Here we concentrate on the second set, of the intrinsic factors in a more genuinely "hereditary" transmission, which was inalterable and largely inevitable yet, paradoxically, also random and unpredictable. The combination of predestination and incertitude ultimately made heredity, more than contagiousness, a more insidious and deep-seated source of prejudice in official measures, if not in popular attitudes. Caution is in order, however, in interpreting the term *hereditary* as it occurred in treatises and certificates because its meaning was not quite congruous with our own usage. Discussions of "inherited" leprosy, which were rarely as detailed as the observations on contagion, generally blurred the now basic distinction between hereditary and congenital disease transmission.[38] Most important, genetics, which is at the center of current thought about heredity, was terra incognita for premodern authors.

Early thoughts about links between leprosy and generation, if not entirely groundless, were speculative. They evolved, however, in an unfavorable direction. The speculation became increasingly categorical and determinist, and likely to fuel suspicions. As an admittedly superficial indication of the widening assumption, we may compare Avicenna's qualified suggestion that the disease "*sometimes occurs* as a result of heredity" with Bocellin's more sweeping claim that "this malady is *commonly called hereditary.*" Johannes Platearius believed the cause "*sometimes* lies with generation itself," while two centuries later Theodoric concluded a list of causes with the straightforward assertion "and *it is* a hereditary disease." Bernard de Gordon, with somewhat inverted logic, argued for a determinist conclusion: "[T]he material of *lepra* is so thoroughly noxious that it cannot be corrected in the womb, and hence it is a hereditary disease, in the same way as gout." Niccolò Bertucci further boosted the importance of heredity by asserting, "[T]his disease occurs *most often* from birth [*ex generatione*] and it is rarely acquired [*per acquisitionem*]." In a similar generalization, Jodocus Lommius claimed that the disease "usually crosses [*transire solet*] from parents to children." Ambroise Paré was more categorical in his assertion and explanation:

> [O]ne can with certainty say that it is a hereditary disease, for a *ladre* begets a *ladre*, because the seed or offspring [*geniture*] comes forth from all the parts of the body. Once the principal parts are spoiled and the substance of the blood is altered, corrupted, and infected, it is necessary that the seed, too, is infected.

Du Laurens revealed how much the assumption had escalated by his time when he counted heredity "among the most certain of causes."[39]

Du Laurens and Bertucci, among others, underscored the certitude of hereditary transmission by pointing out that leprosy has a tendency to remain latent for a long time. Bertucci warned that it "will gain strength around the end of youth and the advance of age, until it surfaces in the body, like the leaven after it has lain hidden in the dough day and night." Du Laurens extended the threat further: "[E]ven if this evil does not erupt in the first years, it manifests itself eventually, and it is propagated into the third and fourth generation." The conviction that leprosy was not only inherited but also certain to become active could lead to prejudicial measures. This happened in Colmar, to a young man whose mother and sister had been found infected: when he developed pustules in the face, surgeons readily deemed him leprous and forced him to leave town; when the pustules disappeared and he showed no other symptoms, the surgeons persisted in their verdict. The man spent five years in isolation before he was rescued in 1589 by the famous Basel doctor Felix Platter, whose riveting story we recounted earlier, when we examined medical judgments and appeals (Chap. 2).[40]

In 1595, Guillaume des Innocens subjugated observation to belief when he recalled an anecdote from his practice which had impressed him as illustrating the power of heredity.

> This is how experience taught me. I recall seeing, in some nice town in Languedoc, an honest and rich family, in which the very beautiful and handsome sons had no mark of this disease in all their life (as I learned from their own avowal and from the neighbors, persons of quality and honor). The girls, however, at the age of twelve or sixteen, and about the time of their menarche, little by little became, from beautiful and white as they had been until then, first extremely red in the face, then covered with nodes and hard spots with tuberosities in the most beautiful areas of their faces, in short, so visibly affected that they no longer

dared or wished to show themselves in public. Indeed, their father had finished his days, with his eldest daughter, [in seclusion] in a beautiful house that he owned in the country; in addition, a grandparent had been buried in a leprosarium. Their mother, nevertheless, was a very beautiful woman (for her age).

When des Innocens introduced his discussion of heredity with this reminiscence, he intended to add a sense of reality to the proposition that the children of a leprous parent carried the seed of the disease or, "what our physicians call in Latin, *potentiam.*" By equating facial deformity with leprosy and dramatizing the turn from beauty to ugliness, the anecdote could only have deepened negative attitudes toward the offspring of leprous parents.[41]

Parentage and family history provided the most direct pretext for insisting, even with limited evidence, that a suspected patient *had* to be leprous. The assumption that a leprous parent necessarily bore leprous offspring—although signs might not appear for years—seems to have been applied primarily to the male ("filius leprosi").[42] If medical authors contributed to the deepening and prejudicial belief in heredity, they also sought rational understanding. They endeavored to identify the source, critical moment, and locus of the transmission between the begetter and the begotten. In the forceful proposition of Du Laurens, "one should not doubt that, if there is any contagion or root of *lepra* in the seed of the parents, whatever is generated from it will be leprous." A similar junction between contagion and heredity underlay the expression of Guy de Chauliac, who counted "a stain in the generation" (*macula generationis*) among the primary causes of *lepra*. Similarly, but in a setting that was quite different from a textbook presentation, Doctor Heinrich Losser characterized patients who inherited the disease as "born of infected seed" (*unreynem samen*). Some authors, searching for more tangible explanations, located the corruption specifically in "the material of the seed," though they were aware that this suggestion raised a further question, namely, how a leprous man, with presumably corrupted seed, was able to reproduce—as observation must have proved. Paré based his certainty about the heredity of leprosy on the premise that

the seed or generating principle [*semence ou geniture*] comes from all the body's parts. As the principal parts are ruined and the mass of the blood is altered, corrupted, and infected, therefore it is necessary that the seed, too, from which one is generated, is infected.

With the eloquence that fueled his widespread influence, Paré managed to bring together not only the notions of contagion and heredity but also speculation and alleged observation in one sentence: "This contagion is so great that it goes to children's children, and even further, as experience makes us believe."[43]

Velasco de Tharanta, the ever helpful compiler, drew together various speculations, and he added some subtle but pertinent touches of his own. Unlike Guy de Chauliac, Paré, and others, Velasco did not rank heredity as one of the primary causes, but he included an inherited melancholic predisposition among the antecedent causes, and leprous parentage among the contributing causes. In the clarifying section of his chapter, he summarized the views on the transmission of leprosy in generation, as they had developed by the early fifteenth century.

> *Lepra* occurs from earlier parents because of the corrupted and infected seed of one of the parents. Therefore it is called a hereditary disease, because together with the law of inheritance it goes on to the offspring, nay, even to posterity after the second and third generation, leaving intermediary [generations] untouched, and this goes beyond rationalization, as we have seen of great princes not so long ago.[44]

With his cryptic comment on princes, Velasco might, at first look, seem to allude to some royal house in which successive generations were afflicted by leprosy. Rather, in a context of ubiquitous dynastic rivalries, he was suggesting the ineluctable transmission of a hereditary disease that, like princely property, overruled rational arguments in passing from forebears to posterity. His suggestion drew additional weight from its place in the midst of causes that, though directly or indirectly related to generation or reproduction, were extrinsic. These causes ran the gamut from dangerous sexual intercourse and ill-timed conception to dietary assaults on the humoral complexion. Their great diversity, together with their many ramifications, can receive due attention only in a separate chapter.

Chapter Six

Causes, Categories,
and Correlations

The Masters in Cologne may have declared Hans free of
the permanent and confirmed leprosy [*bestedige und befestige
ußseczczekeit*], but they could not have cleared him of the
incipient stage [*anhebende ußseczczekeit*]. The latter was, at that
time, visible in his face and verifiable with needles. In addition, we observed an infection in his body which is called
morphea in Latin, and we found that not only the skin but
also the flesh was infected. The masters of the medical art,
including Gordonius in his book *Lilium*, write that, as
ußseczczekeit is in the flesh, so *morphea* is an *ußseczczekeit* of the
skin. People incur *ußseczczekeit* from a bad regimen [*von bosen
regiment*].

When Rupprecht the tailor became disfigured as the
result of a bad regimen, he was given a reprieve and, when
his condition worsened, he was separated. His infection
appeared first on the skin, and then it came into the flesh.
This is why the masters write that *morphea* is a presage of
future *lepra*.

In his defense against the charges by Hans Maderus in
1458, Doctor Heinrich Losser marshaled nearly every
tenet that had governed medical teaching on leprosy for
nearly two centuries. The excerpt quoted here introduces, in part directly and
in part indirectly, the subject matter of this chapter. The "regiment," that is,
the *regimen*, or diet, encompassed the broadest spectrum of causes to which

the physicians attributed leprosy. The distinction between *anhebende* and *befestige*, or incipient and confirmed, stages reflected the evolutionary character of the disease; since this distinction became increasingly crucial in diagnoses, it will receive further attention in the next chapter. Morphea, characterized by Losser as "leprosy of the skin," was one of the ailments that were regularly mentioned together with leprosy, especially in attempts to gain a better understanding by establishing correlations with other diseases.[1]

THE CAUSES OF LEPROSY

The etiology of leprosy received little systematic attention from early authors, even from Constantine the African and the Salernitan masters who generally insisted that the knowledge of causes was paramount to treating diseases. Avicenna, the first to dwell on the causes, divided them, along Aristotelian lines, into efficient, material, and *iuvative*, or contributing. It is easy to schematize the division in the *Canon*, as I have done in Table 6.1. Scholastic authors tended to propose a far greater number of causes, in an effort to cover all the possible explanations—an effort somewhat analogous to the polypharmacy that we will encounter in Chapter 8. They also applied an organization that was both more structured than the one in their sources and more pronounced than the one in sixteenth- and seventeenth-century writings. The Scholastic organization, which may have received its most influential dissemination with Dino del Garbo (d. 1327) at Bologna, moved away from Avicenna, as may be seen in Table 6.2.[2]

Fourteenth-century masters, perhaps influenced by the "New Galen," added the failure of the assimilative faculty as an "immediate," or inherent, cause, which we have considered above, as an aspect of physiology (Chap. 4). They also demoted complexional imbalance from the rank of an efficient to a conjoint cause; inversely, they promoted foods from the status of a material to a basic cause. In addition, Scholastic authors discarded the "contributing" causes—and the perspective that this category afforded. We will follow the Scholastic classification, not only because it framed etiologies for a span of three centuries, but also because it applies a simple logic that lends structure to a variegated and often confusing subject. Conveniently, the categories in this scheme will lead directly into the taxonomies and correlations that constituted the nosology of leprosy.[3]

Basic Causes: The Six Non-naturals

The *causae primitivae*, or basic causes, were not only the roots of leprosy but also most immediately apparent ("evidentes"). Foremost among these was "a bad regimen," in Losser's phrase, which covered a range of aberrations in the diet. The term *diet* (as well as *hygiene*) was more inclusive than it is today, as it referred to the general lifestyle. A balance of the humors was maintained, under the guidance of medical science, by the "governance of health" (*regimen sanitatis*). The matters to be governed were conventionally bundled as the "six non-naturals," that is, factors of health that belonged neither among the "naturals" inherent in the body's constitution (such as the elements, humors, or faculties) nor among the "contra-naturals" involved in the process of disease. With minor variations, the six factors were arranged as ambient air—actually, what we call the environment; food and drink; repletion and evacu-

Table 6.1. Causes of Leprosy According to Avicenna

Efficient causes	Material causes	Contributing causes

—warm and dry imbalance of the liver, burning the blood into black bile, affecting the complexion of the entire body	—melancholic and phlegmatic foods
	—surfeit
		—blocked pores
		—corrupt air
		  –in itself
		  –by the proximity of patients
		—heredity
		—conception in menstruation
		—foods

Table 6.2. Causes of Leprosy According to Scholastic Authors

Basic causes ("primitive")	Antecedent causes	Conjoint causes	Immediate (inherent) cause

—"six non-naturals"
  –ambient air
  –food and drink
  –repletion and evacuation
  –toil and rest
  –sleep and vigil
  –emotions
—conversation with patients
—defective seed
	—humoral predisposition	—complexional imbalance
		—burned black bile
			—failure of the assimilative faculty

ation; exertion and rest; sleep and wakefulness; and "things that befall the soul" (*accidentia animae*), that is, emotions.[4]

The Environment

The surrounding air, exhaled by patients or polluted by the vapors they exuded, was considered the primary means of transmission, as we have seen in the previous chapter. A less direct cause was "corrupt, pestilential air," which might emanate from rotting corpses and the like. The etiology was extended to the environment in general, with extreme temperatures of the air as a contributing factor. One Scholastic author suggested that "prolonged burning in the hot sun or near a fire" might cause leprosy. Examining practitioners in Cologne, however, expressly distinguished burns "from summer heat or a furnace fire" as causing an ailment that was not leprosy. Avicenna had first drawn attention to the role of air temperatures by associating the excessive heat in Alexandria, Egypt, with a higher incidence of the disease. Ambroise Paré believed that greater heat accounted for seeing more *ladres* "in our Languedoc, Provence, and Guyenne than in the rest of France." As Europeans became more aware of other climates, they broadened Avicenna's observation to "Africa" and, in the eighteenth century, to the East Indies, the Caribbean islands, and "the American Colonies—especially the southern ones." Isaac Uddman, however, cautious as always, pointed out, "Hardly anyone will confirm whether the three kinds of *lepra*, that is, the Egyptian, Indian, and American, coincide entirely with each other or with ours."[5]

European authors saw climate from a vantage point that differed from Avicenna's. Bertucci speculated that "prolonged walking in snow and living in the north"—as well as "incarceration between walls"—could cause leprosy by congealing the blood. Paré (who apparently viewed Paris as the ideal latitude) attributed a notable incidence of the disease in northern regions to the fact that "the air becomes thick, sluggish, and frosty in places that are too cold, so that we see many *ladres* in some parts of Germany." In another localized explanation, Paré concentrated on a coastal climate as an environmental factor. He suggested that one might become predisposed by "dwelling in maritime places, where the air is normally thick and foggy." Two generations later, Heinrich Petraeus combined environment and nutrition when he surmised that leprosy might be "widespread in maritime areas, on account of the impure air, salty food, and fish." These speculations are intriguing in view of the lingering reports of leprosy along the Scandinavian coastline, which was the milieu of Isaac Uddman's astute observations and, eventually, of Armauer Hansen's historic discoveries.[6]

Food and Drink

Avicenna extended the geography of leprosy, from climates to cultural differences, when he observed,

> When a hot environment is combined with essentially bad food, whether from the kind of fish, or salted meat, coarse flesh, the meat of donkeys, and lentils, there is no doubt that *lepra* is the result. This is how it is spreading in Alexandria.

This assertion was the source of a steadily growing list of nutritional items that were suspected—some into the twentieth century—of causing leprosy. Among the many versions of the list, I have not yet found two that are identical. While they produce a great and seemingly meaningless profusion of items, the versions yield valuable insights. Quite a few of the suspicions were inspired by vernacular lore, together with underlying associations and cultural preferences. This was the case, for example, for the repeated warning against "regularly eating fish with milk in the same meal," and for misgivings about the meat of the hare (*leporis*). Even though authors often borrowed, presumably without much thought, from other authors, their omissions or additions often suggested personal outlooks. Thus, Theodoric Borgognoni singled out excessive "consumption of cow and ox meat" as synonymous with a "bad regimen": as he was apparently the only one to do so, he may have been influenced by personal distaste. The addition of some condemnations, say, of cabbage or goat meat, was probably due to an author's social bias. Geography played a role in the proscription of some meats, notably fox, bear, and donkey, "because all such animals are eaten in certain regions." From Avicenna to the eighteenth century, donkey meat figured most prominently as prototypical of exotic diets. Isaac Uddman dismissed this idea as obsolete: "[T]he ancients believed that the meat of wild asses, eaten in Africa, triggers this disease." The addition of lion's meat seems to have been solely for the sake of exoticism; however, a Salernitan author gives the impression that it might have found inspiration in the leonine form of leprosy. After all, when authors characterized particular foods and drinks as contributing to leprosy, they usually provided reasons.[7]

Rationales revolved primarily around the principle that, according to Galenic physiology (see Chap. 4), nourishment became part of the body's constitution. This allowed for linking the substance and quality of food and drink to the humoral and qualitative essence of leprosy. Galen established the most general connection by mentioning *lepra* and *elephas* in one sentence with people who "suffer itch" because they "are nourished by foods of a bad humor." In the same sentence, Galen implicated black bile as this bad humor. We will learn more, later in this chapter, about black bile as the principal cause of leprosy; here it will suffice to cite two applications, each of which bears further significance. Nearly every list of dietary causes included lentils, which were branded as "the most melancholic of foods." However, this characterization was a case of post factum rationalization, first, of Avicenna's mention of lentils and, second, of their association with the Levant. The same is suggested, in fact, for the inclusion of donkey among melancholic meats, although the somber temperament of donkeys may have been an additional consideration. In a similar kind of post factum interpretation, slugs were reputed to cause leprosy because they "produce melancholic blood," in contradiction with their usual classification as phlegmatic. This contradiction is one instance, albeit trivial, of the fluid character of humoral etiologies. Indeed, Avicenna juxtaposed phlegmatic and melancholic foods as material causes of leprosy, although he emphasized the role of the latter. Even Galen, who concentrated on black bile as the chief culprit, blamed phlegmatic meat, consumed by eaters of reptiles, for *lepra* and *elephantia*.[8]

In addition to humoral rationalizations, which closely followed the Galenic scheme of digestion, there were qualitative descriptions of suspect foods, which allowed greater latitude. With regard to primary qualities, melancholic ingredients were essentially cold and dry—like black bile itself. Yet Heinrich Losser blamed "a bad regimen of excessively *hot* victuals" for provoking leprosy by "burning and contaminating the blood." Secondary qualities were more open to intuitive associations. A dangerously "hot" or "sharp" diet comprised seasoned dishes, cured meat, and salty fish; the sharpness would be aggravated by aged cheese and old, bitter wine. Salt itself became a suspect substance for Paracelsus and his followers. Traditional authors unanimously disapproved of "thick" and "coarse" foods in general, and legumes in particular, because these thickened the blood. In addition, an entire category of foodstuffs, from mud-feeding aquatic animals to, more understandably, rotten fish or putrid meat, raised suspicions of affinity with the "corruption" of leprosy.[9]

In view of ancient and widespread dietary misgivings about the tendency of pork to decompose and to carry disease, it is surprising that Velasco de Tharanta seems to have been the only Continental author to mention "the use of leprous pork, which has glandular flesh [*carnes glandulosos*]." It is possible that he heard the English speak of the "measled hog" when he served as municipal physician in Bordeaux.[10] In any event, an additional reason to frown upon the consumption of pork was because it was most likely to fall in the harmful category of salted foods. The strongest objection to pigs, however, was voiced by Hildegard of Bingen, who warned that the consumption of pork "readily excites a person's libido and feeds *lepra*." In addition to mixing medical and moral criteria, Abbess Hildegard also combined spiritual and mundane advice—for example, by declaring that "the disease often arises from meats, milk, and strong wine, but *not* from bread, vegetables, or beer." Even more unusual than Hildegard's dietary differentiation was a proposition by the Basel doctor Johann Huggelin. In the only consideration of nutritional deficiency I have encountered, Huggelin claimed that leprosy was common in England on account of a harmful "lack of olive trees and nut trees"; he attributed the information to the "learned physician and philosopher Hieronimus Cardanus," better known as Girolamo Cardano (1501–1576).[11]

"Repletion and Evacuation"; Menstruation

A third factor in the category of the six non-naturals lay between surfeit and lack. It consisted of "repletion and evacuation" (*repletio et inanitio*). The point between plethora and emptiness determined the amount of the body's humors, the balance or "temperament" among them, and the timely evacuation of their corrupt portions. For this category, too, preoccupations with harmful excess overshadowed concerns with deficiency. This slant is surprising, not only for a world in which dearth was commonplace, but, specifically, because we found ample evidence of poverty in the records of *iudicia leprosorum* (see Chap. 2). It is indeed difficult to find an author who associated leprosy with hunger and thirst as causes. Nutritional deficiencies were conventionally held responsible not for morbid growths and corruption but, rather, for chronic wasting, as in hectic fever and marasmus. Avicenna emphasized the role of "nauseous satiety" and "eating beyond satiety" in the overproduction and thickening of black bile that caused leprosy. This emphasis was consistent with Galenic physiology, and it lent itself to moralization. According to Hildegard, "the man who lives in gluttony and drunkenness, often becomes *leprosus* and contorted in his limbs; but he who is temperate in food and drink, will have good blood and a healthy body."[12]

The quantity of intake, however, was etiologically less significant than the elimination of superfluities, inert as well as noxious. Many of these were expelled invisibly, by "transpiration" through the skin. Therefore, as Avicenna taught, blockage or "oppilation of the pores causes the innate warmth to be suffocated, and the blood to be cooled and thickened," which was a contributing cause of leprosy. Galen alluded, though more indirectly, to the same problem when he suggested that "dirty and unwashed people" were prone, if not to *elephas,* then at least to *lepra,* in which the offending humor "settles primarily in the skin." Proper washing was imperative for keeping the pores open. Baths, preferably warm, were routinely recommended for leprosy, both as prophylactic and as treatment, contrary to modern canards about the hostility to bathing in bygone days. Cornelius Celsus was an early and lone exception, with his advice that the *elephantia* patient should bathe infrequently—advice that was inferentially based on the need to "conserve strength." If there was any doubt about the usefulness of bathing, it was because fresh water cleans only the skin. "But in *lepra,* the material lies deeper," Bernard de Gordon argued, so that one would wash "with stronger means, such as a stew [*stupha*] with herbs." Heating the water was the main challenge, so that there is some irony in seeing Doctor Losser blame excessively hot baths for Rupprecht's burned blood and his predisposition to leprosy. Similarly, Peter Lynscydt of Bonn was found predisposed, at least in part, because of unusual paleness, which "resulted from the long use of the hot baths in Aachen, which he had frequented continuously for an entire month."[13]

One of the gravest threats of leprosy arose from a weakening of the "expulsive faculty," whose task it was to dispose of the wastes of digestion, by sending them primarily "into the veins of the anus and uterus." This weakness could lead to a dangerously voluminous "retention of melancholic superfluities," particularly in hemorrhoids, menstrua, and a host of less obvious impurities. Amenorrhea, or suppression of the menses, appeared with some frequency in reports of examinations. As late as 1697, when examiners tended to widen their definition of "leprosy," a physician and a barber-surgeon certified in Waldsee declared,

[W]e have examined Ursula Weberin, forty years old. From the stopping of her monthly period, her blood has fallen into corruption. We have found her afflicted by a scurfy, runny scab, which makes it imperative that she be separated from the healthy.

One may argue that it was logical to assume a nexus between retained cata-menia and corrupt blood as a root cause of leprosy. Nevertheless, amenor-rhea loomed disproportionately large in premodern medicine, both as a cause and as a consequence of ills, and as an illness in itself rather than a symp-tom. Natural philosophers saw the root of the problem in women whose "corrupted superfluities are held in their wombs" because this made them "bear offspring which are lame, blind, suffering from leprosy, epilepsy, and laboring under other diseases as well."[14]

The misunderstanding of menstruation, together with deep-seated fears of its presumed impurity and the impulsive abhorrence of blood, underlay what was arguably the most stunning notion in the etiology of leprosy. This was the proposition that the disease could be contracted by an embryo con-ceived during menstruation. The explanation surfaced in treatises, but, unlike the role of amenorrhea, it was invisible in medical certificates. The notion re-sulted, in a process analogous to the mislabeling of *lepra*, from the fusion of primitive legacies and of their eventual permeation into medicine. While it is impossible to reconstruct the entire process, we are able to trace the out-lines. One root was biblical, again in Mosaic law. Leviticus 15:19 labeled a menstruating woman unclean and untouchable, and 15.24 decreed, "[I]f any man lie with her at all, and her flowers be upon him, he shall be unclean seven days." Talmudic commentaries appear to have reinforced the injunction with a warning that it was dangerous to ignore the ritual impurity.

It is not clear when or where the danger was transferred, from the man who violated the ritual ban, to the product of the transgression. It is likely, how-ever, that the next escalation took place in the early centuries of Christian-ity, and that it was led by ascetic champions of sexual restraint. Saint Jerome (ca. 331–420), remembered as the translator who produced the Vulgate Bible, threatened men who had intercourse with women during "the flow of their impure blood" that the semen would be spoiled, "as it is said, so that from this conception *leprosi* and *elephantiaci* would be born, and loathsome bodies of either sex." About a half century later, another theologian, Theodoret, bishop of Cyr, left the consequences ambiguous enough to be a menace both to father and offspring. He commented on Leviticus, "[I]f a woman is called impure, the purpose is that no one will have intercourse with her. It is said, in fact, that a union of this kind results in injury and *lepra*."[15]

Both Jerome and Theodoret indicated the secondhand character of the idea by employing the phrase "it is said." The rumor had a more likely source in popular imagination than in traditional medicine or natural philosophy. Hippocrates, as well as Aristotle, held that conception was impossible during menstruation. Nevertheless, at some point the notion seeped into learned medicine. While we are able to follow a continued transmission among the theologians and preachers, we can only surmise that it took more time for physicians to adopt a tenet that did not fit in their scientific framework. It is possible that they were swayed by misinterpretations of birth defects, or by observing infections contracted during parturition. Their expressions, however, suggested little more than hearsay—with some attempts at rationalization and, occasionally, equivocation. The idea probably drew its greatest impetus from Avicenna's authority. Ironically, Avicenna was rather circumspect in stating that *one possible* cause of leprosy was "the complexion that the embryo acquired in the womb because of the latter's disposition, *as would be the case if* the conception should occur during the course of menstruation." He could have inherited the suggestion from any of the various traditions that he compiled, but it is tempting to speculate that it had migrated eastward from Antioch in the intellectual baggage of Nestorianism: it is interesting to note that Theodoret wrote in Antioch, and that he has been linked to the Nestorians. In the West, we are faced with something of a puzzle before the thirteenth-century dissemination of Avicenna's *Canon of Medicine.* The idea seems to have found entry in Salerno by the mid-twelfth century. However, it did not appear in the Salernitan *Questions,* of which several dealt with menstruation and others with the sex-leprosy connection. Nor do we find it in the works of Constantine the African, who had much to say about corrupt blood in his *Liber de Elephancia* and who wrote an entire treatise on coitus. An alumnus of Salerno, Gilles de Corbeil, asserted around 1200, with some poetic flourish, that a leprous child would be "conceived from joining at illicit times when nature seeks to pay her monthly dues."[16]

The belief seeped into Latin etiologies of *lepra* in the course of the thirteenth century, but with uneven visibility, and as a postulate rather than a developed theory. This is suggested by Theodoric's terse and matter-of-fact proposition "Item, it is produced in a womb filled with unclean menstrua, even if the seeds are clean." Bernard de Gordon began his etiology by stating that, if *lepra* occurred "from the start in the womb [*ab utero*], this is because [the subject] is conceived during menstruation [*quia generatus est in tempore menstruorum*]." However, Bernard did not elaborate this bald statement, in contrast with his presentation of the other causes. Henri de Mondeville, claiming that causes were not his concern as a surgeon, simply applied the postulate to an intriguing impression: perhaps aware of the Talmudic injunction, he explained that "few Jews are leprous, because Jews rarely have intercourse during menstruation." No academic physician propounded the notion more categorically than Niccolò Bertucci, who taught that "the *most important cause of all* [*causa omnium potissima*] is coitus with a menstruating woman, which produces a leprous child." In contrast, several subsequent authors, including Guy de Chauliac and Velasco de Tharanta, who habitually cited numerous sources, did not mention the idea.[17]

Someone who wrote an essay titled "De lepra et primo de interpretatione," most likely a fifteenth-century bachelor in medicine, clearly remembered the thesis. What is even more curious is that he attempted an explanation that was reminiscent of compilations on reproduction, such as the *Secrets of Women*, which circulated widely at the time. He noted,

> [T]hese days, leprous children are born especially from conception during menstruation. At the time of her menses a woman has greater pleasure in intercourse, because her blood stings and causes a tickle. Thus, when the child is born in the flow of menstrua, he will no doubt incur *lepra* or scabies.

This "interpretation" demonstrates that, even when couched in scientific language, blunt assertions could voice feverish imaginations, while repeating and amplifying hearsay without reference to experience. This nefarious combination survived into the sixteenth century and beyond. By means of qualifiers, however, medical teachers distanced themselves from the certainty of Bertucci's position. Around 1600, Du Laurens sealed the fusion of ritual injunction, ascetic threat, scholastic rationalization, and pervasive gynephobia; yet he injected a note of conjecture. He taught, "[I]f a woman happens to conceive during her monthly flow, the child will be born sickly, *perhaps even* leprous, because the blood in a menstruating woman has some kind of venom." Certainty was downgraded to possibility when medical authors, instead of assuring that the offspring *would be* leprous, speculated that the baby *might* be born with a *predisposition* to leprosy. They further weakened the belief by presenting it as a subjective opinion. A few years after Du Laurens, Heinrich Petraeus signaled the end of certainty by framing the idea as a relic *and* reviving the earlier circumspection. He mused, "Avicenna thinks that, if a child is conceived while the menses are flowing or imminent, he will be leprous, that is, inclined to *lepra*."[18]

Exertion and Rest; Sex

The rise and decline of the myth on conception *fluentibus mensibus* was roughly paralleled by a curve in the intensity with which medical authors connected sex and leprosy. The apprehension about sexual intercourse, which overlapped with the belief in contagiousness and emerged around the same time, peaked notably in the thirteenth and fourteenth centuries and waned—in much of medicine, if not in culture at large—in the sixteenth century. Avicenna's *Canon of Medicine*, Constantine the African's *Book on Elephancia*, and several Salernitan masters did not identify sex as a cause, notwithstanding their attention to the role of proximity to patients, corrupt seed, and menstruation. Intellectual preoccupations with various aspects of reproduction found their first expression in speculations about nature, particularly in *quaestiones naturales*, in the schools of Salerno and at the early universities.[19] These preoccupations, together with the fear of contagion, were intensified by scriptural commentaries, canon law casuistry, and social tensions. In some writings, the intensity bordered on obsession, combining a prurience and phobia that rarely surfaced in official documents. Nevertheless, before examining these writings, it is useful to note that the combination, in its various male manifestations, was less pervasive in medicine than in other disciplines. Thirteenth- and fourteenth-century medical writers often gave the impression that the fear of sexual transmission was marginal to their etiology by paying little or no attention to coitus when surveying the causes of leprosy. Paradoxically, the attention seemed to intensify in the later centuries, when the disease itself was eclipsed by other threats to private and public health, ranging from syphilis to famine.

In the dietetic framework of the six non-naturals, sexual activity sat astride two categories, that of repletion and evacuation and that of exertion and rest. It was logical that excessive sex would be held responsible for many diseases, including leprosy, because it depleted the body's precious fluids and thereby its strength, or *virtus*, and, more critical, because it overheated the blood. Thus, as Hildegard explained in her own graphic style, some men do not have or seek temperance in their libido. As a result, their blood is frequently and immoderately roiled, like when a pot on the fire is neither fully boiling nor completely cold, so that it retains the dregs because it does not have sufficient force to eject them. And when these men are burning with lust to such an extent that their blood is repeatedly roiled to excess, so that it is neither true blood nor true water nor true foam: then it turns into a bad, spoiled blemish that corrupts the man's flesh and skin.

Other authors believed that, while too much sex burned the blood, the opposite was hardly better because "the total removal of coitus" led to leprosy by congealing the blood.[20]

Intercourse with a leprous woman was singled out as a particularly predictable cause of leprosy. The impression that the female, *mulier leprosa*, infected the male partner whereas the reverse did not occur was the staple of *questiones* in natural history. In one typical instance, it was argued that

a healthy man is infected sooner by a leprous woman than a woman by a leprous man. The cause is twofold, one because woman has narrow veins so that the infectious agent is not able to enter so quickly or deeply; second, because she has wide meatuses by which the sperm of the leprous male comes out faster, whereas a man has wide veins and arteries in his penis, and the pores of the penis are opened by the friction, so that the infection of the womb, or the corrupt fumes issuing from the womb, enter the man's penis, and so he is infected.

The unknown author of this fantastic—though far from unique—speculation ends on a surprisingly perceptive note, "[T]herefore, I say that a person is not infected in one instance, nor is a woman, but either gender is infected by prolonged use, and a man sooner than a woman: we have seen the reason." Aside from gynephobia, certain misinterpretations of reality may have contributed to such impressions and speculations. Premodern clues (as well as modern statistics for Hansen's disease) indicate a higher incidence among men. Historians have been tempted to propose other diseases that affect male sex partners and which, hypothetically, could have been mistaken for leprosy; their hypotheses, however, raise as many questions as they answer.[21]

Whatever their underlying causes, speculations and rationalizations about the female agency in the transmission of leprosy became doctrine in scholastic medicine by the end of the thirteenth century; paradoxically, they did not become dogma. We may recall Bernard de Gordon's pithy story, in the discussion of *lepra* which was cited for three centuries, about his patient, the countess, who infected a bachelor in medicine while being impregnated by him. The bachelor's behavior, however, suggests that fears of sexual transmission were hardly absolute—among academics, at least. This suggestion is enhanced by the fact that Bernard felt the need to tell the anecdote for an expressly cautionary purpose, so that "anyone may beware of bedding a leprous woman." A similar paradox emerges from Bernard's alarm about the potential for a woman to transmit the disease indirectly:

> [I]f a healthy man should lie with a woman who has lain with a leprous man, while the semen is still remaining in the womb, he will necessarily be leprous, because in a male the pores are so loose that the infection moves immediately to the whole body.[22]

In spite of the rationalizing and categorical tone, and even though it led up to a dire finale, Bernard's warning turned into an implicit admission of double defeat. He conceded the limited effectiveness of a "Just Say No" philosophy. Second, he acknowledged that there were many ways to elude his dire prediction—by means of efforts that, as he suggested, represented only a portion of known contraceptive methods.

> Therefore one should take the most extraordinary precaution and, if compelled by some unfortunate or bad circumstance, contrive to have the semen come out of the womb, by jumping, sneezing, bathing, and flushing the womb with cleansing waters. Also, as much time as possible should intervene [after the leprous partner]. There are many other methods, which need not be told, for expelling the received semen. Without them, one should prepare oneself for the beggar's cup, the stars, and everlasting opprobrium.

Bernard's contemporary Henri de Mondeville also implied that libido prevailed over fear, and actuality over theory. He assured that, if one has intercourse "with a leprous woman, or with a woman who was with a leprous man, and is aware of it right away, he should immediately rinse his penis with vinegar, and he will not be infected." The normal pattern was amazingly reversed in 1313, in Bernard's own hometown: a man was sent to the stake in Gourdon after a prostitute accused him of coming to her without revealing that he had been leprous for six or seven years.[23]

The diversity of early fourteenth-century statements indicates the extent and boundaries of doctrines about the sexual transmission of leprosy. Nevertheless, with their prevailing focus on the danger of "the first intercourse with a woman after she has been with a leprous man," they bare a potent combination of gynephobia, sexual phobias, *and* the fear of *leprosi.* The potency of this triple fear is palpable (at least in Latin) in the poetry of Gilles de Corbeil:

> *cum subit amplexus coiens et dulcia vota*
> *consummat cum qua carnali gaudia nexu*
> *leprosus complere solet*

"[W]hen, in having intercourse, he undergoes the embraces and consummates the sweet bonds with a woman with whom a *leprosus* is wont to enjoy carnal congress," a man will contract the disease.[24]

Further explanations, while more prosaic, reflected prevalent attitudes toward the "secrets of women." They revolved around arguments that the uterus is so cold, compact, and hard that it retains the spoiled semen without being infected by it. Inversely, and expressing a complementary fear, Ambroise Paré argued that "the man is prone to receiving the leprous venom, because the male organ is very spongy, so that it readily absorbs the poison [*virus*] that arises from the vapors of spermatic matter." Even in the early seventeenth century, when medical interests had shifted to other pathological aspects of sex, Heinrich Petraeus offered the sweeping summation that leprosy

> is contracted by contagion from the very effervescence of the seed [*balsami*] in the intercourse of a healthy male with a leprous female, a leprous male with a healthy woman, or of a healthy male with a healthy woman with whom a leprous male had sex shortly before, if the semen did not slip out.[25]

Other "Non-naturals"

The remaining dietary factors, or non-naturals, paled in comparison with the prominence of nutrition and sex among the elementary causes of leprosy. Some authors surmised that too much sleep and indolence made the blood sluggish and prone to corruption. Most, however, concentrated on the opposite, namely, insomnia and excessive activity, as overheating the humors and contributing to the burning of black bile. When examiners declared Conrad

Pryntz of Bonn "clean" in 1548, they found him dangerously warm, and therefore predisposed to leprosy, due to "the immoderate toil of his entire body": they recommended medical supervision "for eliminating the abundant bad humors," and they ordered him to return six months later for a second examination. The emotions, or *accidencia animae*, while receiving attention as effects and signs of leprosy, often appeared among the causes for the sake of completeness. Velasco de Tharanta, for example, casually indicated, "[O]ther externally originating causes can be the emotions: anger, sadness, fear, and faintheartedness if it persists long, for these increase the melancholic humor." Occasionally, an author added some detail. According to Hildegard, the blood was disturbed by "anger, so that it drops to the liver" and, with thick impurities, "spreads through the whole body, disturbs the flesh and skin; as a result, the skin cracks, and the nostrils thicken and swell while cracking." Du Laurens suggested that he knew the sequence from personal observation: "[A]nxiety and sadness perturb the blood, and we have seen many people who, from too much fear and dread, first became melancholic and then leprous."[26]

DIVISIONS AND DISTINCTIONS

Stages

After dwelling on the "basic" causes of leprosy, the authors tended to treat the other types of causes more cursorily. This was understandable because they discussed the more essential factors under other rubrics. The "antecedents," that is, the humors that predisposed the body, guided most of the internal division of the disease. The "conjoint" role of a complexional imbalance and burned black bile, and the "immediate" cause in a failure of the assimilative faculty, were at the center of the definition and physiological explanation of leprosy (Chap. 4). Most of these causes also figured in the nosological correlation with other diseases, which we will consider shortly. Leprosy itself was divided into levels and types, along diverse lines. Medical theorists recognized four phases in its progress, in accordance with the standard Galenic division of the course of all diseases (though primarily the chronic ones) into four "times" (*tempora morborum*), or stages: a beginning, or onset; an increase, or rise; a plateau (*status*); and a waning (*declinatio*) to the final outcome, of recovery or death. Practical needs dictated a simplification, not only for diagnosis but also for treatment, as we will see in later chapters. Simplified schemes placed the emphasis on the first two stages, with the assumption that, once leprosy reached *status,* further mapping of the course was largely irrelevant. At this point, the firm and patent presence, or *confirmatio,* of the disease created a radically new situation.[27]

A sharp line was drawn, in principle if not always easily in practice, between a confirmed stage and an unconfirmed—or, to be more accurate, non-confirmed (*not yet* confirmed)—stage of leprosy. The distinction was vital because it was the medical—rather than popular or institutional—criterion for determining the patient's social prospects and hope for treatment. In the setting of this chapter, the differentiation was conceptually important because the "confirmed" category represented *real* leprosy. It is significant that the distinction was stressed far more by medical practitioners, such as Doctor Heinrich Losser, than by lay examiners, such as the wardens or inmates of leprosaria. In addition, physicians subdivided this overarching distinction into three more "existential" categories. They differentiated, again more explicitly than lay examiners, between a predisposition, which called for caution; an incipient stage, which allowed for correction but also for ambiguity; and a confirmed status, which was total and irreversible. Possibly the earliest application by medical examiners of this differentiation was recorded around 1335, when a Bavarian judge determined that physicians had found a suspected individual "clean of the stain of *lepra*" but "somewhat disposed toward" the disease. Although the elastic notion of predisposition entailed problems, some of which will become apparent in the chapters on diagnosis and therapeutics, it was spelled out with increasing precision. Thus, by 1455, Hermann Schedel could judge that someone was "not infected" but that "some sufficiently known presages [*praeambula quaedam satis famosa*] had been found, which are wont to precede said disease."[28] Seen by themselves, the differentiations may seem to attest to the advance of "Scholastic hairsplitting" in the medical discourse. They become more understandable, however, if we consider the likelihood that the erratic development and indeterminate forms of the disease eluded classification (as is still the case, to some extent, for Hansen's disease).

Types

Practical interest as well as academic reasoning led to attempts to identify distinct forms of leprosy. The most popular subdivision rested on the application of humoral etiology to selected manifestations of the disease in the body, face, skin, and hair. Leprosy in general was explained by the conjunction of a disproportionately cold and dry complexion with an excess of "unnatural" black bile. This excess, in turn, was due to one of the four humors, which had become overheated and burned into ash (*adustus*) or spoiled (*putrefactus*). The offending humor, as "antecedent" cause, shaped a corresponding category of the disease. Thus, a fourfold scheme arose out of various appearances that the earliest reports had compared to the typical features of certain animals. The Greeks, as European authors noted until the early seventeenth century, had posited only two kinds, of which one resulted directly from burned black bile and the other, more malignant, from yellow bile. Arabic sources concentrated on three kinds. They largely ignored the possibility that phlegm, by definition the cold and wet humor, would burn and thereby change into bad black bile. The Salernitans recognized four types that matched the humors: when burned, yellow bile caused a leonine form; blood caused alopecia, or fox mange; black bile, *elephantia*; and phlegm, *tyria*, named for a snake. A Near Eastern origin of the fourth type may have been imagined by the Fleming Jan Yperman when he explained, in an aside, that *tyria* derived its name from "a serpent that lives in Jericho." At the same time, the fourfold division was presented as a recent insight, formulated "according to the moderns." Notwithstanding the precedents, Guy de Chauliac and Pierre Bocellin gave the credit for the division to "our general school of Montpellier," not entirely without justice.[29]

Few did as much as the Montpellier master Bernard de Gordon to develop and disseminate the classification.[30] Always concerned to streamline teaching with logical consistency, Bernard extended the humoral typology from etiology to a register of distinctive features. The result is schematized in Table 6.3, which incorporates a few details added by other authors. As is obvious from this table, the construct rested more on logic than on methodical observation, and it offered few benefits for differential diagnosis. However, it made such "good sense" that it became commonplace. Each type was distinguished in definition, appearance, and course by the character of the responsible humor, which, in turn, was embodied in the eponymous animal. The scheme appealed not only by the simplicity of its sequences but also by its selective correspondence with reportedly observed facts. Thus, hot and dry yellow bile accounted for the appearance of *leonina* as we have seen it described by Aretaeus and Avicenna (Chap. 3), wrinkling the patient's face by its dry astringency, turning it yellow, and making it look angry or furious (the phrase *facies leonina* remains associated with some patients of Hansen's disease).

The neat distinctions in the medical writings, together with their sche-

Table 6.3. Four Types of Leprosy

Species	Animal	Humor	Qualities	Signs	Prognosis
elephantia	elephant	black bile	cold and dry	nodes and tuberosities thick, cracking skin blackish color	slow increase slow cure
leonina	lion	yellow bile	hot and dry	loss of eyebrows protuberant forehead hoarse voice yellow skin and urine	swift increase
tyria	snake	phlegm	cold and moist	white scaliness whitish face and urine	intermediate
alopecia	fox	blood	hot and moist	hair loss, in patches red face and eyes	safest

matic representation in Table 6.3, risk conveying the impression that mark-edly different ailments, such as fox mange, were being lumped together, or even confused, with leprosy. This impression might be reinforced by the for-mulaic claims of examiners. Practitioners at Marseille, for instance, claimed in 1464 that they had observed "the canons of the doctors of medicine" by examining a suspect "with regard to the four species of *lepra* which exist on account of the four complexions of the humors." In order to avoid erro-neous inferences, it is important to realize that most authors treated the four forms as subspecies of the same disease. Occasionally, the unity of leprosy was underscored by the comment that the forms "occur seldom by them-selves, but they are mostly combined." Rather than suggesting a muddled taxonomy, this admission may very well have reflected the elusive nature and the polymorphic presentation of the disease. In another effort to subdistin-guish while maintaining nosological unity, Du Laurens replaced the four humoral types with three "kinds" (*genera*), which were closer to the stages yet also hinted at the forms of the disease. The first kind was defined qualita-tively, as a dual "distemper" (*distemperatura*), namely, an imbalance of the ex-cessively "warm liver, whence the blood is burned," and an indisposition of the body's parts, "which are uniformly cold and dry." The second genus, pertaining to the body's constitution, was "a bad conformation," which consisted in growths and in "disfigurement by swellings, nodes, and scabies." In the last kind, "separation of the continuum appears in fissures, ulcers, and so on."[31]

If most authors oriented their taxonomies toward the overall unity of leprosy, a few espoused an opposite trend, toward greater compartmentalization. Petraeus collated no fewer than three elaborate divisions from the *Volumen paramirum* of Paracelsus, "who was evidently more diligent and felicitous on the subject of this disease." Paracelsus envisioned a "quadruple" leprosy on the basis of the elements ("matrices").

> If the *air* is infected, [patients] have heavy respiration and discoloration of the entire body; and it begins in the extremities; if the *earth*, it appears in the face, they feel no prickings, and the breath in the mouth is heavy; if *fire*, tiny ulcers occupy the throat and gums, and these disappear and return; if *water* is a cause, the feet swell and become cold and numb, the pudenda and groin swell, ulcers follow that are difficult to heal, and the patients feel prickings.

Another Paracelsian scheme, based on the principal organs, comprised an "octuple" division of leprosy, with affected lungs responsible for a hoarse voice, the bladder for fetid urine, and so on. In yet a third construction, in his book *De vita longa*, Paracelsus turned the classical four types (leonine, alopecia, *tyria*, and *elephantia*) into six by adding morphea and *undimia*, or edema. Even if the categories reflected actual and observed differences, this does not diminish the irony that they were so readily multiplied by the most fiery critic of Aristotelian categories and Scholastic subdistinctions.[32]

It is hardly surprising that the amplifications, pushed to an extreme by the Paracelsians, ultimately provoked a skeptical reaction. Isaac Uddman, ever the vigilant critic, questioned the relevance of conventional subdivisions of leprosy:

> Many species are listed by the authors, but when we look more closely at their descriptions, and when we collect and compare their symptoms or diagnostic signs, we find nothing accurate and consistently distinctive, by which these species would be differentiated. One who has observed this disease just a little more closely, understands fully that a division into a dry and moist *lepra* or *elephantiasis* is of very little importance.

Rather, Uddman continued, "the hoarse and nasal voice, the disfigured face," and other symptoms "barely differ, except for the stage and degree" of the disease. Therefore, he preferred to rely on the certitude of physicians who "recall and describe from observation," rather than on "the rambling attempts of commenting authors."[33]

Allocation in General Nosology and Humoral Taxonomy

Uddman was convinced that the kinds of diseases "should be defined as accurately as possible," though without unnecessarily multiplying their genera.[34] Thereby he may have proved to be an exemplary student of Carl Linnaeus, under whose aegis he was writing. He was also, perhaps unwittingly, encapsulating centuries-long endeavors to classify leprosy in the universe of diseases, in order to gain a fuller understanding, to sharpen the differential diagnosis, and to design the appropriate treatment. Some classifications were implicit, primarily in the place or rank assigned to leprosy in the overall medical discourse. In quite a number of comprehensive treatises, the table of contents holds clues to the standpoint of the compiler—whatever his or her avowed taxonomy. By the arrangement of their chapters, the authors revealed whether they approached the disease as a systemic or a topical problem; to which area of medical practice they assigned the principal response; and even whether they were concerned primarily with the morbid process or with its cosmetic effects. In several cases, the organization of the subject matter indicated not only whether the author was more interested in teaching or in practice but also to what extent the work was addressed to physicians, surgeons, apothecaries, or lay readers. In general, arrangements differed primarily in the area of therapeutic practice, which will be considered in a later chapter (Chap. 8). Suffice it here to cite some illustrative differences, without pursuing the implications.

The Greek sense of *lepra* and Galen's defining phrase, "a change in the skin," exerted a lingering influence on the thematic allocation of leprosy. At least partially as a result of this influence, the disease was designated by some authors as a superficial condition, most notably by Constantine the African. In the thirteenth and fourteenth centuries, Italian authors discussed leprosy between freckles and tuberosities (Theodoric Borgognoni), juxtaposed it with "other foulnesses of the skin" (Bartolomeo da Montano), or included it in a "treatise on beauty care" (*de decoratione*) (Niccolò Bertucci). In the seventeenth century, Samuel Hafenreffer covered *lepra*—in both the Greek and the Arabic sense—in the chapters on scabies in his textbook *Nosodochium,* on "all affections of the skin" (see Plate 5 in Chap. 4). The authors of earlier Latin manuals allocated leprosy quite differently, in the wake of the *Canon.* Avicenna's influential compendium featured *lepra* in the section devoted to apostemes or tumors (bk. 4, *fen* iii). This *fen,* or section, was preceded, somewhat oddly, by a section on prognostication, and followed by a section on wounds. By not assigning his chapters on leprosy to a part on sundry diseases, Avicenna confirmed his view of leprosy as a "universal" disease that pervaded the entire body. This view also underscored the *dispersion* of burned black bile as the efficient cause, whereas Constantine—and most of the authors who concentrated on the superficial character of leprosy—emphasized the intrinsic *malitia* of the humor. Notwithstanding their differences, in emphasis and dependence on Galen, both Avicenna and Constantine built their pathology and nosology on the Galenic tenet that *elephas,* as well as *lepra* in the Greek sense, "is produced by melancholic humor."[35]

Avicenna owed much of his influence in the Latin West to the sense of order with which he integrated seemingly disparate medical knowledge, including the profuse and often diffuse teaching of Galen. This general observation applies seamlessly to the subject of leprosy, on which Avicenna conflated discrete Galenic statements into an orderly synopsis. The synopsis stimulated further organization of the subject matter, which eventually became grafted into an entire "tree of diseases." A simple parsing of the first 130 words of the *Canon*'s "tractatus de lepra" yields a clear image of the nosological location and connections of leprosy, as may be seen in Figure 6.1.

In the course of three centuries, commentators and compilers endeavored to delineate this image with greater precision. To be sure, it is natural for the human mind to seek understanding (and comfort) by identifying or creating connections, regardless of concrete implications. Furthermore, medical authors tended to be more concerned than practitioners with delineations—and medical practitioners tended to be more precise than lay examiners. Nosology, however, was not merely an intellectual game. It offered benefits, at least with regard to leprosy, for more precise definitions or "pathologies," differential diagnoses, and rationally justified treatment. These benefits are most obvious in the contrast between medical practitioners and lay examiners in their use of nosological categories (aside from the differences in their diagnostic criteria, which will be discussed in the following chapter). While physicians strove for precision, nonmedical jurors, apparently biased toward finding a suspect leprous, applied broad and undefined labels. This was par-

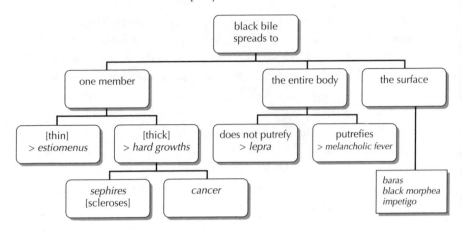

Figure 6.1. Taxonomy of *lepra* in Avicenna's *Canon*

ticularly striking in the few documented cases in which a jury consisted of leprosarium residents or *leprosi* at large. Suffice it to illustrate the contrast by an admittedly selective comparison between academic writings and lay certificates from one region. The archives of the Pas-de-Calais in France show that, unlike the medical authors, the examining *ladres* mentioned a "liepre courant," a "running" and, inferentially, acute variety of leprosy; they distinguished a "grand" type from the "petit" or "menu mal"; they classified a form as "white evil" (*blanq mal*) without identifying it—although several candidates might be suggested (including "white morphea," about which more below); and they introduced vernacular categories such as "pouacre" and "luffre." In Arras in 1489, a physician and a surgeon found Flourens Polart not leprous, "in spite of some blockage in the nose which can result from another cause than *lepre.*" Two years later, however, six *malades* declared Polart "infected with said disease, in various places, and with several of the diseases such as *brun mal, gro-mal,* and *mormal.*" In casting a wide net, the examining *malades* drew neither a connection nor a distinction between the "brown" and "grand" (*gros*) evil; furthermore, they extended their positive verdict to "mormal," or *malum mortuum,* that is, gangrene, a sequela of leprosy which no medical author considered synonymous with leprosy itself.[36]

Medical classifications, and the degree of precision in distinctions, affected pathology and etiology as well as diagnostics. The classifications grew out of Avicenna's outline, and they found a rational foundation in the broader humoral taxonomy, which is illustrated in Figure 6.2. The reader will note that the ramifications of an excess of blood are omitted from this chart. There are several reasons for this omission: the consequences of an imbalance in the most vital humor were believed to be unique and complex; they would make the chart too large for fitting on a single page; and, with regard to our

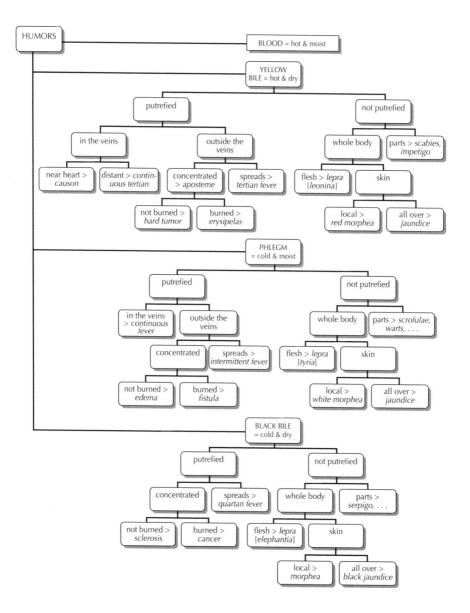

Figure 6.2. Humoral taxonomy of diseases

subject, they bore on the mildest type of *lepra*, namely, "alopecia." The chart is drawn primarily from a tightly organized treatise by Bernard de Gordon on prognostication. However, it represents the structure that framed or underlay nosology for all the authors who attributed diseases to a humoral imbalance, or, to be more precise, to the excessive dominance of one humor.[37]

Whatever their simplifications and contradictions, redundancies and lacunae, the schematizations aimed at making diseases intelligible, logical, and open to analysis and orderly reconstruction—as in a sort of "algebra." Thus, the puzzling syndrome of jaundice recurred with axiomatic predictability under each humor, as a generalized skin condition. The feverish skin eruption of erysipelas was logical for the hot and dry complexion of yellow bile, as the soft watery swelling of edema was for the cold and moist complexion of phlegm. There was symmetry in the correspondence between hard and soft swellings, the parallel between sclerosis and cancer, and the contrast between edema and fistula. In sum, this architecture was ideally suited for making sense of the most baffling afflictions. This is why *lepra* (under three of its manifestations) figured prominently in this outline, notwithstanding the alien context: by its chronic nature, the disease fell outside the purview of Bernard's *Liber pronosticorum* or *De crisi et creticis diebus*, which was devoted to acute diseases. It is particularly instructive to note that *lepra* appears in this chart (not merely by my design: compare Figure 6.5) in the middle of diseases that were often clustered, not only as prominent effects of excessive black bile on the body, but also on account of other correlations, as we will see shortly.

Connections and Parallels
"The Tree of Diseases"

More ambitious efforts were made to integrate all of nosology into comprehensive frameworks. The results were less symmetrical, reflecting the organic jungle of reality. At least one of the outlines was called a "tree of diseases" (and attributed to Bernard de Gordon in a single extant manuscript). This outline represents—with allowance for some variations—the medical taxonomy that prevailed until sixteenth-century authors moved away from Galenic categories and conceptual pathology, as we have seen with regard to definitions (Chap. 4). The "tree" sheds light on the place that was assigned to leprosy in the network of pathological identities, causes, and manifestations. In order to appreciate this place, as well as the correlations with other diseases, it is useful to present the context in diagram form. Even though the subsequent charts may seem extensive, they cover only the "universal"—that is, generalized or systemic—diseases, which accounted for just a portion of the nosological "tree," as may be gathered from Figure 6.3.[38]

The ramifications of the first principal branch, comprising fevers, are charted in Figure 6.4. The relevance of this chart may not be immediately apparent, since premodern nosology contained no direct associations between fevers and leprosy. In fact, the two were mutually exclusive, for the former consisted in an unnatural warming while the latter was accompanied by a drastic cooling—as the "ash" of burned humors spread through the body. Nevertheless, two indirect connections helped further to circumscribe the taxonomy, and even the terminology. First, leprosy had a counterpart in quartan fever, an intermittent fever with attacks every third day: both were attributed to the diffusion of black bile, not putrefied in the former, putrefied in the latter. In general, the four forms of *lepra* were due to the spreading of a corresponding humor that was corrupted by "burning" rather than "rotting"; when the same humor was "rotten" while spreading through the body, it caused one of the major fevers (which, in premodern medicine, were not viewed as symptoms but as morbid entities).

In a second elucidating connection, *lepra* had a nosological antithesis in hectic fever. This was a chronic and continuous condition ("hectic" was derived from *ethos*, habit) and, contrary to what modern expressions such as "a hectic pace" might suggest, it progressed slowly. Like *lepra* in the standard definitions, it affected the entire substance of the body. Unlike *lepra*, however, hectic fever was due not to the defect of a particular humor but to a general drying of the life-sustaining or "radical" moisture. It did not consist of uncontrolled growth but of steady consumption (and, indeed, it was usually

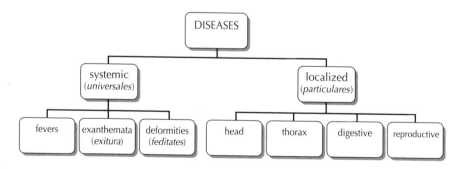

Figure 6.3. Tree of diseases: principal branches

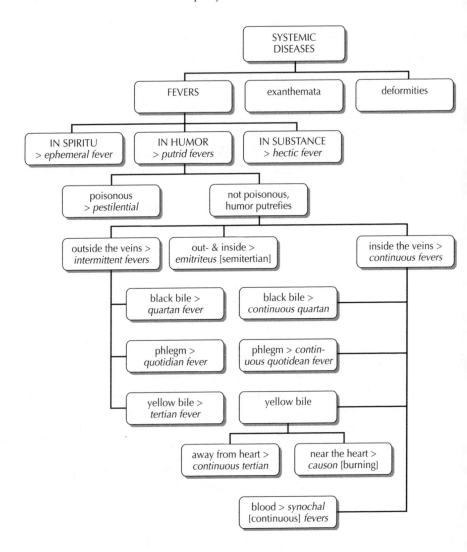

Figure 6.4. Tree of diseases: fevers

associated with pulmonary consumption, or *phthisis*). Consumptive emaciation was "not a differential characteristic of *lepra*," according to Bernard de Gordon. Bernard observed an additional contrast:

> It is remarkable that in *ethica*, the patient has a sense of the external but not of his or her own illness, while the opposite happens in *lepra*, where there is no

complete sense of the external—indeed, the patient will not feel when pricked with a needle—and yet, in some way, she or he has a sense of her or his own suffering.[39]

This observation, and the contrast with hectic fever, raised poignant and perennial paradoxes about leprosy, not only of deepened internal feeling and external loss of sensation, but also of the patient's inner "consumption" and the uncontrolled outer growth of *exitura*, or eruptions. The context of exanthemata, as schematized in the tree of diseases, is diagrammed in Figure 6.5.

Before examining the place of *lepra* in Figure 6.5, it is helpful to realize that the apparent order and symmetry of nosology were offset by imprecisions and contradictions. In our tree of diseases, the logical allocation of *lepra*, among the effects of black bile and among the large eruptions called "apostemes," was followed by an odd but revealing appearance in the section on "foul deformities" (*feditates*), which is diagrammed in Figure 6.6. In the inclusion and branching of items, this catchall section followed less exacting criteria. It is strange to find *lepra* on the same taxonomic level as perspiration, and matched with rabies or, to be more precise, hydrophobia. It is more significant, however, that only two diseases were identified as contagious—constituting a separate branch of "foul deformities"—and that *lepra* was one of the two. With rabies as the other, it is obvious that the notion of "contagion" was applied with great, if not unusual, latitude. At the same time, by not classifying scabies (which was always feared as highly communicable) or even variola (which was recognized as an epidemic sequela) *as* contagious, the schematization highlighted contagiousness as a defining characteristic of *lepra*.

Correlates

If the branch and further ramifications of "deformities" revealed the logical weaknesses of nosology, the diagrammed correlation of leprosy with other diseases attested to structural coherence. As seen in Figures 6.1, 6.2, and 6.5, black bile, the efficient or "conjoint" root cause of leprosy was responsible for at least a half dozen other problems, which showed "family resemblance" as well as distinguishing features. Like leprosy, scrofula, warts, and the condition called *herpes estiomenus* or *herpes ustiomenus* ("burning herpes") represented disfiguring, uncontrolled, and sometimes fatal growth; unlike leprosy, they were localized. Lupus, on the other hand, was a baffling and generalized combination of a neurological malfunction with skin manifestations; however, it

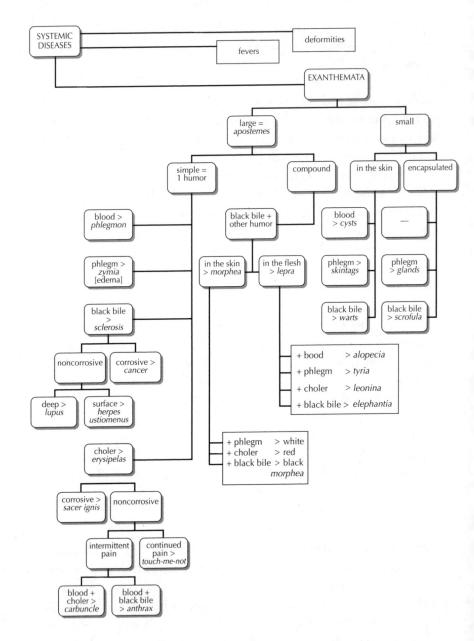

Figure 6.5. Tree of diseases: exanthemata

represented neither growth nor corrosion. No diseases were correlated more closely or constantly with leprosy than cancer and morphea. The latter was routinely paired with leprosy, not only in learned discussions but also in notarial acts, and even in pictorial presentations. In Plate 7, which presents a detail from the "Disease Woman," a woodcut in the now famous *Fasciculus medicinae* (printed in Venice in 1491), the pair appears among the diseases of the entire body. Cancer, on the other hand, marks the breast in this pictorial inventory, as it did in most of the written nosologies; it marked the face in

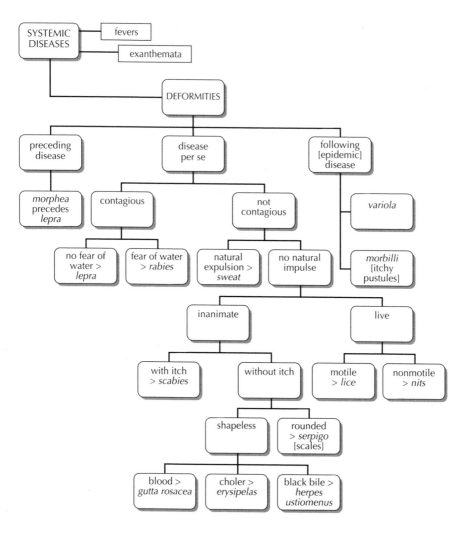

Figure 6.6. Tree of diseases: deformities

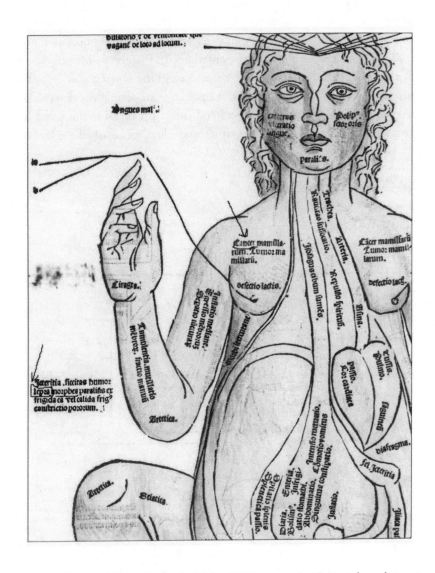

PLATE 7. "Disease Woman": woodcut (detail) illustration in *The Fasciculus medicinae of Johannes de Ketham Alemanus. Facsimile of the First (Venetian) Edition of 1491.* With an introduction by Karl Sudhoff, translated and adapted by Charles Singer (Milan: R. Lier and Co., 1924), p. [1–10]. At the National Library of Medicine, Bethesda, Maryland.

some diagnoses, as we will see. In any event, it stood out as a disease that by definition affected a specific part and thereby stood in direct contrast with leprosy; in every other way, both diseases shared the same essence. These two correlates require special attention, not only on account of their close noso-logical connection but also because of the practical implications.

From the second to the eighteenth century, parallels were drawn between cancer and leprosy. Rufus of Efesus pondered that *elephantiasis*

> appears to be a superficial disease, since it makes its appearance on the skin; but the difficulty of curing it, a difficulty which comes close to impossibility, sug-gests to us that it has a deeper origin, an origin not easy to penetrate; it is even as deep as that of carcinoma, by common opinion; in truth, Praxagoras [Kos, ca. 330 BCE] accepts a deep origin above all for carcinoma.

In his *Histoire de l'Éléphantiasis,* François Raymond considered the analogous corrosive effects of both diseases in his description of the ultimate disfigure-ment "by horrible ulcers that represent a universal cancer and which eat even the bone structure." Raymond also adopted the implicit comparison with hidden cancer by Aretaeus, who characterized incipient *elephantiasis* as "a fire that, after lurking in the viscera as if in the dark netherworld, flares up in the end and, having vanquished the body's insides, bursts out to the surface." The parallel between leprosy and cancer was promoted above all, as we have seen earlier (especially in Chap. 4), by Avicenna, who also bequeathed a full elab-oration. Both diseases resulted from burned black bile; were extremely diffi-cult to be known in their early stage; behaved duplicitously on the surface and deep inside the body; manifested themselves first in hard nodes that turned livid; corrupted the complexion, form, and figure; corroded the flesh and often broke the integrity of parts; sometimes ulcerated and sometimes not; and defied treatment once they entered an advanced stage, although some patients lived longer than others. It would be surprising if the shared attri-butions of sinister features such as, in particular, a melancholic origin, a stealthy presence, a corrosive or "eating" progress, and an unstoppable "rot-ting" did not cross-fertilize the extraordinary popular fears toward these two diseases.[40]

The paramount difference between leprosy and cancer was that one af-fected the entire body, the other a single member. This theme recurred con-stantly, not only in academic discussions but also in evidence much closer to practice. It even seems to have overruled Galenic influence on learned dis-

course: to my surprise, I have not encountered echoes of a potentially contradictory proposition by Galen, that "both scabies and *lepra* are melancholic
afflictions of the skin only, for if these occur in the veins and flesh, it is called
cancer." The practical implications of the conventional distinction between
the general and localized disease were not overlooked. In a very concrete application of the difference, the surgeon Hans von Gersdorff explained,

> [T]his disease is called "a universal cancer," contrary to "particular" as for a can-
> cer that settles in only one member. When a disease is seated in only one mem-
> ber, then it is possible that it may be cured, because you may perhaps be able to
> separate the member from the body, such as a foot, hand, finger, and the like, as
> I have often done with my own hand.

Practitioners remembered the Avicennan pair when they examined suspected
patients. At La Seu d'Urgell in 1372, as we have seen, the *Canon* was summoned
for determining that Ramona Isern's facial ailment was "a kind of cancer,
which is particular to one organ," and that she was free of leprosy, which,
"on the contrary, is general, because it proceeds to the hands, feet, nails and
skin." Similarly, in 1457, Heinrich Losser cited Avicenna when he certified
that widow Fransbergerinn suffered from a facial cancer rather than from the
disease that "originates in a weakness of the liver, is specifically called *lepra
universalis*, and can infect the entire body." However, even the express differentiation, that "cancer is leprosy of a single member," left room for ambiguity. Losser proposed that the widow's cancer had "something in common with
true general leprosy," and even that it was just cause for her separation, "on
account of its predisposition to leprosy."[41]

In the nosology of leprosy, only morphea appeared as a closer correlate
than cancer. When stemmas, such as the one illustrated in Figure 6.5, presented cancer as a "second cousin" of leprosy under the parentage of black
bile, morphea was an immediate sibling. The medical literature did not pursue detailed differentiations between leprosy and morphea, beyond the contrast between their manifest effects. While the former affected the solid substance of the body, the latter manifested itself on the surface, in scattered
light or dark spots and hair loss. The authors, in fact, gave the impression
that the close correlation between both diseases extended to pathology and
etiology: for example, both were due to a failure of the assimilative faculty.
As we may recall, Paracelsus even included morphea as a type of leprosy.
Learned practitioners, in general, adhered to the conventional taxonomy, and

they treated the simplified distinction as an ironclad premise. Doctor Losser was typical when, as we saw at the opening of this chapter, he referred to "the masters of the medical art," and to Bernard de Gordon by name, for the maxim that, as leprosy or "*uβseczczekeit* is in the flesh, so *morphea* is an *uβseczczekeit* of the skin." Avicenna's *Canon* and other Arabic writings appear to have boosted this direct juxtaposition of leprosy with morphea. Galen was quoted as pairing the two when he illustrated the effects of a disturbed "digestive faculty," which "alters the nature [*speciem*] of a member, as we see in morphea and *lepra*"; however, he did not link morphea with *lepra* or *elephantia* in his basic discussions of humors and complexional imbalances. Figure 6.7 presents a diagram of the Galenic taxonomy, which Arnau de Vilanova constructed out of these discussions.[42]

The number of correlates increased with time, to include other "loathsome" skin problems such as impetigo, vitiligo, and psora, or psoriasis. The correlation with leprosy, and the taxonomy itself, were often not clearly delineated. The most confusing addition to the group, from Arabic sources,

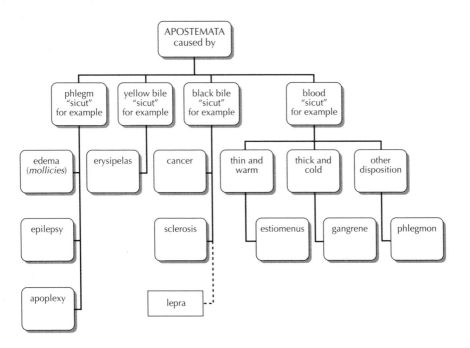

Figure 6.7. Correlate effects of complexional imbalance according to Galen, and schematized by Arnau de Vilanova

was *baras* or *albaras,* an unstable term in the Latin nomenclature (Chap. 3). While many authors applied the label to leprosy, some identified it, rather, as "white morphea." One practitioner, who may have possessed limited book learning, set this disease in direct contrast with leprosy. In 1501, after having participated in examinations, the Béthune barber Pol Roze was reported by fellow barbers to be infected: he contradicted the report by claiming that he had "a white *morfée,*" and that "as long as he had this *morfée,* he would not be leprous." Around 1517, the surgeon Gersdorff both inverted the coloring and blurred the distinction between both diseases by stating that, while an evenly malignant infection was *lepra,* it was "called *morphee alberas* if it is black"; his attribution of this view to "the common school" proves difficult to verify. Actually, Gersdorff observed elsewhere that "the old masters are unclear in their distinctions, for some call morphea '*albarasa,*' while others call it '*lepra.*'" In some cases, white morphea may have overlapped with *tyria* (see Table 6.3). It may also have coincided with the "white sickness" distinguished by Sister Barbara Sagers and her Brugge successors (Chap. 2), or with the *blanq mal* of examiners in Pas-de-Calais (earlier in this chapter). In Saint-Omer in 1550, physicians and surgeons, "in full and unanimous accord" with *ladres* from two leprosaria, declared that eighty-six-year-old Nicolas Bernard was "infected with the disease of *lèpre* called 'white morphea,'" and that he needed to be separated because "this kind of *lèpre* is the most dangerous and contagious."[43]

Regardless of the garbled interpretations, a practical reason for pinpointing the relationship between morphea and leprosy was that, as Figure 6.6 illustrates, the former was generally believed to be a precursor to the latter. In 1458, as we saw in the beginning of this chapter, Losser summarized conventional teaching by observing, "[T]he masters write that *morphea* is a presage of future *lepra.*" In 1501, a week after Pol Roze made his claim, two physicians and two surgeons found him free of *leppre* but infected with "a white *morfée,* which is one of the sixteen equivocal signs preceding the said disease." At Saint-Omer in 1532, physicians and surgeons found, together with several *ladres,* that Leurens Waultier was "infected with morphea of the head"; they determined that, while he appeared "predisposed to the disease of *lèpre,*" he "should not yet be separated from the conversation of healthy people": in this determination, they disagreed with the *ladres* who considered him leprous. In an alarming—albeit ambiguous—correlation, medical examiners in Lille in 1559 found

Boudewyn Cupre presently not infected by the disease of *lepre* or *ladrie,* but closely disposed to it, to the extent that we find it necessary to separate him for some time from intercourse and conversation with the healthy, otherwise he might infect others with the same disease that torments him. His present disease, which comes from salty phlegms and the burning of blood, is called *impetigo* by the Latins and *morphea* by the people, and it is the road and path to real *ladrie.*

These and other archival documents demonstrate that taxonomy was not a mere Scholastic creation, even if it faded when the medical emphasis shifted from the concept to the description of leprosy in the sixteenth century. Categories and correlations played a role in the encounters between medical practitioners, leprosy, and patients. In the second chapter of this study we surveyed the most dramatic moment in these encounters, the *iudicium.* The two remaining chapters are devoted to the most vital episodes, namely, the diagnosis and treatment of the disease.[44]

CHAPTER SEVEN

Diagnosis
Signs and Symptoms

March 22, 1574
In re the visitation of Adrien de Bellevalet,
licentiate in the laws, solicitor at the Council of Arras,
who was suspected of *lèpre.*

Said Bellevalet has been examined, after swearing to tell the truth in the interrogation by the physicians and surgeons. He has been bled from the right arm, and the blood from this letting has been inspected, together with his urine, in order to judge about the inside with better knowledge. Then, after examining him diligently for the unequivocal as well as the equivocal signs, and after carefully considering everything the physicians and surgeons have found Bellevalet not to be confirmed in the said disease. They have reached this conclusion because, according to the ancient teachers [*docteurs*], *leppre* is a very great error of the assimilative faculty, by which the entire form is corrupted. They have not found this in Bellevalet, given the integrity of all his limbs, without any spot, weakness, or deformity in his entire body, even though his face shows disfigurement of the nose and mouth: these signs are unequivocal, but insufficient for separating him at this time.

Bellevalet has requested a record of this, which we are granting him in these presents, of which he may rightfully avail himself.

I ssued by the municipal attorney of Arras, in northern France, this certificate offers an elegant and valuable synopsis of a medical "visitation." As such, it introduces the present chapter, which is focused on diagnosis, while it also recalls several subjects of earlier chapters, ranging from the *iudicium leprosorum* to the definition and division of leprosy. The certificate is somewhat unusual because it is more orderly and comprehensive than most archival records. Nevertheless, it represents the conventional diagnostic techniques, criteria, and conclusions, which we find documented across the Continent. It illuminates the features of a formal examination which stand out in academic treatises, particularly the concentration on unmistakable symptoms, the methodical procedures for obtaining evidence, and the theoretical justification in interpreting signs. At the same time, the legal document reveals—even better than the treatises—the variability of subjects, circumstances, and outcomes.[1]

THE VISITATION
Protocol

Adrien de Bellevalet was hardly the prototypical subject of a visitation. If he initiated the request, as we may infer, he represented only a portion of the suspected patients; the staging at his own house was most unusual; and the practitioners apparently proceeded with uncommon care. In essence, however, the examination followed the normal protocol for a medical intervention—which, as we may infer from the certificates, was both more uniform and more detailed than appearances before lay examiners. It was common to have several practitioners present, sometimes as many as seven; physicians were routinely accompanied by surgeons, and frequently by barbers; and each type of practitioner had a designated task, at least in principle. The presiding examiner performed the opening formalities, which included an interrogation of the examinee. Members of the team applied their respective knowledge and skill in the physical examination that followed and which, for Bellevalet and most other individuals, consisted mainly of two parts: a thorough inspection of the body's exterior and a careful "reading" of the inside. After carefully assessing all the evidence, the practitioners determined whether the subject was free of leprosy, suffering from an advanced and "confirmed" state, or somewhere in between. They justified their finding by matching their observation with authoritative teaching on the definition or pathology, etiology, and semiology of leprosy. Bridging medical opinion and legal action, the concluding verdict translated the diagnosis into a decision about the patient's future.

Subjects requested an official record of the decision, either as proof that they were "clean" or as voucher for their admission to a hospital. For the bearers, the certificates opened the door to freedom or succor; for historians, they open windows on actual medical responses to leprosy. We have already seen how these archival records illuminate and complement the image conveyed in academic writings. In this chapter, the certificates will prove invaluable for reconstructing—and assessing, at least tentatively—the premodern diagnosis of leprosy. One, out of the hundreds that I have found remarkably well preserved in archives, is reproduced in Plate 9 below. The certificates also open a much broader perspective, for they alert us to various nonmedical factors (ranging from the status of the subject to the local situation) and human dimensions that were so poignantly evident in the "judgment" of people suspected to have leprosy (Chap. 2). In this chapter, we will not dwell on those aspects, which are arguably extraneous to the diagnosis. For example, common protocol included an invocation and consolatory words to the patient, in addition to an oath such as that imposed on Bellevalet. These opening formalities, which were rooted in nonmedical antecedents, pertained more to the judicial arena than to a clinical procedure.

After the preliminaries, the examination itself began with the taking of the patient's history, in keeping with the timeless and (supposedly) universal practice for any doctor's visit. The recurrent assurance in notarized records that the examinee had "answered the questions in the usual fashion [*consueto more*]" suggests that an interrogation was an integral part of the *visitatio leprosorum*. The oath to tell the truth probably received a different degree of attention when it was administered as part of the interrogation itself rather than as one of the preliminary rituals. In either case, combined with the customary presence of witnesses, the oath added a judicial weight that must have made the event considerably more stressful for the subject than a mere medical visit or consultation. For the examining practitioner, too, the obligation to make the right inquiries was made heavier by the implications of a wrong decision. Academic authors considered a thorough background check indispensable for preventing irrational misperceptions and inconclusive examinations. The physician-poet Gilles de Corbeil urged examiners to "consult the circumstances of place and time, and the complexional vigor and personality of the patient, so that reason may stand firm without nodding or wavering."[2]

Authors from the fourteenth to the seventeenth century elaborated an out-look that was similarly holistic—as well as, paradoxically, differentiating—when they drew up detailed agendas for questioning a potential patient. First, the examiner should unearth "the things that dispose to *lepra*." These began with the suspect's parentage and consanguinity, as it was important to know "whether anyone among the relatives is known to be infected with *lepra*: for the disease is often propagated into the third and fourth generation, and not only in children but also in nephews and kindred." Had the subject lived among *leprosi* or kept company with them, and "perhaps contracted some-thing by contagion"? A probe of the person's lifestyle would also reveal pre-disposing factors that could give the examiner "a handle on the burning of black bile." These factors ranged from "the frequent consumption of thick and melancholic foods, salted meats, old and rotten fish, or thick and full-bodied wine" to emotional troubles, including "some sudden fright or sad-ness." Acquaintances might provide information about the individual's "wits [*astucia*], behavior, and dreams and desires." The assessment of predisposition concluded with the "medical history" proper: "whether some regular evacu-ation, such as the menstrual or hemorrhoidal flow, had been suppressed; which illness had been more troublesome than others; and whether she or he had ever suffered from quartan fever, depression, mania, morphea, or a similar melancholic disease." The number and thrust of the questions may some-times have moved toward the inquisitorial, as if driven by preset expectations. André Du Laurens ended his instructions for the interrogation on a some-what ominous note by suggesting that "by these inquiries the physician or surgeon should be able to obtain some conjecture that the patient's complex-ion or constitution inclines him toward this disease in such a way that, if he is not yet leprous, he will soon become."[3]

Authors eventually extended the interrogation to "the condition of the natural, vital, and psychic functions" in order to check for "internal symp-toms." Under the "natural actions," Samuel Hafenreffer itemized the ques-tions of "whether stomach eructation is constant, the stool unusually dry, the urine frequently resembling that of beasts of burden—although many call this a misleading sign—, the sex drive compulsive, and the sweat and breath fetid." Problems with the "vital" functions were to be revealed by ques-tions about pulse and respiration. For the "animal" or psychic faculties, the inquiry should address a variety of items. These ranged from "numbness of the senses" and "anesthesia of the skin" to "timidity." Hafenreffer added "ulcerous lassitude" (*lassitudo ulcerosa*), a puzzling phrase by which he may have meant feverish melancholy. His verbal examination was so detailed that it may have pushed the concern with the patient's medical history to the point of impracticality. Even in the unlikely event that answers would have been forthcoming, let alone reliable, the questions covered many of the signs that, in any event, were to be determined by actual inspection.[4]

The Physical Examination

One of the earliest certificates of a medical examination preserves a clear outline of the nucleus of a methodical inspection. On April 2, 1289, at Pistoia, "in the house of the heirs of Spina Philippus" and in the presence of a notary and four male "witnesses invited to this procedure," Master Laurentius, *medicus*, dictated his finding on "the Lady Contessa, daughter of the late Moro de Cavinana."

> Having seen the Lady Contessa, having diligently inspected the signs of her illness, and having fully deliberated on them, by what he saw and knew, after invoking the name of God, Master Laurentius said and pronounced his judgment that she is not infected, and that she does not suffer from the disease of *lepra* at this time.

This outline was so programmatic and ubiquitous that, as simple as it appears, it deserves to be spelled out and expanded with details from treatises and records. While "seeing" referred to the procedure in general—and emphasized the visual component—the "diligent inspection" entailed several discrete activities. In numerous certificates, examiners vowed that they had looked, "in detail" (*exquisite, umbständlich*), at "every single part of the body, from the sole of the feet to the top of the head"—this sequence, customary in certificates, inverted not only the usual head-to-toe order of medical manuals but also the priority stated in guidelines for leprosy examinations.[5]

The requirement that the patient needed to be naked for this complete "scan" could cause problems, such as the misunderstanding between Doctor Losser and the allegedly menstruating Gerusse Guldenlewe (Chap. 1). No difficulties with dignity or decorum seem to have arisen on other occasions—for example, when assembled members of the Cologne Faculty of Medicine reported having "seen naked, and felt by touch [*tactu contrectata*]," female patients. Not only examiners but also officials, such as the sergeant at arms of Saint-Omer in 1535, reported that the patient had been touched ("visité par palpacion") as part of a diagnosis that was performed "diligently" or properly ("bien et deuement"). The basic visual and tactual inspections were supplemented by several methods, which were not indicated in the 1289 certificate and will unfold gradually in our survey of the entire examination. The Pistoia certificate, however, offered further insight by centering, not on the person of the Contessa, but on "the signs of her illness" as the target of the inspection—a target that lay examiners usually identified with the same "clinical" precision.[6]

Physicians and surgeons devoted great attention to these signs, "noting all of them carefully" (*omnia bene notando*), as will become clearer shortly. Moreover, they would not proceed to a decision unless they had "fully deliberated on them." This deliberation was guided, in the phrase of Master Laurentius in Pistoia, by "what he saw *and knew*"; alternately, in the expression of Hermann Schedel at Nuremberg in 1481, by the "the canons of the medical art"; or, in the discriminating phrase of the Cologne Faculty of Medicine in 1550, by "the true rules of medicine." A physician and a surgeon in Lille in 1559 emphasized that their "pertinent and requisite duties" revolved around an inseparable pair, namely, "not only the seeing but also the considering" of all the signs. Their colleagues in Ieper indicated that the dual activity, "to observe and deliberate maturely," was a customary and integral part of a formal examination or "solemnity" (*solemnitheijt*). The insistence on deliberating, and the habit of doing so in a collegial fashion, constituted a major difference between lay and medical examinations. It is instructive to compare two certificates from Arras. In 1489, a physician and a surgeon certified not only that they had "visited and examined Flourens Polart" in the "usual and adequate" way but also that, "after this examination was completed, we have deliberated together"; they deemed Flourens clean. Two years later, six *malades* from two leprosaria simply testified that they had performed "the probe in the customary manner" before judging Flourens infected.[7]

The reports of practitioners as well as the instructions of professors indicate that the physical inspections proceeded methodically and in deliberate stages, and the examiners applied the senses of sight, touch, and smell "with care" (*diligenter*). Variations in the order, however, suggest that priorities may have changed. According to Guy de Chauliac, as soon as the interrogation was complete, the physician "should feel the pulse, and then have blood drawn for him." After inspecting and testing the blood, the physician should "observe the patient's appearance, and tell him to retire and to bring his urine in the morning." The night gave the examiner a chance to "think about what he had seen and needed to see." In the morning, the patient was to "come before the physician, who should first of all see the urine and consider whether it would signify anything about a disposition of *lepra*." Then, after devoting special attention to the face, the examiner should make the patient "undress completely" and inspect the body from head to toe, not only visually but also by palpation and tests; he needed to check the face one more time before weighing his findings and applying them to a verdict.

Du Laurens, while avowedly following Guy de Chauliac as "our author," structured the inspection differently. He allotted a first stage to "the signs from the overall presentation of the body," and a second to "individual parts." By an examination of the entire body, he understood the checking of the body's appearance by means of sight, touch, and tests; an "additional" diagnosis covered the pulse and urine; and "lastly, look at the blood." When moving on to the parts, Du Laurens devoted more than half of his guidelines to the head. Hafenreffer, too, began with an inspection of the head, "after the thorough interrogation, and when all the parts have been stripped naked," but he followed a different sequence and ranking. The difference is attributable, at least in part, to the fact that Hafenreffer's book was dedicated to skin diseases. Indeed, he did not reiterate the ubiquitous maxims about the chief importance of the head, and he devoted proportionately greater attention to dermatological symptoms. It is telling, nevertheless, that Hafenreffer completely omitted uroscopy and reading of the pulse. For him, too, the examination of drawn blood was a last step, "if doubt remains, after all these things have been investigated."[8]

By his emphasis on the interrogatory phase, and by the generally impersonal tone of his instructions, Hafenreffer gave the impression that he normally supervised the physical phase of the examination rather than carrying it out. In fact, he guided the examiner by listing some twenty questions, rather than by describing procedures for the collection of signs. A similar impression of remote involvement may be conveyed by the use of commands by other authors of guidelines. While Ambroise Paré did sketch the procedure in his own twenty-step checklist, he seemed to take a similarly supervisory stance. He began his instructions with the declaration, "According to the doctrine of the ancients, one should examine the entire head"; and for subsequent steps he reiterated the phrase "one should." Impressions of aloofness find a corrective, however, in the realization that Paré happened to be a surgeon, who operated far from the academic milieu of Hafenreffer and exemplified the true "hands-on" practitioner. There is no reason to doubt

Paré's claim of extensive personal experience, "In truth, I have often been involved in the examination [*l'espreuve*] of *ladres,*"when he recommended one
test that, "among all those worth special attention, was routine for me." Even
among the academic physicians, who were so often portrayed in pontificat-
ing poses and who seemed to shun personal contact, several interacted
directly with leprous patients. Two of these physicians, Bernard de Gordon
and Niccolò Bertucci, will give proof of such interaction, in various capac-
ities, further in this chapter.[9]

A Model Inspection

The information on the distribution of tasks, which we have drawn from
written sources until now, is extended by the remarkable print in Plate 8. The
"Inspection of Leprous Patients" (*Besehung der uszsetzigen*) preserves an ideal
image of a medical examination. The woodcut, of which slightly different
versions exist in two printings of Hans von Gersdorff's *Feldtbuch der Wundtartzney,* is attributed to Hans Wächtlin of Basel. Wächtlin, or Wechtlin, was
an accomplished engraver of medical illustrations, and he may have been
trained by Hans Baldung Grien. This picture, created shortly before 1517,
marks a near-perfect halfway point between the guidelines of Guy de Chauliac (1363) and Samuel Hafenreffer (1660). I should emphasize that it documents a *medical* examination rather than just any "visitation," at a time when
many were still performed by juries consisting of leprosarium residents or
officials. By the same token, the woodcut depicts an idealized examination,
rather than following the directions or sequence dictated by Gersdorff or any
individual author. It presents the crucial moment of each of three discrete
stages as if they were occurring simultaneously, in a device that is often applied by illustrators and which may be compared to multiple-exposure pho-
tography. Of course, as we may note in passing, the iconography is typical
of the time in certain conventions, including the assumption—thoroughly
embedded in the treatises, though not in the archives—that the whole world
was male. Nevertheless, it is a welcome mise-en-scène of the academic discussion as well as the archival testimony of the medical diagnosis of leprosy.

The event is staged in a room devoid of decoration, which suggested a
public place rather than the home of a patient or a practitioner. The picture
invites reading from right to left. A physician is performing uroscopy, in the
prototypical stance of the learned practitioner. By this time, the urine flask
(*matula*) had long been the physician's trademark, comparable to the stetho-

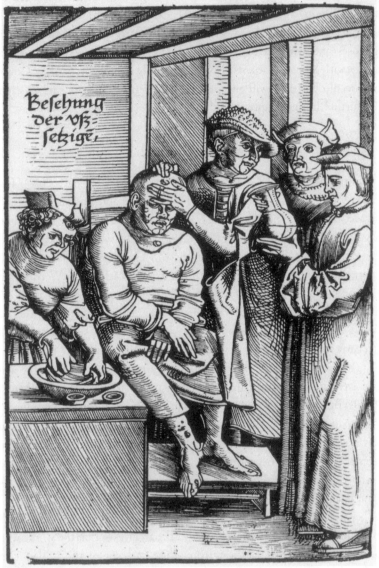

PLATE 8. "Inspection of Leprous Patients," woodcut in Hans von Gersdorff, *Feldtbuch der Wundtartzney* (Strasbourg: Joannes Schott, 1517), p. 67. At the National Library of Medicine, Bethesda, Maryland.

scope and other diagnostic instruments in later ages. With an authoritative expression, the doctor is applying his privileged knowledge, which he had acquired not through mere experience but in his university education. He had learned the science of interpreting the various colors and consistencies of urines and studied the treatises, particularly by Isaac Judeus and Gilles de Corbeil, which became part of the *Ars medicine,* or *Articella,* the core of the curriculum. The specimen in the flask was most likely brought by the patient, since the first morning urine was considered the most reliable for diagnosis. We will see shortly what it was supposed to reveal. The observant person in the background may be a second physician, but signatures and remarks on certificates lead me to surmise that he was a witness—perhaps the notary or official who would record the outcome.

Considering the status of the learned physician in most guidelines for examinations (and in sixteenth-century society in general), it seems surprising that the central actor—and more surprising that the best-dressed man— in the scene is a surgeon. He may be featured so prominently because his peers were working hard to make their craft more respectable and, after all, the woodcut was produced for a surgical manual. In any event, as a medical practitioner who, by definition, works with the hands, χειρουργός or *chirurgus,* the surgeon is palpating the patient, or, to be more precise, feeling the head. As Ambroise Paré insisted, the examiner should "feel with the finger" every part of the head, especially the eyebrows and behind the ears. Almost every protocol treated the head as the most significant part of the body for the diagnosis of leprosy, "and, above all, the face, in which the specific and most truthful signs appear, because the face is soft and delicate," according to Paré. While bracing the subject's head with one hand, the surgeon in the woodcut is pressing the upper forehead and pointing to a lesion with the other. He may be wearing a protective apron, which would be rather customary, although the drawing is not entirely clear. With a facial expression that suggests a reaction to fetor, he is set to report his findings to the doctor. The patient, covered by some kind of undershirt, either before or after he has been undressed, is seated impassively on an elevated chair. His crossed hands, reminiscent of the "Man of Sorrows" in the art of the time (or, even, of Albrecht Dürer's famous self-portrait of 1522), underscore his dependence on this medical intervention. Nodules on his cheeks, legs, and feet allude to the single most common characterization of leprosy, although it was singled out more sharply in iconography than in medicine (in the face: see Plate 10 in Chap. 8). The lesions were intended more to symbolize the disease than to depict clinical details, and thus we should not infer too much from the arguably disfigured lips, nose, and ears of the patient.[10]

The participant who is portrayed on the left, on the lowest level, also occupied the least dignified—though not the least popular—position among the medical practitioners. He is a barber, possibly a barber-surgeon, whose primary association with medicine consisted in the drawing of blood, usually called phlebotomy or venesection. He is dressed rather plainly (except for the biretta) as a layman, and with his sleeves pulled up. He has drawn blood, normally "from the right arm," as was stated in the Arras certificate at the beginning of this chapter, and presumably "from a wide opening," as was recommended for leprosy diagnoses. Now the barber is washing a cloth that has been soaked in the patient's blood, so that the filtrate and the residue in the basin may be inspected. The two small dishes contained salt and vinegar for mixing with the blood, and in some cases with the urine, of the patient. These further tests could be done to make the inquiry more complete, or to add weight to the examination, in the words of Guy de Chauliac, "for the sake of solemnity." While we will consider tests at the end of this chapter, we may recall their importance for the judicial aspects of the diagnosis (Chap. 2).[11]

Wächtlin's illustration is riveting and effective. His solemn yet vivid portrayal of the "visitation" makes the viewer wonder, not only about these procedures and the evidence that each of them would produce, but also about the extent of a complete diagnosis, the spectrum and assessment of signs, and the consequences of the medical finding. A few rudimentary clues for answers are offered in the folksy verses of the caption above the woodcut,

> Blood, urine, nodule, glands, foul limbs,
> Fetor of the breath, and many signs:
> Forsooth, I pronounce the verdict
> That this is a leprous man.[12]

These verses correspond closely to the highlights of the 1574 certificate that opened this chapter, although the inspection staged by Wächtlin and the examination recorded by the Arras attorney had opposite outcomes. In both documents, gauging the body's internal conditions by hematoscopy and uroscopy preceded the recognition of external signs. The priority of these diagnostic techniques reflected authoritative teaching, namely, that the disease lurked deep inside before coming to the surface (Aretaeus), and that it was basically caused by a complexional imbalance (Galen). When it came to external manifestations, the woodcut and the certificate also shared an emphasis on deformity or nodosity and glands, and on the corruption or foulness of limbs. On the other hand, the documents diverge in certain features; the divergence is due to the format rather than to chronology. Most significant, the certificate, unlike the caption, drew an explicit distinction between unequivocal and equivocal signs; inferentially, it also differentiated symptoms as indicating or not indicating a confirmed stage of leprosy, and criteria as justifying or not justifying sequestration of the patient.

SIGNS

It took a battery of signs ("zeychen vil") to declare a patient leprous. An elaborate catalog, as well as a selection of prime symptoms, took shape in a centuries-long and uneven development. In an admittedly simplified overview of this development, it becomes apparent that there was little change— let alone "progress"—in the identification and interpretation of the signs themselves. Rather, attempts to improve the diagnosis consisted in compiling more symptoms, describing them with greater precision, differentiating and organizing them methodically, and evaluating their reliability. These attempts encountered many obstacles. First, as Isaac Uddman observed, no disease is more variable. The polymorphic character of leprosy, together with its latent and uneven progression, baffled diagnosticians. Additional challenges ranged from the assumptions on which diagnostic techniques were based to the fluidity of language. Visual, tactile, and olfactory perceptions of external appearance were susceptible to the interference of personal feelings and cultural settings. The reading of internal conditions in urine, blood, and pulse often served to confirm an impression that had already been formed, with a misleading sense of certainty anchored in theoretical foundation.

Even language contributed obstacles to the development and teaching of a reliable semiology. These obstacles were similar to those that complicated the naming and definition of leprosy (as we saw in Chap. 3), but their consequences were more immediate. It was difficult to teach the recognition of a disease when supposedly differential names for symptoms were as imprecise and fluid as, for example, "glands" (*drüszen* in the caption of Wächtlin's woodcut), which also referred to buboes, goiters, and the neck growths of scrofula. The difficulty was compounded by a broader and long-term process, the forging of a standard medical vocabulary. Some of the foreign terms, introduced into Latin by translators, remained opaque: this, as we saw in the previous chapters, was the case for the Arabic word *albaras*, which neither became integrated as a neologism nor translated into one single European equivalent. Moreover, the transmission of Latin descriptions to vernacular versions and nonacademic milieus was uneven and incomplete. Authors of vernacular texts did not always transfer the diagnostic nuances of their Latin sources: thus, in Middle Dutch, Jan Yperman abstracted various ocular symptoms into the casual statement that leprous patients "see with different eyes." The linguistic perspective reveals a further problem in structuring the diagnosis of leprosy, namely, the frequent disjunction between depictive impressionism and anatomical precision, which may be sampled in various descriptions of ocular symptoms. Avicenna mentioned "a darkening of the eyes, tending to reddishness"; Gariopontus, a "reddening of the eyes"; and Bertucci, "a disturbance of the whites of the eyes, accompanied by lividness or darkening, and with a great number of red veins in the corners of the eyes." Subsequent authors did not integrate these traits. Nor did they correlate them with two classical descriptions, one by Galen, of "the whites of the eyes becoming more livid," and the other by Aretaeus, of "dark" or "misty [*caliginosi*] eyes, the color of bronze."[13]

In some way, the second-century report by Aretaeus, who identified at least forty signs of *elephantia*, remained unsurpassed in its perceptiveness. Even though he mentioned them discursively rather than in an organized list, he recognized the progressive character of the disease and the distinctiveness of at least three successive stages. He indicated the escalation, for example, from the onset with a general emergence of tumors, to the advancing "eruption" with disfigurement of the face, and to the ultimate affliction when, "unless death prevents, the nose, fingers, entire hands, feet, and genitals fall off." The inclusion of genitals was fanciful, as was Aretaeus's assertion that patients "have an appetite for sex." His apparent liberties, however, were balanced by details that suggested keen observation and which were absent from dozens of later descriptions. When observing that, at the first advance of *elephantia*, "hairs over the entire body fall prematurely," he remarked that, "if a few hairs should remain, they are more unsightly than those that fell out": most authors, while mentioning the loss of hair, omitted this fine point. Even when Aretaeus was rediscovered in the Renaissance, several of his finer brush strokes were ignored. For instance, hardly anyone noted the graphic precision of his sketch of the symptomatic frownlike swelling by which the forehead "covers the eyes, as happens in people when they become angry, or in lions: whence this disease is called leonine." Behavioral profiles, which were part and parcel of an examination, lacked the poignancy with which Aretaeus observed the listlessness of patients in the last stage, when "they enjoy neither bathing nor being dirty, neither food nor fast, and neither movement nor rest."[14]

This kind of psychological observation by Aretaeus, together with his statement that *elephantia* was also called *satyriasis* because it manifested itself by an "insatiable and shameless lust for sexual intercourse," shows that a person's conduct was counted among the "signs" since the early centuries. The belief that "an intense eagerness for coitus" was one of the symptoms of leprosy persisted well into the seventeenth century. Over the centuries, however, this belief often became just one item in a more inclusive behavioral profile that was rife with connotations. This development was boosted by the proposition, in Avicenna's *Canon,* that erratic behavior accompanied the incipient stage of leprosy. In the Latin version, the attribution of "mores melancolici et mali et dolosi" (melancholic, bad, and cunning habits) opened the door to suggestions that the disease manifested itself in the area of "morals" and on the level of "evil." Inversely, and without depending on Avicenna, Hildegard of Bingen attributed certain manifestations to cardinal sins or vices when she proposed that *lepra,* "from gluttony and drunkenness, produces red tumors and worm-shaped eruptions," and "from lust, broad-surfaced ulcers, like crusts, with red flesh underneath." The more down-to-earth Guy de Chauliac expanded Avicenna's dictum into the declaration that leprous persons "are clever, cunning, and hot-tempered, and they want too much to thrust themselves onto people." The sociopsychological speculation was elaborated most fully by a surgeon who often proved to be close to popular perceptions. Concluding his catalog of symptoms, Ambroise Paré claimed that *ladres*

> are nearly all timid, deceiving, and suspicious, because they distrust themselves on account of the melancholy. They also greatly desire the company of women, especially during the increase and stasis of their disease, on account of the warmth that burns inside them; in the [last stage, of] decline, however, they abhor this kind of diversion.[15]

It is easy to hear, in these characterizations, undertones of fear and suspicion; it is tempting to speculate about the deep roots of these feelings; and it is reasonable to assume that medical practitioners were susceptible to and, at the same time, contributed to a climate of moral censure of leprous patients. It is useful, nevertheless, to allow for some distance between treatises and everyday life, and to wonder whether the characterizations played a role in actual examinations. While authors included unusual conduct in their lists of symptoms, they did not suggest methods for investigating it when they drew up diagnostic protocols. We have seen that this investigation fell under the interrogation about predisposing factors, and that examiners were supposed to reach out to acquaintances. Hence, the result would depend on the practitioner's subjective evaluation and on potentially prejudiced testimony. However, I have not come across any record of an examination that entailed a diagnosis of "depression, mania, or a similar melancholic disease," in the words of Du Laurens. Certificates, which preserved many clues to physical appearances, were silent on the mental traits of the examinees. In sum, it is impossible to estimate the impact of this unstable "moral" category on the outcome of inspections. At the same time, it is difficult to ignore, in both prescriptive writings and official records, the efforts to gain objective knowledge from secure, tangible, and somatic evidence. Only the most skeptical reader of the sources would doubt that medical examiners were concerned to secure this knowledge with the scientific literature.[16]

From early in the fourteenth century, the didactic literature on leprosy was dominated by Bernard de Gordon, as may have become apparent in the preceding chapters. His influence was most pervasive in the area of diagnostics, where it eclipsed the impact of Theodoric Borgognoni and other compilers of Latin summae. In fact, Bernard even emended Avicenna's teaching on the diagnosis of leprosy, especially by making it more practical. He brought order to the signs that were cataloged somewhat randomly in the *Canon;* he oriented them expressly toward the judgment they should support; and he extended their significance to the prognosis of the final outcome. In some way, Bernard's influence seems disproportionate because he devoted a relatively small portion of his chapter on *lepra* to diagnosis and prognosis (9 percent, though this ratio jumps to more than 23 percent if we include his substantial *clarificacio* on facial symptoms—about which more below); moreover, the second half of his section "on signs" was adapted from Avicenna. Nevertheless, the chapter in the *Lilium* offered the first enunciation of several notions that framed the medical examination of people suspected of having leprosy.[17]

Differentiating Signs

The most consequential notions were shaped in three overlapping propositions by Bernard. First, he proposed that certain signs were "occult" and others "manifest." Second, he drew a sharp line between signs that reliably indicated that leprosy was present, and those—far more numerous—that were ambiguous. Last, he stressed that even reliable indicators were most convincing when appearing in the face. These ideas were adopted by authors, and applied by examiners, for more than three centuries. Montpellier masters played the principal role in elaborating these notions, but the resulting diagnostic framework spread across the Continent. In the first proposition, which was an intensified version of the long-standing recognition that leprosy developed slowly, Bernard intended not so much to warn about the occult signs as to concentrate on the manifest signs that revealed that the disease had become established or "confirmed." I should clarify here that the phrase "occult signs" was not an oxymoron but pointed to a hidden presence of incipient *lepra*. The clarification is important because at least one author applied a more literal interpretation. Ambroise Paré drew a startling and suggestive contrast between patients with confirmed leprosy and those who showed no outward signs,

> such as the white *ladres*, called Cachots, Cagots, and Capots, who are found in lower Brittany and in Guyenne in the direction of Bordeaux, where people call them Gabets. Even though few or none of the described signs appear in their faces, their bodies emit a strange heat and burning, as I have witnessed personally: sometimes one of them kept in the house, for an hour or so, a fresh apple, which then appeared as arid and wrinkled as if it had been in the sun for a full eight days. Now, such *ladres* are white and beautiful, almost like the rest of us.[18]

Except for this unusual excursus, Paré and his colleagues shared the view, formulated by Bernard de Gordon, that the examiners' primary challenge lay in distinguishing the manifestly established disease from an earlier stage, at which the patient did not have to be sequestered and might be cured. Bernard carried sovereign authority, on a par with Galen and Avicenna, in Frankfurt in 1458, when Doctor Heinrich Losser argued,

> It is necessary to know that there are two kinds of examinations [*czweyerley ist besehen*]. One bears on the separation of people, as they are infected with the complete, confirmed, and permanent *uſseczczekeit*, which is called *lepra confirmata* according to the masters: this is what Gordonius writes in the special chapter of his *Lilium*. The other examination is for an *uſseczczekeit* that is incipient and not yet complete, confirmed or permanent, and which is called *lepra incipiens nondum*

confirmata: it is curable, as the masters write, including Avicenna, the *liliator* [Bernard de Gordon], and Galen.[19]

The decisive role of Bernard and his successors can be more fully appreciated if we first survey the tradition, which preceded them, on recognizing confirmed leprosy, from Aretaeus and Avicenna to Salerno and Theodoric Borgognoni.

Substantial numbers of signs and symptoms were assembled by Aretaeus (more than forty) and Avicenna (twenty-nine), in catalogs that hinted at the gradual advance of *elephantia* or *lepra.* Aretaeus concentrated on manifestations of the "growing disease," which he distinguished from a preceding occult presence, when "the physician, unaware of the patients' calamity, does not apply his art to the very feeble beginnings of the disease." After thirteen general indicators of the initial growth, a subtle transition introduced the next fifteen signs, which would appear unless the disease jumped directly "from the insides to the extremities." Aretaeus intimated two further stages by brief conditional clauses, "if the disease increases further," and ultimately, "as the disease has a long life—like the elephant." To the limited extent that Avicenna recognized developmental stages in his catalog, he inserted markers that were even more inconspicuous than those of Aretaeus. After listing eleven signs as visible "when *lepra* begins," he introduced three subsequent groups tersely, by repeating the adverb *deinde,* that is, "thereafter," or "and then." Even an author of an influential surgical manual, such as Theodoric, who wrote shortly before 1300, might pay little attention to chronological categories, except for mentioning "the signs of incipient leprosy." The schools of Salerno divided manifestations primarily according to the four humoral types— a division that became a regular feature in both diagnostic discourse and procedure—and they barely alluded to distinct stages. The minimal allusions ranged from the characterization, by Constantine the African, of some signs as "subsequent," to the association, by Gariopontus, of a facial change with "the incipient disease."[20]

Salernitan treatises may have evinced a growing interest in recognizing leprosy, as is suggested by a recurring *incipit,* or opening sentence, which might be translated as "We catch [*apprehendimus*] *leprosi* and recognize them first from macules that resemble impetigo, and which appear in the beginning." Nevertheless, the Salernitan masters shared a significant trait with Aretaeus and Avicenna, namely, the intention for their catalogs of signs to be primarily descriptive. Bernard and his Montpellier successors, on the other hand, methodically developed the lists into diagnostic tools, in other words, from descriptions into checklists. It stands to reason that mounting social pressures sharpened the need to recognize the signals of actual or "confirmed" leprosy. Bernard referred to the confirmed/unconfirmed dichotomy in 1299, when he differentiated between medications that were appropriate for confirmed and for unconfirmed *lepra*: by mentioning this differentiation without comment, he gave the impression that it was already established several years before he wrote his chapter on *lepra*.[21]

In the *Lilium*, Bernard offered a variant, tripartite distinction. A first group consisted of the "occult signs," which indicated that the disease was still "in a beginning stage" (*in principio*), and that the suspect was "not to be condemned [*iudicandus*] to separation, but to be admonished very forcefully"; the phrase *in principio* may also be translated as "in principle," or "potentially present," or "in a predisposition." To be sure, the line between predisposition and onset was so vague, in definition as well as fact, that it would occasion hesitant or disputed judgments—or, viewed differently, would offer physicians an opportunity to give the examinee a reprieve. On October 5, 1532, the master physicians and surgeons of Saint-Omer determined that Leurens Waultier, while "infected with morphea in the head and having nodules on the nostrils," was "predisposed" but not "completely leprous"; the *ladres* of the La Magdelaine leprosarium, on the contrary, deemed him leprous and due to be sequestered. By April 24, 1535, however, Leurens was brought back to La Magdelaine, where he was "duly examined in many parts of his body, by palpation and the drawing of blood": this time, the medical practitioners and the *ladres* "judged and declared, unanimously and in common accord," that Leurens was "infected with the malady of *lepre* in the beginning of the form [? *phangie*] which is called 'the brown evil,' for which he must be separated from the healthy."[22]

The crucial transition, in Bernard de Gordon's diagnostic scheme, was marked by the "manifest signs, upon the appearance of which the patient is to be sequestered from the people." While these signs came second chronologically, Bernard presented them first because he considered them of the highest importance and in the greatest need of his elucidation, which we will consider shortly. In the third group, which pertained less to recognizing the presence of the disease than to anticipating the final turn, he listed the extreme phenomena that announced "shipwreck and the approaching end." In spite of Bernard's logical skills and practical concerns, his spectrum of symptoms fell short of being organized so that it could readily be applied in

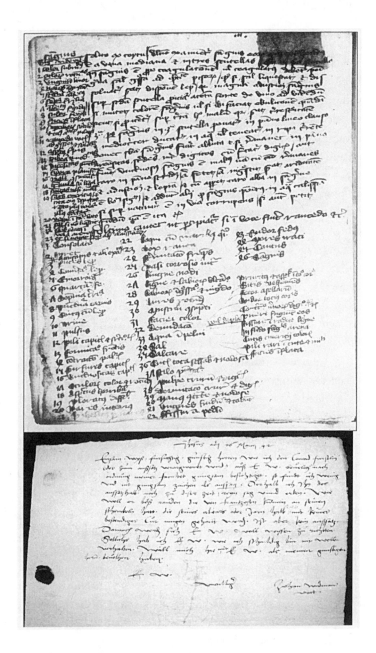

PLATE 9. (*Top*) Marginal gloss on "The Signs of Leprosy," following a text *De cautela preservandi a contagio leprosorum* in Vatican, © Biblioteca Apostolica Vaticana, Palatine MS Latin 1207, fol. 19v; courtesy Biblioteca Apostolica Vaticana. (*Bottom*) Certificate of a medical judgment of Conrad Fursthin, signed by Johan Widman, Doctor, in Nördlingen on May 26, 1542; Nördlingen Stadtarchiv, R39 IV F5, Nr. 37 ("Ärztliche Atteste, 1503–1694").

actual diagnosis. His younger colleague Henri de Mondeville avoided the problem altogether by limiting himself to the "manifest" signs, which would suffice for those occasions "when some people come to us with questions about the signs and judgment of *leprosi*."[23]

Checklists

There was interest at Montpellier, however, in drawing up a practical check-list that would be convenient for consulting and, at the same time, secured by medical science. The interest is suggested by some versions of a brief text, variously called *De examine leprosorum* and *Signa lepre*, which were attributed to Arnau de Vilanova. In one of these, phenomena were roughly divided into two categories, with this instruction: "[S]o that the examination may be done properly, take a tablet and write the good signs on one side and the bad signs on the other, and thus you will not fail." It is impossible to know how widely this method was applied, but one instance is worth noting, from a report by the Delft physician Pieter van Foreest. During his practice in France, after thoroughly inspecting a man who wished "to know from us whether or not he was afflicted by true *lepra*," he and a surgeon took "a tablet in which we wrote the good and bad signs side by side, so that we would not fail in the judgment." A different kind of tabulation was introduced, rather casually, in a marginal annotation on another version of the Montpellier treatise. As may be seen in the gloss, reproduced in the upper half of Plate 9, the man-uscript page was kept "current" or "alive" with successive additions. Glos-sators inserted two numbered lists of twenty-three and forty-six signs and checks, respectively, and then eleven unnumbered items. While the numbers were intended to summarize the treatise, and presumably to mark steps in the procedure, they added up to a rather incoherent synopsis, and they skipped randomly from symptom to cause to technique, in the second set, for exam-ple, from (8) "tingling of the flesh" to (9) "sex with a leprous woman" to (10) "urines."[24]

Efforts to produce a more coherent diagnostic guide were made at Mont-pellier around 1320. Jordanus de Turre divided his *Notes on Lepra* into twenty-one rules that, while primarily concerned with therapeutics and deferring to Avicenna for diagnostics, drew attention to some of the prominent symp-toms, including hair loss and hoarseness. Far from misunderstanding the sense of "occult," Jordanus believed that even the most subtle sign was "man-ifest to one who inquires carefully." It is plausible that he also drew, or at least inspired, a remarkable diagram that was inserted into blank spaces in a manuscript of his *Notes*. One half of the diagram presented the signs of *lepra* in general; the other, the signs of each humoral form. Anchoring diagnosis solidly in medical theory, the general synopsis rested on a Galenic framework. The examiner who followed this outline would proceed methodically—though, presumably, with a degree of comfort and competence which would depend on his academic standing—by checking three areas. A first group of symptoms consisted of aberrations in the operation of the vital, natural, and animal or psychic faculties, such as, respectively, breathing problems, unusual libido, and loss of sensation. A second category contained indications in the "output" of urine and blood. The third group presented the largest number of signs (thirty-one, out of fifty-two in all), which were "qualitative changes" in the body, ranging from discoloration of the skin to the loss of extremities.[25]

For the sake of perspective, this synopsis may be correlated with the parallel approach that Niccolò Bertucci took, probably not more than a decade after Jordanus de Turre. Bertucci cited Galen for the proposition that "the general signs of *lepra* are drawn from three points, just as there are three kinds of symptom [*accidens*]." Even though his semiology was cast in a discursive rather than a schematic format, it was more lucid than the synopsis in Plate 9. In particular, he correlated symptoms anatomically, identified each one concisely yet completely, and limited the overall number to less than half of most lists. The form and content of Bertucci's summary show that, in the fourteenth century at least, diagnostic guidelines could reflect the personality of their author, and that they evolved organically, rather than simply duplicating the authoritative chapter in Avicenna's *Canon*. The summary has the additional benefit of broadening our purview beyond Montpellier, to Bologna, even if we are unable to discern, in this case, the direction taken in the regular communication between the two medical universities. As it happens, Guy de Chauliac, who offered the most lucid diagnostic taxonomy overall, was both a student of Bertucci's and an alumnus of Montpellier.[26]

Guy opened his section on semiology by revealing who had straightened out Bernard de Gordon's inchoate schematization of the emergence of full-blown leprosy. "With regard to the signs and judgments of *lepra*, it should be understood that *lepra* consists of disposition and actuality [*habet disposicionem et actum*] according to Master Jordanus in Montpellier." Supplementing the synopses of his teachers—and, it is fair to speculate, under pressure to provide clearer categories—Guy next streamlined the distinctions by which one should recognize the phase of the disease and decide the fate of the patient.

His first dividing line was no longer between incipient and confirmed lep-rosy, or between occult and manifest signs, but between a preparatory con-dition and the actual disease, which gradually became more manifest. He set aside "a preparedness or predisposition to *lepra*" as a potentially dangerous "property of the body." Actual *lepra*, on the contrary, consisted in "damage to the said [assimilative] faculty which results from the spread of black bile through the body." Then, for a more concrete classification of symptoms, Guy superimposed the concept of "confirmed" leprosy onto a standard Ga-lenic grid.

> This actuality is said to have four times: a beginning, increase, stasis, and decline—at least toward death. The beginning is when the damage touches the internal organs: at that time, weaker signs appear.... The increase is when it appears on the outside: then, the signs grow in magnitude and number. The sta-sis is when the members begin to ulcerate: then, the signs are manifest. The decline, when members fall off: at that time, the signs are obvious to everyone [*popularia*].

In view of his cogent thinking and limpid style—and of his well-known re-port about a personal bout with bubonic plague—Guy disappoints histori-ans in this passage by not providing as much as an inkling of actual or expe-rienced situations.[27]

His teacher, in contrast, sounded more "alive." Bertucci's *Collectorium*, a voluminous summa "of nearly all of medicine," attested to the symbiosis of theory and practice in the "scholastic" discourse on the diagnosis of leprosy. Neither the rigidly logical presentation of the symptoms nor the Galenic philosophical framework precluded the insertion of practical instructions, precise descriptions, and echoes of personal observation. A few vignettes may exemplify these features. When Bertucci presented "the appearance of the tongue" as one of the "qualitative" symptoms, he combined clinical detail and direct address: "[I]f you depress the tongue and look inside toward the root of the tongue where the uvula hangs, and similarly at the palate, you will see tiny cracks and millet-like grains of various colors." Sharing his tactile sense with the reader, he described a symptom of the entire body as "a nodulous-ness that is found under the skin when it is squeezed between the fingers." For another change in body appearance, he pointed out that, if any hairs grow back in bald patches, "they are so short and thin that you will barely see them unless you look at them toward the sun." Bertucci paid special attention to

Table 7.1. Signs Common to All Forms of Leprosy According to Guy de Chauliac

I. Signs of Predisposition
   hideous color
   morphea
   scabies
   fetid excretions
   predisposing factors
II. Signs of Actual Leprosy
   A. Unequivocal (always signify leprosy, intensely or remissively)
      1. roundness of eyes and ears; eyebrows thick or tuberous, loss of hair
      2. nostrils widen externally, constrict internally
      3. lips turn hideous
      4. voice hoarse, nasal
      5. fetor of breath, of entire person
      6. fixed and terrible stare
   B. Equivocal (also found in other diseases)
      1. hard and tuberous flesh, especially in joints and extremities
      2. dark and variegated color
      3. loss of hair and growth of thin hairs
      4. consumption of muscles, especially of the thumb
      5. anesthesia and cramp in extremities
      6. scabies, impetigos, gutta rosa, ulcerations
      7. grains under the tongue, eyelashes, and behind the ears
      8. formication
      9. goose bumps when skin is disposed to air
      10. skin appears oily when water is sprinkled onto it
      11. rarely run a fever
      12. [deviant behavior]
      13. oppressive dreams
      14. weak pulse
      15. blood black, dark and leaden, ashen, sandy, clotty
      16. urine livid, white, thin, ashy

these manifestations because they were among the signs that he called "more certain than the others, so that, when they occur together, they do not lead one into error."[28]

The abundance of misleading signs compounded the daunting task of examiners, who were charged with distinguishing not only between the presence of leprosy and other diseases but also between a predisposition and actual infection, and between an incipient and established condition. Bernard de Gordon framed this challenge when he selected six "infallible signs" out of the relatively amorphous inventories in the *Canon*, the Salernitan summae, and other sources. He warned, nevertheless, "[W]e should not judge by one sign but by two, three, or more, because they often are equivocal. These, however, are the most certain, although there are many others." Henri de Mondeville placed the same six, almost verbatim, at the head of his own list, but he indiscriminately added six more, from those that Bernard considered less than infallible; and, where Bernard moved on to Avicenna's extensive catalog, Henri concluded, "[T]he rest may be found in the authors."[29]

Guy de Chauliac, who became a leading authority on this matter, reconstructed the core of Bernard's diagnostics along straighter logical lines, as may be seen in Table 7.1. He turned Bernard's tentative dichotomy of reliability into a symmetrical pair, grouping signs of actual leprosy as "unequivocal, which always indicate and follow *lepra*, and of which there are six," and "equivocal, which are also found in other diseases so that they do not always indicate *lepra*, and of which there are sixteen." It is interesting to note that Guy counted the items but did not number them, unlike his teachers. Jordanus, as we have seen, numbered his twenty-one notes on *lepra*. Bertucci, while adhering to the triple scheme of Galenic semiology rather than to the new dichotomy, numbered the symptoms within each section. If the attention of these authors to itemization makes them look pedantic or influenced by numerology, we should consider that it may have been inspired by their wish to make the guidelines easier to memorize or copy, and more convenient for consulting in the field. In fact, when Ambroise Paré cataloged twenty signs of confirmed leprosy, he numbered them as steps in the overall procedure, beginning with the head and ending with changes in mood (16, nightmares, and 17, behavior), (18) urine, (19) blood, and (20) pulse.[30]

Assessing the Signs: Semiology

The numbered lists, as well as the diagnostic categories, were important tools for applying the knowledge of signs to actual examinations. This practical relevance is amply evident in the archives. Let us, for the sake of continuity, return to two vividly documented cases that have been introduced earlier in this study. In Marseille in 1464, the physician Jacobus Buadelli and the surgeon Johannes Grivilhoni were commissioned "to see, palpate, examine, and probe thoroughly [*exprobandum*] Bartholomeus Giraudus." They reported, "We have performed a diligent examination, also by testing the blood [*per experitionem sanguinis*], straining and inspecting it as the teachers of the canons of medicine prescribe, and on the four kinds of *lepra* which exist on account of the four complexions of the humors." After identifying the four kinds and respective humors, the examiners assured that they had checked the patient

Table 7.2. Signs Common to All Forms of Leprosy

	A	B	C	D	E	F	G	H	I	J	K	L
1 nodules: forehead, chin, cheeks		X		X	X		X			X	Xx	X
2 face livid, darkens			X	X	X	X	X		X			
3 frowning, jutting brow		X			X	X		X		X		X
4 hairs, eyebrows/lashes thin, fall		X	X	X	X	X	X	X	X	X	Xx	X
5 unmoving, terrifying stare				X	X	X	X			X	Xx	X
6 eyes become round					X	X	X	X		X		X
7 eyeballs darken, bloodshot	X	X	X				X		X	X		X
8 nostrils dilate	X	X		X	X	X	X	X			Xx	
9 nasal tightening/voice			X	X	X	X	X	X				
10 nasal cartilage corrodes				X	X		X			X		X
11 strained breathing	X			X	X		X			X		X
12 hoarse voice			X	X	X		X	X	X	X	Xx	X
13 granules under tongue		X					X	X	X	X	Xx	X
14 lips thicken, turn livid		X		X	X		X	X	X	X	X	X
15 ear[lobe]s contract		X	X		X	X	X			X		
16 tuberous, rough, thick, uneven skin	X	X	X	X	X	X	X	X	X	X	Xx	
17 skin oily: water runs off			X				X	X	X		X	
18 body becomes darker				X	X		X	X	X	X		
19 scabs, pustules, morphea			X		X	X	X	X		X		X
20 fetid breath, body, sweat	X	X	X	X	X		X	X		X	Xx	X
21 hideous appearance (*figura, forma*)				X	X		X			X	X	
22 melancholic, bad conduct				X	X	X	X	X		X	X	
23 high libido		X					X			X	X	X
24 insomnia, nightmares		X		X	X		X	X		X	X	X
25 anesthesia of extremities, etc.			X		X	X	X	X	X	X	X	
26 shrinking muscles, esp. thumb/index					X	X	X	X		X	X	X
27 extremities swell, fissure, drop	X	X	X	X	X		X			X		
28 urine thick, turbid		X	X				X	X		X	X	
29 slow, weak pulse		X		X	X		X	X	X		X	
30 clotty, sandy, ashy blood				X	X	X	X	X	X		X	X
	A	B	C	D	E	F	G	H	I	J	K	L

Note: A = Galen; B = Aretaeus; C = Salerno (Gariopontus, Constantine, Gilles); D = Avicenna; E = Bernard de Gordon; F = Henri de Mondeville; G = Jordanus de Turre; H = Guy de Chauliac; I = Niccolò Bertucci; J = Jodocus Lommius; K = Ambroise Paré (X = treatise; x = 1583 letter); L = Samuel Hafenreffer for "the unequivocal signs, of which there are six, as well as the equivocal, which are sixteen, as the said canons of the teachers of medicine determine." They found that Bartholomeus displayed "certain congruent equivocal signs, not unequivocal but equivocal; to wit, in a numbness [*in dormitione carnis*] or lack of sensation in certain places of his person; scabies and *albarasis* [whiteness?] and tuberosity of the flesh; and as if he had morphea." Alarming as these symptoms might have been, in themselves and in their simultaneity, they were emphatically dismissed as equivocal.

Next, the commissioners even deemed it necessary to qualify the presence of a more specific phenomenon.

> By the appearance of his eyes he seems to fall into an unequivocal sign that, however, is not yet entirely confirmed [*formatum*] but only tentative [*dispositus*]: if this were confirmed, such a person would have to be considered leprous, because the unequivocal and equivocal signs would be in agreement, and he would not be able to be cured from this disease except only by God.

At this point, the report demonstrated that a patient's fate could hang on the thin thread of a subdistinction, in this case, of a single symptom, namely, the appearance of the eyes, which most authors ranked among the most reliable.

> Because this unequivocal sign is not confirmed in him, as stated above, but only tentative, and even though the other equivocal signs are confirmed in him, we state, know, and declare that Bartholomeus will not have to be completely separated from the healthy.

These learned practitioners of Marseille will amuse some, as parading the kind of sophistry that would inspire Molière; they will make others wonder whether this judgment was "fixed." In their defense, we should keep in mind that they faced a triple challenge, of distinguishing—in one decision—between congruent and disparate phenomena, between unequivocal and equivocal signs, and between a potential and confirmed condition. The distinction was all the more challenging because so many signs needed to be weighed, as may be gathered from the selection in Table 7.2, which is limited to the top thirty signs mentioned by at least five authors.[31]

A century after the finespun verdict by the Marseille examiners, Ganges doctor Pierre Serieys and surgeon Pierre Brunier tackled their challenges with similar circumlocution in their testimony.

> Following our art, we have pursued all the signs, unequivocal and equivocal, which indicate and demonstrate to us whether the disease is present as a predisposition or a condition [*en disposition ou habitude*]. We have found that said Guilhaume Vézye is not predisposed to falling into this disease of *lèpre*, but quite in a confirmed condition. Therefore, following our duty and conscience, having considered everything as above, we declare him to be completely *ladre* and admonish him, in the name of God, to withdraw from the community and company of healthy persons.

While reaching different conclusions, the examiners in Marseille and Ganges behaved typically in several respects. They hung their decision on a "reading" of the unequivocal and equivocal signs, without much indication about the content of these categories. Some of their expressions suggest that they were overwhelmed by the complexity of the literature on leprosy and by the challenge of applying theory to practice. Even when they felt compelled to find against the patient, medical practitioners preferred to judge conservatively, as we have seen in Chapter 2.[32]

Equivocal and Unequivocal Signs; Focus on the Face

Numerous certificates from the fifteenth and sixteenth centuries, including the affidavit at the beginning of this chapter, as well as the two quoted above, show that the differentiation between unequivocal and equivocal signs routinely governed a medical examination. This bifurcation was far from simple, however, because the former category was neither defined nor applied uniformly, and the latter was vast and not without consequence. One might expect that, once certain symptoms were rated "infallible," they would coalesce into a solid category, but even a glance at a few early lists should dispel such assumptions. Rather than a detailed collation that, though worth pursuing, might distract from the larger picture, a few samples will suffice. The "more certain signs," as Niccolò Bertucci characterized them, did not include the "external dilation and internal constriction of the nostrils," which Bernard de Gordon and Guy de Chauliac (and many others, into modern times) counted among the prime indicators of leprosy. Granules on the tongue and palate constituted a symptom that had been noted by Aretaeus but ignored by Avicenna, the Salernitans, and Bernard de Gordon; Bertucci viewed this symptom as a certain sign, while Chauliac deemed it only equivocal. Bernard believed that swollen and livid lips marked a terminal stage, which was beyond ambiguity; Bertucci, on the other hand, ranked "unnaturally prominent and leaden lips" third in importance among the specific symptoms, as Chauliac did for "hideousness of the lips."[33]

With such variety in the authoritative guidelines—and the lack of progressive debate among their proponents—it is difficult to draw a composite picture of the examiners' overriding criteria. The difficulty is compounded by the fact that most of the certificates recorded only the decision, not the justification. Some contained general references, to "a great number of symptoms," or "bad signs," or "sufficient evidence." A few, affording an inference from negatives, identified critical signs that should have been manifest or combined for finding a suspect leprous. Thus, according to the affidavit at the beginning of this chapter, the disfigurement of Bellevalet's nose and mouth was "unequivocal but insufficient" for a positive finding, in the absence of the clinching sign, which would have been the corruption of his "entire form." Even when medical examiners accounted for their decision, their reasons were sketchy when compared with the precision of their protocols and with the detailed descriptions of phenomena that they discounted as symptoms. One such relatively sketchy affidavit was signed in 1347 by Petrus de Urbana, sworn physician of Saintes-Maries-de-la-Mer (Provence). Master Petrus decided that Petrus Bernardi needed to be separated because of "manifestly present signs, *such as* deformity of the face in the eyes, nostrils and eyelashes, as well as the consumption of the muscles of the hands and feet, and the *many other signs* that make for a complete disease of leprosy [*et aliorum multorum morbum lepralem complentium*]."[34]

One certificate of an actual examination supplied a more precise justification. In 1530, in a letter to the diocesan court of Besançon, the *commissarii* certified that they had inspected Nicolaus Chapu of Montbéliard,

in every member, from the soles of his feet to the top of his head, and following the signs of *lepra*, unequivocal as well as equivocal, and taking good note of each. We find him to have, rather than the due formation [*plasmacione*] of his members, a breakdown of his form and figure, white morphea, nodules, fetid ulcers, serpigo and impetigo. Whereas these are symptoms [*accidencia*] of the disease, we state that he should be separated from the company of the healthy, and that he is leprous.

Unfortunately, the commissioners were not identified, so that it is possible that they were not medical practitioners, notwithstanding their use of conventional technical expressions, and even though the letter wound up in a miscellaneous medical manuscript. In any event, this certificate was exceptional in specifying the manifestations that eliminated doubt about the presence of leprosy.[35]

An affidavit that is far more typical, but interesting for a different reason, is reproduced in the second half of Plate 9. On May 26, 1542, Doctor Johan Widman certified to the magistrate of Nördlingen,

> At your command and in accordance with my faculties I have diligently inspected Conrad Fursthin who was suspected of *aussatz*, and I find few and insufficient signs of *aussatz*. Therefore, at this time, I still find and declare him clean with regard to *aussatz*. He does have bad lesions on the thighs, which come from the French disease, and which cannot be cured permanently because of his age. This is not *aussatz*, however, as I am informing you in accordance with my duty.

The matter-of-fact reference to "the French disease" indicates, as do some other certificates, that confusion between leprosy and syphilis was not commonplace. In contrast with Widman's precision on the syphilitic lesions that he dismissed as not indicating *aussatz*, we note the total absence of information on the "few signs" that he did find, even though these seemed to make his judgment provisional, "at this time."[36]

Notwithstanding the customary taciturnity of examiners and the variant rankings by authors, some general observations can be made about the criteria that physicians considered unequivocal. In the fourteenth century, at the peak of academic concern with leprosy, most authors emphasized facial signs, while classifying abnormalities in the rest of the body as "general *or* equivocal." Latin authors singled out and expanded the facial signs that Avicenna mentioned, without special emphasis, in his chapter on diagnosis. François-Olivier Touati draws our attention to the fact that the concentration on facial symptoms declined after antiquity, in the early Middle Ages, to reemerge from the twelfth century on. He raises the stimulating question of whether this curve might reflect mutations in the disease itself. Even if this was the case, social and medical factors undoubtedly played a role, as well. The awareness of facial features and interest in cosmetics, while as old as humanity, seem to have intensified during a period of adaptation to the rise of towns. Urban life brought closer proximity, greater attention for visual detail, and—

arguably—a keener sense of stigma. Bernard de Gordon remarked that "people are more solicitous" about the face than about the rest of the body. While he may have been thinking of human nature in general, his outlook was no doubt shaped by the thriving urban world around him. His solicitude about facial symptoms evidently arose from a keen personal experience; it resonated with other authors and practitioners, although in varying waves, as we will see.[37]

Bernard was the first to state emphatically, "[T]he signs that matter to me, are the ones that are found in the face, as I will clarify; for when these manifest signs appear, the patient is to be separated from the people." He returned to this emphasis in the most elaborate of the twelve "clarifications" at the end of his chapter on *lepra*, with the formal question of "whether it would be possible for anyone to be completely leprous with no sign of *lepra* appearing in the face." At first sight, this *questio* may look like a mere pretext for a pedagogical exercise in dialectic, especially since the answer opened with the formula of Scholastic *sic et non* debates, "It would appear so" (*Videtur quod sic*). The issue was "existential," however, and his doubt proved that even so-called unequivocal evidence did not guarantee certainty. After confiding that this question "has occurred to me on account of what I saw in a certain man," Bernard switched to a narrative mode:

> This is what happened. There was someone whose toes and fingers were all so corrupted in shape and appearance and breakdown of integrity, that they were only one inch long instead of three, and lacking nails; there was utter deformity and hideousness, yet, with all this disfigurement, nothing at all showed up in his face. From this example, it appears that one may be confirmed leprous without having any visible sign in the face, even after the feet and hands become misshapen and fall off.

Galen, Avicenna, "and all the others" appeared to support the opposite viewpoint, for "when they give the signs of *lepra*, they begin with the signs observed in the face."[38]

Actually, in this appeal to authoritative teaching, Bernard interpreted Avicenna selectively, and he projected this interpretation back into Galen. The chapter on signs in the Latin *Canon* opened with the statement "When *lepra* begins, the color turns red with a tendency to blackness." Avicenna did not declare here that facial symptoms carried diagnostic priority, but that they were the first to appear—possibly, of any in the entire body. Galen, on the other hand, only once linked *lepra* and *elephas* to the face, and more often to the skin; on the whole, he "took the signs from the corruption of form and figure," as Bernard quoted him. This quotation, together with the conclusion "that no one should be judged leprous unless the corruption of the figure is manifest," points to the fluidity of the expression *corrupcio figure*. The phrase could refer to disfigurement in general, or to a complete breakdown of the appearance and particularly of facial traits, as Bernard seemed to understand it here. The absence of facial abnormalities led him to conclude, "with the conjecture closest to the truth," that the observed case "was not *lepra* but some kind of arthritis [*arthetica*], a certain leprosity [*leprositas*—a term I have not encountered elsewhere], a scabies, and an ulcer of the nails by which the feet and hands became disfigured."[39]

There is much to disconcert us in this passage, which exposes some of the weaknesses in early medical teaching. Bernard geared his entire outlook to a single incident. Moreover, he settled his diagnostic doubt on the written tradition rather than on deeper probing and broader experience. What is worse, he cited the canonical authors inaccurately, and he steered ambiguous arguments to an inconclusive conclusion. Nevertheless, it makes sense to view his ruminations in a more positive light. The patient's fate was the real source of his agonizing:

> I tell you, I wanted to absolve him, and I questioned him repeatedly if any sign had appeared in his face; he had remained like this for nearly twenty years and he still lives with this hideousness of the extremities, but nothing shows in his face.

Bernard felt that he had reason to be concerned "because it appears to me that *leprosi* are judged very poorly in our day." Unlike those who ignored the difficulty of diagnosing true leprosy, he showed honesty and intellectual humility by reiterating the admission "God knows the truth, I do not know." Suggesting that he, too, may once have been more aggressive in his diagnoses, he claimed, "[A]lthough I once held another opinion, after laboring diligently in this work, I think differently, and now I would not judge someone [like that man?] leprous."[40]

Bernard's change of mind may reflect the variability in the choice of ultimately decisive signs. His soul-searching, furthermore, reveals some of the basic dilemmas in the diagnosis of leprosy. This disease forced medical teachers and practitioners, more consistently than most other diseases, to choose between personal and professional response, between the authority of doctrine and the persuasiveness of appearances, and between a tightened or widened scope of symptoms. Within the narrower spectrum of unequivocal signs, priorities fluctuated between the face and the entire body. It may have been easier to reach a balance in academic discourse than in actual practice.

The persistence of the diagnostic concentration on the face, as well as the hold of textual authority on practitioners, is illustrated in a document of 1541. This was a template, drawn up by Claude Chastellet (Claudius Chastelleti), *medicus* in Constance, for the certificates that he would issue as "commissioner deputed by Lord Carolus, duke of Savoy, to visit and palpate *leprosos* through the entire country." The template was closely based on Bernard's criteria, as may be seen from the juxtaposition in Table 7.3. In order to demonstrate the closeness, the table presents the original Latin texts, rather than translations; italics mark the matching words and phrases; and the last two columns add the pertinent excerpts from two other sources that Chastellet might have consulted but which place less emphasis on the face. It is clear that Chastellet not only followed the *Lilium* in his itemization of symptoms but also made Bernard's standpoint his own in professing, "[T]hose that are found in the face are the ones that matter to me, as commissioner."[41]

This document should save us from relying exclusively on the overall impression that the sources project. In general, it appears that the concentration on facial symptoms gradually became less lopsided, first in academic treatises and eventually in health certificates. It is difficult to determine to what extent this trend was due to changes in the semiology itself, in general perceptions, or in the nature or incidence of the disease. In any event, the trend was uneven, and the face remained important in the diagnosis of leprosy. This continued importance makes it surprising that the protuberant forehead, lack of eyelashes and brows, round eyes, and flared nostrils were neither equated with the lion's features, after the image introduced by Aretaeus and mentioned by Avicenna, nor bundled into a syndrome (*facies leonina* is today associated with the presentation of lepromatous Hansen's disease).

The gradual but uneven loosening of the focus on the face may be gathered from a brief look at a series of authors who generally treated Bernard as a major authority. Henri de Mondeville and Jordanus de Turre, while close to Bernard in time and place, did not mention him with regard to the diagnosis of leprosy; more significant, both attributed special importance to facial symptoms but placed them on a broader diagnostic spectrum. Half a century later, Guy de Chauliac subscribed to Bernard's priority by recom-

Table 7.3. A Practitioner and His Authorities

CLAUDE CHASTELLET *Attestatio leprosorum* Constance, 1541	BERNARD DE GORDON *Lilium medicine* Montpellier, 1303	GUY DE CHAULIAC *Inventarium sive* *Chirurgia magna* Avignon, 1363	AVICENNA *Canon medicine*
reperi leprosos in *depillatione superciliorum et grossitie eorum* atque *rotonditate oculorum,* [*coartacione*] *interi*[*us*] *cum difficultate anhelitus et quasi* sic quod *cum naribus loquerentur;* quorum *color facierum est lucidus, mortifficatus ad [fusce]dinem;* quorum etiam *facierum* aspectus est terribilis *cum [fixo intuitu] et cum gracilitate* palpebrarum [!?] *aurium et potissime illius qui est inter policem et indicem et fissure [et infectiones]* pedum ac manuum. Necnon inventa est grossities labrorum; quorum etiam oculi sunt rubei et palpebre inversantur. Que omnia *sunt signa detecta et maniffesta quibus pacientes sunt a populo sequestrandi* habentque multas in toto nodositates. *Mihi* tamen commissario *sufficiunt* ea *que in facie inventa sunt,* sicut postule, excrescentie, morphea, sic de aliis. Ad quorum rei veritatem et fidele testimonium presentes litteras signavi ego · dictus medicus.	Signa infallibilia sunt ista: *depilacio superciliorum et grossicies eorum,* et *rotunditas oculorum,* et dilatacio narium exterius et *coartacio* interius cum difficultate anhelitus, et quasi si cum naribus loqueretur; et *color faciei lucidus* vergens ad *fuscedinem mortificatam;* et terribilis aspectus *faciei cum fixo intuitu, et cum gracilitate et* contractione *pulparum aurium.* Et per unum signum non debemus iudicare, quia frequenter sunt equivoca, sed duo vel tria vel plura. Ista tamen sunt certiora. Et sunt multa alia, *sicut* sunt *pustule et excrescentie* et consumpcio musculorum *et potissime illius qui est inter pollicem et indicem,* et insensibilitas extremitatum, *et fissure et infectiones cutis*...Et multa alia valde que ponunt auctores. *Mihi* autem *sufficiunt illa que in facie inventa sunt.* Ista autem *sunt signa detecta manifesta quibus* apparentibus paciens est a populo sequestrandus.	consideret faciem suam: *supercilia si sunt depilata,* si sunt inflata et tuberosa; *oculos si rotundos* ; de *naso,* si est tortus, grossus, ulceratus intus; de *auribus,* si rotundantur et curtantur; de voce, si *loquetur* rauce et *cum naribus;* de labiis et lingua, si sanguinantur et ulcerantur et si habet grana; si *hanhelitus est difficilis* et fetidus; et si eius forma est diversa et horribilis. Et consideret bene ista, quia *signa faciei* sunt cerciora.	et apparet in *oculo* obfuscatio ad rubedinem declivis, et apparet in *anhelitu* strictura...et apparet in *nasu* gaveto, et fortasse fit strictura in *naribus* et canna...et *facies* fit *terribilis* et color denigratur...et additur constrictio *anhelitus* donec pervenit ad *difficultatem* vehementem et disniam magnam...et ingrossantur labia et denigrantur...et corroditur cartilago *nasi*...

Source: Borradori, *Mourir au Monde,* pp. 223–224. Bernard de Gordon, *Lilium,* fol. 18ra. Guy de Chauliac, *Chirurgia,* p. 285. Avicenna, *Canon medicinae,* sig. FF 7rb.

Note: I have corrected—and indicated with square brackets—obvious errors of orthography, although I was unable to determine whether these slipped in at the first recording of the attest or in the modern transcription of the archival document.

mending that the examiner "carefully consider the signs of the face, because these are more certain"; of the six signs that he identified as unequivocal, four were strictly facial features; and he instructed the practitioner to "return to the consideration of the face" after completing the examination. However, Guy's instructions for examining the face were very succinct, not only in proportion to his entire discussion of diagnosis but also when compared with Bernard's detail (as may be seen in Table 7.3); on the other hand, he covered the rest of the body methodically, whereas Bernard had dismissed, almost casually, "the great many other signs that the authors mention." Velasco de Tharanta, who relied on Bernard for several aspects of leprosy, largely ignored him for the diagnosis; and he mainly repeated the *Canon*'s inventory of signs, together with a note of contingency, "Avicenna says that, *sometimes*, pustules and large tuberosities appear in the face and around the nose."[42]

In the sixteenth century, the diagnostic weight of facial symptoms was surpassed or at least matched by consideration of the entire body. Ambroise Paré insisted, as we have seen, that the face presented "the specific and most truthful signs." Nevertheless, he devoted three-quarters of his diagnostic instructions to other manifestations, ranging from loss of sensation to behavioral changes. Jean de Varanda fully recognized the equivalence among the primary signs of leprosy. He declared (with an interesting juxtaposition of neurological and dermatological symptoms) that certain phenomena, particularly "the loss of sensation in the extremes and between the shoulder blades, are no less specific and pathognomonic of this affliction than those observed in the face." He added, however, with rather uncharacteristic conformism, "[L]et us state the latter in the usual order, lest we be seen to introduce novelties."[43]

If we expect not only professors but also practitioners to have been wary of novelties, and to follow authoritative tradition as faithfully as Claude Chastellet did in 1541, we may be surprised by the tenor of some certificates. Consider, for example, the affidavit that was quoted at the beginning of this chapter. In 1574, medical examiners in Arras deferred to "the ancient teachers" for the definition of *leppre* on which they based their decision, but then they turned the traditional primacy of facial symptoms upside down. They pronounced Adrien de Bellevalet free of confirmed leprosy—and free to go—because the absence of symptoms in his *body* outweighed the presence of facial signs, however unmistakable:

[G]iven the integrity of all his limbs, without any spot, weakness, or deformity in his entire body, even though his face shows disfigurement of the nose and mouth: these signs are unequivocal, but insufficient for separating him at this time.

Exactly one century later, in 1674, physicians and surgeons in Ravensburg decided the fate of Jacob Loben on a comprehensive basis, without singling out the face for special mention. They certified that Jacob needed to be separated because he was "afflicted with ample marks of *Aussatz* on his whole body" (*mit genuegsamen kennzeichen an seinem gantzen Leib behafftet*). It is of interest to note that the shift in emphasis, from facial to all-body symptoms, paralleled the loosening of the boundaries of "leprosy" by examiners. Both developments were no doubt susceptible to social pressures.[44]

Diagnosis in Perspective: Social and Conceptual Frameworks

Official documents preserve many details that, though scattered and often marginal, shed light on social and conceptual frameworks of the diagnosis of leprosy. Direct and lively expressions, for example, reveal the depth of popular anxieties about facial blemishes. The record of Ramona Isern's examination at La Seu d'Urgell in 1372 told us, poignantly, that suspicions were aroused when she had "a grave lesion in the nose, and some people say that it is *meselia*." More indirectly, many certificates indicated what compelled someone to seek a medical examination when they dismissed phenomena that, inferentially, had stirred the initial alarm. Thus, examiners regularly cleared a person of leprosy "even though" (*combien que, wiewol, al eist dat*) the disease seemed manifest, or "notwithstanding" (*non obstante*) troubling signs that could range from "pustules over the entire body" to "a red nose given by a parent." These waivers, scattered through the archives, deserve analysis beyond what this chapter can accommodate. They illuminate a vast range of social facets, in addition to aspects of diagnosis. Among others, they point to the prevalence of scabies, which matched the prominence of facial abnormalities among "false alarms." The sifting of symptoms also demonstrates the examiners' reliance on interrogation, the nexus between semiology and etiology, and, above all, the diversity of "equivocal" signs.[45]

Two examples will suffice to give an idea of the rich contents of the waivers. Both cases date from 1548 and are extant in the Deanery Books of the Faculty of Medicine in Cologne. The assessment of Petrus Lynscydt's appearance, in conjunction with his—and his brother's—explanations, led two doctors to declare him "clean" of leprosy,

> notwithstanding the unevenness of his skin, with soft, dry, reddish pustules on
> the eyebrows, the upper part of the eyelashes, and the cheekbone near the eyes.
> His brother confirms that he has contracted this from the parents. Also, his nose
> has become flat by a fall from a horse, so that the right nostril appears bigger and
> thicker, but without corruption of the face. His color happens to be paler from
> the prolonged use of the thermal waters in Aachen, which he has frequented con-
> tinuously for an entire month.

Nevertheless, the doctors ordered Petrus "to come back, on account of tuberosities on the forehead and nose, hairlessness of the eyebrows, and macules on the arms and legs." The same two members of the faculty examined Conrad Pryntz, "former baker of the canons of the collegial church in Bonn." They found him

> clean, notwithstanding a certain darkness and protuberances in the face, because
> these have a different cause, namely the washing, with extremely cold water, of
> his face when it was burning both from the summer heat and the furnace fire and,
> in addition, from the immoderate toil of his entire body. He has testified under
> oath that, in the same way, he gradually contracted an ugly color of the face and
> hands from his continuous toils.

Wishing, again, to be on the safe side, the doctors recommended medical supervision "for purging the superfluous bad humors," and they scheduled another visit for the fall.[46]

These two cases, and numerous others, indicate the extent to which the examiners relied on causal explanations for reaching a differential diagnosis. While this reliance was crucial for screening out misleading signs, it entailed the danger of prejudgment when it was applied to symptoms. Causes served to rationalize not only diagnostic guidelines in treatises but also the conclusions of examinations. The rationalization might be relatively innocuous, as when the examiners cleared Bellevalet because of "the integrity of all his limbs," whereas authoritative teaching held that "*leppre* is a very great error of the assimilative faculty, by which the form is entirely corrupted." On a few occasions, however, physiological notions and teleology—"the way things are supposed to be"—seemed to eclipse physical appearances and actual observation. This distraction may have played a role, for example, when Jacques Mahieu was judged leprous at Lille in 1560. The examining doctor and surgeon, while basing their conclusion on the unequivocal and equivocal signs, certified that they were "taking into account the extreme warmth of the liver, and the violent and excessive inflammation and incineration of all the blood, which were followed by a drastic consumption of all his muscles." In different situations, when the signs did not point to true leprosy, causes might be summoned for a positive finding. Thus, in 1559, the commissioners of Ieper testified that Joos Gauweloos should be hospitalized even though he was not "infected by any kind of *laderie*": they argued that he would "remain permanently unclean, simply from the superfluous humors in the head, and this is just as incurable as every kind of *laderye*."[47]

Semiology, etiology, and pathology were still garbled, with far-reaching consequence, in a considerably later affidavit. At Waldsee in 1719, a doctor and two surgeons observed in Catharina Maucherin

> a corrosive lesion in the face and neck, particularly on the right side, as well as a sizable tumor on the elbow and another under the left knee; also, a large bluish stain on the right foot: this does not properly fall under the label of *Aussatz*. Yet, considering that it has been there for fifteen years, we should take it for incurable, since it takes its origin and nourishment in totally corrupted blood; by the same token, there is more than a little danger of contagion, just as there would be for *lepra* itself. In order to prevent this, it is necessary to separate her from human company and to commit her to a leprosarium.

It is plausible that this had been Catharina's wish all along, and that the examiners were influenced by social circumstances in making their decision. At the same time, they demonstrated the influence of preconceived notions by hinging their decision largely on the corruption of the blood (which they may have determined after venesection).[48]

To some extent, the interference of preconceptions was facilitated by some learned authors who injected theory into diagnosis. Velasco de Tharanta, for example, added explanations to the signs he adopted from the *Canon*: "[T]he first origin of *lepra* is in the viscera, as Avicenna says, and *therefore* the lung is injured, its pipe is narrowed, and respiration is impeded"; further, "breathing is rapid *on account of* smoke ascending to the membranes of the brain, and then [the patient] speaks as if through the nose." Ambroise Paré's interest in tying signs to physiology led him to reinforce a stereotype by explaining that *ladres* "greatly desire female company, especially during the onset and stasis of their disease, *on account of* the unnatural warmth that burns inside them." Most authors, however, observed certain boundaries between semiology and etiology. Some separated leprosy in general, for which they focused on symptoms, from the four types, for which they identified signs as effects of an "offending humor": thus, while *lepra* as such manifested itself in a livid face, corrupt blood turned the darkness reddish in *alopecia,* choler gave it a yellowish tint in *leonina,* and so on. Others implied etiology only when, instead of a humoral scheme, they chose the Galenic framework of faculties and based their diagnosis on "the condition of the natural, vital, and animal functions."[49]

Gradually, and in a trend that paralleled the abandonment of conceptual categories (Chap. 6), physicians freed diagnosis of rationalization, to concentrate on the symptoms in themselves. This trend, which Lommius may have been the first to cultivate, characterized later clinical reports such as that drawn up by Isaac Uddman.

> In the individual whom we have observed in Ostrobothnia, these were the principal signs: protuberances that are mobile, numb, dark reddish, and rooted in the solid flesh, particularly on the forehead . . . in all of them, as many as I have seen, the protuberances lacked sensation, while the adjoining areas were itchy.

Uddman exemplified a parallel development in semiology, namely, the diminishing recourse to a multiplicity of signs. Academic and practical interest in casting the widest possible net, by proposing a multitude of phenomena, began to fade in the sixteenth century, for several reasons. After surveying half of the conventional symptoms, Jean de Varanda asserted, "[T]he remaining signs are of little importance, either because they are not apparent by themselves but have to be discerned from the patient's report, or because they are quite common to other diseases." One factor, no doubt, was the growing appreciation of direct observation and induction. Another was the paradoxical realization that the more numerous the signs, the more a judgment would depend on the examiners' discretion. The Marburg professor Heinrich Petraeus observed that "the internal and external symptoms, which are almost infinite in this monstrous disease, each have their specific indications, and these need to be discerned by the physician's particular prudence."[50]

A question arises here: Why did it become so important, in the fourteenth century, to post as many signs as possible? More than fifty were entered in the marginal gloss reproduced in Plate 9, and a more organized list presented forty-five signs that a scribe "collected from all the authors of medicine." After all, premodern physicians apparently knew (and modern leprologists observe for Hansen's disease) that the disease could be identified from two or three out of a few unequivocal signs, above all, livid macules, dispersed nodules, and loss of sensation. It is illogical—although not foreign to popular docudrama—to interpret the proliferation of diagnostic criteria as evidence that leprosy was becoming ubiquitous. A more likely explanation is that, with Scholasticism at its height, physicians employed as many subdistinctions as possible in valiant attempts to identify a disease in the tangle of individual variations, multiple forms, and uneven stages—like multiplying the dots to be connected, or the numbers by which to paint. An additional source of the profuse symptomatology lay in Avicenna's diagnostic inventory, which, unlike the temporarily eclipsed description by Aretaeus, lent itself to enumeration. With time, moreover, popular perceptions, impressions, and alarms enhanced the accumulation of signs, many of which became classified as "equivocal." The ultimate impetus for applying a broad spectrum of criteria to the diagnosis of leprosy—broader than for any other disease—came from social pressures on the diagnosticians. Medical examiners were expected to deliver a decisive report. Just as learned practitioners were establishing credibility, as more trustworthy than lay healers not only in the treatment but also in the recognition of disease, they stepped from the bedside into the public forum of the *iudicium*, as we have seen (Chap. 2). Here, instead of ministering to a patient, they decided the future of a "suspect"; in addition to applying privileged knowledge, they issued judicial verdicts; and, even when deferring to religious and secular authorities, they almost wielded the power of the Levite priests.[51]

It is impossible to ignore certain parallels between the aggregate of medical diagnoses and the biblical semiology of *zara'ath*, or, to stay within the European ambit and the Vulgate tradition, *lepra* in Leviticus 13. Like the medical examiners, the priest was credited with an ability to differentiate which placed him above the lay people. He, too, could translate this knowledge into a determination (*arbitrium*) by which a reportedly infected individual would be exonerated and freed, condemned and separated, or provisionally quarantined. Unlike the practitioners, however, the Levite priest issued not only the diagnostic determination but also the corresponding order. In giving him the option of isolating a suspect temporarily, "for seven days," Levite law anticipated the practice of allowing signs to become less ambiguous with the passing of time, which was common in European examinations (most often for six months). The Mosaic ordinances, like the Latin treatises—though less explicitly—recognized the developmental character of the disease, in emerging, growing, and advancing stages. They also reflected differential recognition by distinguishing *lepra* from scabies, burn lesions, and baldness. The three critical signs were color change and whiteness of the skin, depressed lesions, and yellow hair; these have, understandably, stimulated speculation that there was, if not complete identity between biblical *lepra* and Hansen's disease, at least an overlap with tuberculoid leprosy. The significance of this possibility, however, may be offset by some sobering considerations, even if one is not averse to making retrospective diagnoses. The symptoms of *zara'ath*, limited in number and couched in terms that allow for widely varying interpretations, are applicable to a host of diseases.

The relative vagueness with which symptoms are described in Leviticus stands in contrast to the agonizing preoccupation with precision among medical authors and practitioners. Most important in the context of this study, the Levite priest, with direct power over the fate of the suspect, pronounced a ritual verdict on purity rather than a decision on public health. With a mandate secured by status, he did not have to worry about credibility. Divine scripture provided the criteria for his judgments. Medical examiners, who lacked these powerful credentials, sought security in utmost caution; their safeguards undoubtedly limited the number of erroneous decisions. We have seen that they were keenly aware of the consequences of a *iudicium* that erred in either direction, and of the overall impact on their credibility. The awareness that, for the sake of justice, "great attention should be paid to the examination and judgment of *leprosi*," which was expressed by Guy de Chauliac and other authors in various formulations, pervaded the formulation of certificates and oaths. Concerns with credibility were palpable, above all, when the "solemnity" of tests was added to examinations.[52]

Diagnostic Tests and *Experimenta*

Diagnostic tests, as they were described in guidelines, anecdotes, and affidavits, reflected a fluid mixture of inquiry and display, empiricism and rationalism, forensics and gambling. Paradoxically, even if they were conducted primarily for the sake of diagnostic thoroughness and comprehensiveness, they often became rituals, performed "according to custom" or because they were "prescribed" in the textbooks. Even while admitting that eminent authors, including Jean Fernel, considered the results of blood tests "utterly uncertain and misleading," Foreest carried them out as currently fashionable ("iuxta neotericos"). He had the blood, drawn by the surgeon,

> divided into three small dishes, placing in the first, when the blood was almost congealed, three grains of coarse salt the size of peas, to see whether these would dissolve; in the second, very strong vinegar, to see whether it would change the color of the blood; the remaining part, from the third dish, we placed in a linen cloth that we closed and immersed in water.

This is a perfect caption for the far left portion of the woodcut in Plate 8. It is ironic, however, that this woodcut illuminated the textbook by Gersdorff, who warned, "[T]he doctors write that the signs in the blood are not certain, so we should not attach too much faith to them."[53]

Even when diagnostic tests stayed close to otherwise recognized symptoms and to the written tradition, they often took on an element of ceremony or, one might argue, theater. It is instructive, for example, to view Foreest's procedure in the perspective of tradition. By the early thirteenth century, Latin authors presented as verifiable certain hematological changes that had been posited by medical authorities such as Avicenna. Thus, Gilles de Corbeil assured that the blood drawn from someone with *elephas* was "notably black or dark." Theodoric Borgognoni sketched two simple tests:

> [I]f three grains of salt are placed upon the blood of the patient they are immediately dissolved. But if the blood is not infected, they are not dissolved. Likewise another common sign is, if blood is taken and rubbed in the palm of the hand and it squeaks or is too greasy, it is a sign of infection and corruption.

Jordanus de Turre offered more detailed instructions, which allow us to appreciate the variable nature of premodern tests, and their potential for ritualization. His four *experimenta* were alternatives, rather than sequential steps as in the examination by Foreest. In the second option of Jordanus, when a "large lump of salt" (formalized into "three grains of salt" by Theodoric and others) was placed in the strained blood, it would indicate leprosy if it did not dissolve: this proposition was exactly the opposite of Theodoric's. For the third test, Jordanus specified that strong vinegar would foam when poured into the blood of a leprous patient: this differed from the instruction of another diagnostic guide, to "join an equal amount of blood and vinegar, and let stand: if they do not mix, the subject is leprous." In the fourth option, urine poured onto the blood would indicate leprosy "if it sinks and mixes": this criterion appears to have been commonplace. In principle, however, this test was akin to the tossing of a coin, and to some "trials" or "ordeals" by water. This similarity was suggested further in the assertion by Velasco de Taranta that, if the blood "be put in very clear water and floats, it bodes ill for the subject."[54]

Guy de Chauliac, a most perceptive witness on this matter—as on many other matters—pulled together several threads. He considered the results of straining the blood in a cloth "very important," but, like Jordanus, he added the other tests as optional. "If the physician wishes to test in the other dish whether salt melts quickly [in the blood] or whether vinegar and urine mix in it quickly, and whether [blood] descends like flour in a basin full of water, he may do this for the sake of completeness [*gracia solempnitatis*]." If the medical examiners sought diagnostic certainty in the tests, the "solemnity" of the procedure must have carried more weight for the witnesses, who often numbered six or more, and who lacked the knowledge for verifying the results. One *experimentum,* with a diagnostic relevance that remains difficult to evaluate, had the trappings of a traveling magic show. In 1438, the Jewish physician Astrug Abraham and the barber Antoine Crote examined Hugues Rémusat in a garden in Tourves (Provence). They immersed one egg in the drawn blood of Hugues and set another one aside; after a moment, both eggs were broken and displayed to everyone present. The chosen witnesses compared the eggs and declared that the tested one was unchanged. This culmination of his step-by-step examination allowed the physician to pronounce Hugues healthy. In 1462, also in Provence, the less fortunate Johan de Bonafe presented various ominous signs, including "heat of blood" which boiled two eggs! Bonafe was subjected to another solemn test when the examiners "made him walk on a board covered with spinach grains," after they had looked extensively for anesthetic spots by means of a needle.[55]

The local loss of sensation was generally recognized as a primary and early symptom of leprosy and, therefore, as an imperative target in examinations. Tests to determine this loss, however, often transcended their diagnostic significance. Perhaps the most ritualized *experimentum,* performed in Le Mans at least from 1404 to 1491, was the practice of placing the suspect naked on a slab of marble: while "medicalized" by the occasional presence of a surgeon, and probably determining the patient's tactile response to the cold, this test emanated an almost sacral aura. Even the conventional methods for detecting the loss of sensation displayed telling nuances. We may compare, for example, the terse prescription of Guy de Chauliac with the more elaborate directions of Jordanus de Turre. Guy ordered examiners to check whether the patient "feels it fully when pricked in the heel and thigh and knows, when asked, where and with what." Jordanus almost turned it into some sort of game:

> [M]ake the patient cover his eyes so that he cannot see, and say "Look out, I'm going to prick you!" and do not prick him. Then say, "I pricked you on the foot"; and if he agrees, it is a sign of leprosy. Likewise, prick him with a needle, from the little finger and the flesh next to it up to the arm.

Jordanus was eager to offer a rational explanation for several of the signs and tests, including this one: "The reason for doing it in these fingers rather than others is that they are the weakest and therefore the ones which are first lost to the natural state." Velasco de Tharanta returned to a simpler test when he instructed the patient to "stand erect with closed eyes, and to be pricked lightly with a needle in the heel or the sole of the foot. When he sits down and is unable to place his finger on the place where he was touched, it is a sign of *lepra*." Velasco also greatly simplified the rationale "because the flesh is insensible."[56]

Another nuance in the diagnosis of anesthesia emerges when we contrast Velasco's instruction, that the subject should "be pricked lightly," with later, more aggressive techniques. For example, Johann Huggelin remembered seeing

> a fifteen-year-old boy in Pforzheim. While he was looking out the window, a barber freely pricked him with a needle about half a finger in length, stuck it quite deep, and asked him what was done to him. He answered only that a middle finger was poking him: he was so *maltzig* that he felt almost nothing.

Ambroise Paré may make the reader queasy by boasting that one test on suspects was

> routine for me, namely that, having pricked them with a rather thick and long needle in the big tendon attached to the heel, which is a great deal more sensitive than the others, and seeing that they did not feel anything, even though I had pushed the needle very deep, I concluded that they were truly leprous.

The fact that Paré was a surgeon may have a bearing both on the graphic nature of his report and on his predilection for this drastic probe.[57]

The aggressiveness of the latter two probes is apparent, albeit less pronounced, in other sixteenth-century treatises. I think that it suggests a growing determination to overcome the multiplicity, polyvalence, and deceptiveness of lesions, and to have the ultimate certainty of an infallible criterion. In any event, the treatises demonstrate the concurrence of academic and practical interests in vigorous examination techniques and tests. Authors adjusted the conventional proportions of the discussion of leprosy, and the expansion of diagnostics was generally matched by a reduction in therapeutics, and particularly by the steady decline of polypharmacy. Some, including Paré, Du Laurens, and Lommius, reduced the treatment of patients to little more than an afterthought; others, such as Huggelin and Varanda, limited their section on *cura* to prevention and palliation. In short, at about the same pace, the coverage of remedies tended to shrink, and the efforts of authors to identify patients and to recognize the disease increased—as did, arguably, their knowledge of its nature. As we have seen in this chapter and in our earlier examination of the *iudicium,* the academic preoccupation with diagnosis is richly confirmed in official certificates. In other chapters, this documentary richness has also shed light on subjects that ranged from nomenclature to etiology—and it supplies the substance of this entire study. For our last chapter, however, on the medical discourse on the prospects and treatment of patients, we should be prepared to find a comparative dearth of archival or "factual" evidence.

Chapter Eight

Prognosis, Prevention, and Treatment

 Reverend Father,

Together with Doctor Hartmann Schedel, who was in Nuremberg at the time, I have with all diligence examined the bearer of this letter, Conrad, a former member of your household. We have found him not infected by *lepra*, though having a certain ulceration of the foot, which does not yet portend *lepra*. I understand that Conrad has spent some time in the leprosarium with others who were infected, and I suspect that this stay has considerably affected and indisposed the humors of his body.

The Doctor and I deem it advisable that he be governed by a good regimen, observing a good diet and abstaining from all very salty and acidic foods, sharp victuals, and excessively hot spices. He should also avoid contact of his body with extreme cold, and with intense warmth, especially of hot baths and steam baths and the like, as he has been sufficiently informed. It will further be necessary that in due time his body be suitably and completely purged of bad humors by an expert physician, and that he has bloodlettings administered according to immediate need, and other treatments according to the discretion of the treating physician. In addition, it will be imperative that he have an experienced surgeon to devote utmost care to the ulcerated foot.

If all these things are carried out correctly and if he is obedient, and since he is young and at the age of puberty, there is very good hope that, with God's help, he will recover his original health and, at last, escape clean and whole.

Hermann Schedel made this transcript, probably around 1500, of a certificate that he and his uncle Hartmann had first drawn up in the late 1470s. By concluding with hope of recovery, the Schedels confirmed Conrad's exoneration in the strongest possible terms, since there would have been no prospect of cure if he had been found leprous. Incurability was the ultimate characteristic of leprosy, and its most constant attribute from late antiquity into the twentieth century. It is easy, however, to underestimate the complexity of this attribution and, therefore, to assume that, beyond diagnosis, medicine was largely irrelevant to the disease. In simple terms, from the premise that leprous patients were beyond remedy, the assumption concludes broadly that they were not a medical but a social concern, objects of charity rather than cure, treated by ecclesiastical and secular authorities rather than physicians, and accommodated rather than attended to in hospitals "for the incurable." In this vein, a frequently quoted historian of leprosy in the Middle Ages, who lumps patients together in "une société d'exclus," perpetuates a double disparagement in the terse assertion that they were "mercifully abandoned by medicine."[1]

THE PATIENT'S PROSPECTS
"Mercifully Abandoned"?

There are various indications, scattered but reliable, that people labeled leprous were not abandoned by medicine or, for that matter, automatically excluded from society in the Middle Ages. To be sure, the vast majority of records identify practitioners as participating in examinations rather than attending to patients, but these are certificates and commission reports that date from the sixteenth and seventeenth centuries. Earlier medical involvement in the care of leprous patients is documented more indirectly and more haphazardly. In the earliest instance, admittedly an exceptional and inconclusive one, the *medicus* Aubert, or Olbert, was "taking care of the affairs of the *leprosi*" in the chief leprosarium of Troyes from 1151 to at least 1159: while it is likely that Aubert's care was primarily administrative, it probably included health concerns. When the medical needs of inmates were not mentioned in statutes, we should not infer that they were ignored. There is at least some explicit evidence that sequestered patients, whether or not their confirmed leprosy was still treated, retained access to "professional" medicine for other illnesses. The 1493 statutes of the leprosarium Terzieken of Antwerp provided that "if one of the resident patients suffers from some curable disease and requests a physician, this physician should be provided." It does appear, however, from our limited—and relatively late—evidence that doctors delegated most of the bedside care to surgeons, religious attendants, and relatively able-bodied patients. From 1479 to at least 1632, the city of Saint-Omer commissioned a surgeon with visiting the Madeleine leprosarium "for the examinations" *and* "all the other necessary duties of a surgeon toward the patients." This surgeon drew a higher compensation than the examining physicians. In an instance of less "professional" attendance at Saint-Omer in 1505, care was given to a patient in an apparently advanced stage. A resident of La Madeleine, Jean de Wolcrincove, was paid 25 sous "for having administered steam baths, baths, and ointments to heal [*garir*] Loyette, daughter of Marc Hazebart, and having cleaned her thoroughly and properly for five to six weeks."[2]

There are also passing but compelling hints that there were confirmed patients who were not sequestered—without necessarily being homeless beggars—and who received medical attention. The 1240 statutes of the University of Montpellier ordered bachelors, at their promotion to masters in medicine, to

> swear that they will accept no *leprosus* in their care for longer than eight days in Montpellier between the Lez and Mausson rivers, except with permission of the royal court if the *leprosus* is in the king's territory, or from the episcopal court if he is in the bishop's territory.

We do not know whether Bernard de Gordon observed—or was required to observe—this statute when he took into his care the leprous countess; her illness was presumably far enough advanced to infect the bachelor who broke the Hippocratic Oath with her. Around the same time, a woman in Spoleto who suffered from "a kind of leprosy" also attested to the availability of sustained medical treatment: she testified that prayer to the saintly Chiara of Montefalco made her feel better but that her brother scoffed, "[Y]ou said the same thing the times you were treated by a *medicus*, and every time your infirmity came back."[3]

Recorded agreements for actual treatment preserve mundane traces of medical service to leprous patients. In 1388, the physicians Geoffroi Magenc and Jean Raynaud signed a contract at Aix-en-Provence to cure the leprous Louis Audibert, for a fee of 100 gold florins. In 1448, Montpellier alumnus Jean Guilhelmi agreed in Avignon to cure a priest of Barbentane, for 40 florins; sixteen years later Guilhelmi was called as an expert to evaluate the results of another physician's treatment of leprosy. In 1464, also in Avignon, Doctor Loup de Sandoval took on a contract to cure a leprous student, for 80 gold *écus.* Expense was probably an incentive for going to cheaper practitioners; this meant—at least according to the medical establishment—settling for quackery. However, even engaging an expensive physician did not necessarily guarantee care. These three points were documented in the case of a poor priest, Heinrich of Wald. In August 1519, the Cologne faculty judged him not leprous but ordered him to seek medical counsel and to return in six months. The Reverend Heinrich skipped his scheduled reexamination, but he wound up back with the faculty in September 1520, when he was declared leprous. According to the examiners, "he maintained that, in the interval since the first visit, he had subjected himself to the treatment of certain quacks [*empirici*], under whose hands he almost gave up the ghost, because they applied empirical liniments for suppressing the spots." Heinrich then sought recourse, perhaps at a discount, from a more learned though apparently equally unreliable practitioner: "He also complained about a certain physician of Antwerp who, after accepting considerable gifts, promised health but then failed him."[4]

Documentary information on the charges of physicians for treating leprosy is limited, but it suggests that remunerations could be steep. This suggestion accords with Henri de Mondeville's appreciation of the persuasiveness of "urgent pleas or very high fees." In the absence of these, he sternly warned his literate colleagues that "neither physician nor surgeon should deal" with cases of leprosy, especially when these were in an advanced stage. In fact, Mondeville vowed to "go lightly" over the subject of *lepra* in general, because it was "difficult, and much more medical than surgical." Another surgeon, Guy de Chauliac, similarly believed that the treatment of confirmed *lepra* belonged "more to the lords physicians than to the surgeons, except with regard to judging and manual operation." Ironically, yet another surgeon, Guillaume des Innocens, believed that "it pertains strictly to the vulgar surgeons to undertake the treatment of this kind of disease, whether to preserve, palliate, or heal"; limiting himself to diagnostics, he skipped therapeutics altogether "because it is not my intention to heal *ladrerie,* which everyone considers to be incurable."[5]

Beyond Cure?

The topos of incurability reveals significant contradictions and ambiguities in the premodern discourse on leprosy, in actual responses, and in the interaction between practitioners and society. The ambiguity begins, albeit in a modest way, with the interpretation of the universally employed term *curare*— for example, when Paulus Iuliarius addressed the council of Verona, which had chosen him "ad curandum" the disease of *lepra.* The possible meanings ranged from "to cure" to the more indefinite "to take care of" and "to treat"; the sense of "to cure" could further be extended from "to stop" to the ultimate "to heal"—although the term for the latter, *sanare,* was far less common in medical writings than in miracle stories. The notion of incurability is inflexible in itself. Nevertheless, it was often tempered in academic declarations, administrative policies, and actual applications. In addition, incurability was sometimes applied inversely, not as a presumed characteristic but as a diagnostic criterion of the disease. This was the case in one well-documented and instructive situation that evolved from the sixteenth to the eighteenth century, at a time and place where the notoriety (as well as, supposedly, the incidence) of leprosy was becoming negligible. The situation entailed a stubborn disagreement, with troubling implications, which ran between the medical jury and the management of the leprosarium Terzieken in Antwerp. Terzieken was staffed and managed by women in religious vows; at various junctures, their issue embroiled also the municipal welfare officers and even the town council.[6]

From 1573 on, one mother superior of Terzieken after another complained that her asylum was burdened by the number of nonleprous patients whose admittance was forced upon her by the certificates of examiners and the orders of magistrates. In 1587, the hospital even had to admit several syphilitics, to the great scandal and considerable risk of the young novices who were to take care of them as if they were *ladres.* In 1653, the magistrates abolished the public *proeven,* or examinations, as too expensive, but the firm diplomacy of Mother Anne van Burkroy had them reinstated within the year. Without the existence of examinations (*proeven*) and people judged to be leprous, the nuns would lose their income and even their raison d'être; on the other hand, with existing examinations beyond their control, they regularly found it necessary to challenge the judgment of the examiners. On May 31, 1701, Mother Marie Isabelle de Montagu succeeded in having six persons reexamined and declared "immune of *Elephantiasis* or the disease named after Saint Lazarus," only three weeks after they had all been judged leprous. In 1703, Mother Superior still protested that the sworn physicians and surgeons far too readily declared someone leprous. She successfully demanded that, for each examination, they swear an oath, provide a list of observed signs, exclude patients with other diseases—ranging from scabies to syphilis—and allow her to attend with a few of her senior sisters.[7]

Far from being settled, the controversy culminated in 1727, in an epic three-year legal contest between the municipal almoners (welfare officers) and the nuns. The contest at last revealed the crucial issue—and the reason that this Antwerp episode bears more directly on this chapter than on the two preceding ones. Mother Superior refused to admit the orphan Anna Catharina Spinneloy, whom the medical practitioners, she claimed, had declared leprous without making every attempt to cure her. Only the *failure of every treatment* would determine that a person was incurable *and therefore* leprous. There were Flemish precedents for constructing a practical definition of leprosy on the sole criterion of incurability. In 1559, the medical examiners in Ieper were "unable to find Joos Gauweloos infected by any kind of *laderie*," yet they assured that excess humors in the head would "keep him unclean and in need of alms forever, because this is *just as incurable* as every kind of *laderye*." In 1570, they did not judge Pieter vanden Briele leprous, but they declared that "his miserable illness is not curable"; on this basis, the regent of the leprosarium concluded that Pieter should be admitted "like other burghers who are found *lasere*."[8]

The medical practitioners and the leprosarium directors in Antwerp were unable to share the same flexibility on the boundaries of "leprosy." Therefore, the magistrate tried to resolve the disagreement by allowing an appeal to a second jury, which would consist of six sworn commissioners in addition to the six original examiners (three physicians and three surgeons). Even this did not settle the conflict. In 1733, the commissioners repeatedly returned the orphan Nicolaes van Goordaert to the almoners as free of leprosy. After subjecting the boy to all the "recommended medical treatments," the physicians brought him to the Faculty of Medicine at Louvain, where two professors declared that he was "infected by that kind of *lepra* which the Greeks call *Elephantiasis* and people call 'the disease of Saint Lazarus.'" The same day, Nicolaes was seen by a senior professor, who modified the verdict: "[H]e suffers from *lepra*, but in such a way that I consider him curable by antimonials and decoctions of woods, as long as a suitable diet is followed." Nevertheless, six weeks later the Antwerp medical college, that is, the expanded jury, pronounced the orphan leprous. The decisive factor appears to have been the report of a surgeon. Master Jan Baptist van Deventer testified,

> I have treated Nicolaes van Goordaert to cure his *lepra* if this was possible, and I confirm having administered a complete salivation by mercury, externally as well as internally, for at least thirty days, and then a suitable decoction, while observing a good regimen. All without success.

This declaration was the final proof that the boy's disease was truly leprosy *because* it was impervious to the most advanced treatment.[9]

The surgeon's admission of utter failure confirmed what Aretaeus had observed in antiquity, namely, that leprosy entailed "inevitable demise" because it remained hidden until there was "no hope of recovery" (*salus desperata est*), and it was "by far the most efficient of all diseases in bringing death." It is natural that this outlook was virtually unanimous for more than eighteen centuries. On the other hand, the view was often expressed in relative terms that left room for hope and continued care. We may compare, for example, the gloom of Aretaeus with the nearly contemporary stoicism of Rufus of Efesus, who mused that leprosy seemed to have "a deeper origin" in view of "the difficulty of curing it, a difficulty that *comes close to* impossibility." A similar relativity may be perceived in the contrast that the thirteenth-century preacher Jacques de Vitry drew for his audience in a leprosarium chapel, between the Divine Healer and human *medici:* the latter, admitting their lack of omnipotence, "*sometimes* say, 'We cannot, or we do not want to, or we do not know how to heal your illnesses.'" It is of interest to observe here that medical authors, even those who prescribed religious rituals for other immedicable diseases, ranging from insanity to sterility, did not recommend such remedies for leprosy.[10]

Bernard de Gordon mentioned no supernatural remedies in the elaborate section on therapeutics of his chapter on *lepra*. Yet elsewhere in the *Lilium* he reported that scrofula was healed by the touch of kings, that epilepsy was said to be cured by a certain incantation by a priest, and that herpes, anthrax, and other advanced apostemes "are cured by divine rather than human hand." Also, sixteenth- and seventeenth-century authors viewed recourse to the supernatural as outside the pale of medicine. Jean Fernel and Samuel Hafenreffer reiterated the classical maxim that leprosy is "so stubborn an evil that it simply cannot be cured by any remedies, but the common people are wont to beseech the deities [*divos*] for a cure." Thereby, both the Paris humanist and the Tübingen Protestant suggested a distance between their learned standpoint and the popular outlook. Even in popular piety, however, expectations of the supernatural healing of leprosy faded steadily from the ninth century on. The disease became almost invisible in the voluminous and growing body of miracle stories. In the early eleventh century, there still were shrines, such as that of Saint Vedastus in Arras, where *leprosi* were "cleansed"—note the biblical echo—by the relics on display. Subsequently, pilgrimages to places associated with the patron saints Lazarus and Magdalen—most of which were leprosaria—were made as penance and reminders of death (*memento mori*), rather than in search of a cure. In the mid-thirteenth century, the irreverent Champagne troubadour Rutebeuf had a character mock, while debating a barber, "There is no need to go on a pilgrimage if Saint Lazarus, not giving you a break, has hit you in the face."[11]

From the thirteenth century on, the hagiography presented saints neither as healers of *leprosi* nor as defying the fear of contagion. They were venerated as models of charitable dedication, who suppressed the revulsion toward loathsomeness. Saint Francis (ca. 1182–1226) was portrayed as kissing a leprous wanderer (on the lips, in some iconography) and then hastening to the leprosarium of Assisi, where he distributed alms to the inmates. Saint Elizabeth of Hungary (1207–1231) reportedly nursed a leprous beggar and took him into her bed, and she devoted much of her short life to feeding, washing, clothing, and sheltering the most repulsive outcasts. A fifteenth-century preacher recounted a legend that graphically highlighted the combination of charity and self-control—and the silence on healing and contagion—in describing one of Elizabeth's heroic acts:

> She abhorred the horrible bath of a *leprosus* with natural horror, but she wished to subjugate the domestic enemy, that is, her own body. . . . Right away, with unheard-of fervor, she drew the water of the fetid, dirty, and scabby bath with both hands and drank it. Thus she conquered herself.[12]

The new pattern was both confirmed and broken by one notable yet littleknown saint of the same era. The Provençal Elzéar (1285–1323), Count of Sabran, was eternalized around 1373 in the marvelous marble relief that is now at the Walters Art Museum in Baltimore and which is reproduced in Plate 10.

PLATE 10. Saint Elzéar and leprous patients, ivory relief, Provence, ca. 1373; The Walters Art Museum, Baltimore, Maryland, inv. no. 27.16.

Saint Elzéar, who was canonized in 1369, was named after Saint Lazarus. It is fitting, therefore, that one of the highlights in his life involved leprosy. The sculpture captures Elzéar's encounter with *leprosi* when he was setting out with a hunting party, during a stay in the kingdom of Naples. The artist has combined realism, in the contrast between the saint's smooth visage and the knobby masklike faces of the patients, with euphemism, in the lack of disfigurement and loathsomeness. This dignified portrayal of the poor sick contradicted their description, in a *vita* of the saint, as "so deformed by the disease that it was pitiful to behold them" and with "lips so corroded that their teeth were showing." The most pertinent point is that—pace some captions by art historians—Saint Elzéar is not portrayed as performing a miracle by "Healing Lepers" but, rather, as greeting, comforting, and touching them. The report for the canonization process concentrated on Elzéar's embrace and almsgiving in the leprosarium, and, almost in an epilogue, it added that, after the count left, the patients were "suddenly healed, while a sweet odor filled the house." This sudden recovery was an exception for the time, as leprosy, unlike most other diseases (and with sustained church emphasis on acceptance), became an affliction beyond the reach of miracle as well as medicine. The blessed encounter with Saint Elzéar offered the ideal reminder, not so much that supernatural omnipotence surpassed the power of medicine, but above all that the salvation of the soul far outweighed the health of the body. What better way to drive this reminder home than to emphasize the incurability of a disease of the entire body![13]

Glimmers of Hope; Prevention

Medical authors, too, often sounded unrelenting when they proclaimed that leprosy was incurable. Ambroise Paré asserted forcefully, "*[L]a lèpre* is totally incurable, in the same fashion as plague: it is incurable because it is a general cancer of the whole body; and even a cancer, which is in a single part of the body, accepts no treatment." It is worth pointing out that Paré's statement was in direct contradiction with the suggestion of his principal authority, Avicenna, that leprosy "is kept in check by the administration of strong treatments exactly because, unlike cancer, it spreads through the entire body." Few assertions were more dire than that by Hafenreffer in the mid-seventeenth century: "On account of its desperate obstinacy, the evil of *lepra* is left untouched by the physicians, ancient as well as more recent; and [patients] are relegated to hospitals [*Nosodochia*] because of the communicable *virus*." It almost seemed a token concession to humanity when he added, "[B]ut so that they may have some mitigation, the doctors prescribe baths." A closer look, however, reveals that even the most negative declarations had their counterpoints, and that the apparent medical fatalism had several soft edges.[14]

Paré's contemporary Paracelsus defiantly challenged the notion that leprosy was generally untreatable, with the same verve with which he claimed to discard the entire humoral tradition. He proposed that, if *lepra* and *elephantia* were called incurable, this was the fault of "the physicians who did not understand the healing art": those physicians "stole the power from Nature, which God has so richly endowed that there is no disease for which it holds no cure." With his known penchant for inconsistency, Paracelsus elsewhere admitted, "There is no cure for *Elephantia*," and, once *lepra* is in an advanced stage, "neither treatment nor preservation will help." Nevertheless, he applied his revolutionary outlook on nature to several remedies for leprosy, some of which we will consider later in this chapter. Suffice it here to mention one of his theses, a bold but arguably far-fetched one: that since leprosy was basically a rotting of the body, the most effective therapeutics would employ substances that were known to preserve dead bodies, including the balms of embalmers, essence of juniper, and pulverized pearls.[15]

Closer to the time of Hafenreffer's avowed pessimism, a patient in an advanced stage of the disease could still insist on being treated by a physician. In a reminiscence with timeless echoes, Felix Platter recounted,

[A] countess came to me from France with a large retinue, including a surgeon, and she asked me to cure her. From the first look, I recognized unequivocally confirmed *lepra*, and I knew that, on my part, I was unable to achieve anything here. I communicated the situation first to the surgeon and then to the steward. Both initially refused to believe me, for the surgeon as well as a number of physicians had declared that the patient was not infected, and that they did not find anything suspicious in her blood and urine. In the end I convinced them with striking proofs. When the countess was informed of this, she departed very angrily.

Platter added, "Both the innkeeper and I bore this rather lightly, as her departure saved the house from ill fame, for she already smelled bad and was half rotted [*semiputrida*]." With this rather cantankerous dismissal, which seemed to suggest a lack of compassion, he underscored the firm conviction of his inability "to achieve *anything*."[16]

The gloomy proclamations by relatively late authors such as Paré and Hafenreffer, that leprosy was simply incurable, as well as the confident claims by the likes of Paracelsus and the desperate demands of patients, tend to eclipse the more subtle mitigations of medical defeatism—and the occasional disappointment of a physician when a patient dropped treatment. Foreest disappointedly recalled a young man who promised "to return, after he checked with his friends whether they would deem it necessary to use the medicines that we prescribed; he did not come back, and so no medications were prescribed." Inversely, a quiet hope could mitigate even the most melancholy conclusion, such as that of André Du Laurens that *"lepra* or *Elephantiasis* is an incurable disease in its essence." Allowing for the possible discovery of an effective treatment, he added, "[I]f it requires medication, this will have to be by some specific antidote, something that has *not yet* been invented by human industry." More obliquely, Isaac Uddman left room for hope when he characterized leprosy as "a highly chronic disease, which is hardly ever cured *by the benefit of nature.*"[17]

In the supposedly darker fourteenth century, hope could glimmer in the modification imparted by a simple adverb, a clause, or an afterthought. Even an avowed abandonment of hope precluded neither the belief in the power of medicine nor the will to persist. Bernard de Gordon expressed the paradox of hopelessness and determination:

> We can prognosticate with certainty that, once *lepra* comes to the manifest corruption of form and figure, it will not ever be cured. *Yet* we can prolong life and, with appropriate medicines, prevent the poisonous incinerated melancholic matter from going to the heart and the principal organs.

With a similarly mixed sense of defeat and duty, Alberto de Bologna declared, "[T]his disease cannot be cured once it has gained strength, *but* we should persist so that it stabilizes and does not increase." Guy de Chauliac applied subtle modifiers to an otherwise sweeping statement, "all judge that leprosy is the worst disease, and it is *almost* impossible to eradicate, *especially* when confirmed." Implicit in these paradoxes and modifiers was the admission that predicting the outcome, even "with certainty," was not infallible. Prognostication, which in the Hippocratic tradition focused on "critical" or acute rather than chronic diseases, admittedly relied more on the art of conjecture than on scientific certainty. It may be noted in passing that astrology, which was associated with prognosis for so many diseases, played no role in dealing with leprosy. Even as a chronic disease, leprosy presented special problems: it advanced not only slowly but at an uneven pace; this pace depended on the patient's overall constitution and momentary condition; the progress was often interrupted by partial or apparent remissions; and the increasing interference of various secondary complications clouded the final outlook.[18]

There were two responses to the special problems with the prognosis of leprosy. One consisted of attempts to improve the diagnosis (Chap. 7), in order to recognize a person's predisposition; a second response lay in a greater emphasis on prevention. While there was little development in the descriptions for prevention, efforts were concentrated on differentiating the disease's stages, and adapting remedies to these stages. Avicenna considered leprosy "more susceptible" to treatment at the onset, "more resistant when confirmed," and beyond hope when in full attack. Alberto de Bologna adopted Avicenna's differentiation and applied it to two groups of medications, namely, "evacuative" and "alterative," which would prevent or slow down the progress of the disease. When Alberto urged the practitioner to "persist so that it stabilizes [*stet*]," he was alluding to the *status*, or stasis, the plateau that chronic diseases reached, in the fourfold Galenic scheme, before moving into the final stage.[19]

Physicians generally retained this scheme for tracking the progress of leprosy, but a student of Master Alberto's adapted the scheme to practical needs. Guy de Chauliac reduced the "four times of disease" to the two stages that were decisive for the treatment—as they were for the judgment—of leprosy, and he added a preliminary state. He outlined the three resulting stages in correlation with distinct medical endeavors:

> In the treatment of *lepra*, the doctors commonly have three objectives. The first is to preserve those who are disposed, before the disease comes; the second is to treat those who are actually sick, when the disease has been introduced but not confirmed; and the third is to palliate, once it has been introduced and confirmed.

The first objective came to dominate responses to leprosy, while it received limited consideration for other diseases. At the same time, the complementary notions of predisposition and preservation opened the door to confusion.[20]

When someone was found "not infected but disposed," the criteria for diagnosing a predisposition were imprecise and almost arbitrary. It was easy to entertain adverse presumptions about the suspect's past and future, as we have seen earlier (Chap. 7). A neutral finding meant a deferred judgment about the person's status as clean or infected, and uncertain or tentative recommendations for care. In the excerpt that introduced our chapter on the *iudicium,* Guillaume des Innocens noted that the categories themselves were problematic: "Now, the terms of 'disposition' and 'act' in *lepre,* as Mr. Guy de Chauliac interprets them, are full of ambiguity and reasonable doubt." Neither diagnostics nor therapeutics drew a clear and consistent line between potential and incipient leprosy. Moreover, the verb "to preserve" ambiguously covered prevention, for example, in guarding against plague, as well as maintenance, as in keeping phthisics from total consumption. Jean de Varanda attempted to resolve the ambiguity:

> Let us say that the inclination to this disease is to be avoided by the application of careful precaution, namely with the removal or amelioration of all the causes that can contribute to the emergence of the disease, and this is true prophylaxis. On the other hand, to suppress the disease when it is beginning, by the elimination of internal causes and the correction of the imbalance: this is protective [*curatoria*] prophylaxis, which is consistent with our book.

In other words, Varanda evaded the ambiguity by omitting the truly prophylactic course.[21]

Other authors, however, offered prescriptions for *preservatio* which were unequivocally preventive. The advice of Hermann Schedel and his uncle, recorded in the memorandum that opens this chapter, was clearly intended to protect Conrad from the disease after he was exposed to infection. A much earlier allusion to real prevention occurred in a medical manual attributed to Gil of Portugal (aka Egidius de Valladares, d. 1266), with the chapter "on treating and *avoiding* the disease of *lebrosia,*" which began with a remedy for the person "who is afraid of *lebrosia.*" One physician, Gulielmo de Varignana (ca. 1270–1339), gave the impression that he personally practiced prophylaxis: "[S]ince *lepra* begins with blockage in the spleen, the authors highly recommend drinking a decoction of tamarisk roots with raisins, every morning; and we, who have tried this, add little to this concoction." A fifteenth-century treatise on leprosy by an unknown author contained a "recipe for the physician to protect himself from *lepra:* take toast dipped in vinegar and eat in the morning"; indeed, the same protective recipe was recommended as "very good against other contagious diseases, and against putrid air where there are many swamps." The majority of the "preservative" recommendations, however, were not preventive but therapeutic, for they aimed at maintaining stability in bodies that had already crossed the threshold from predisposition to onset.[22]

The development of leprosy therapeutics, while generally unremarkable, reflected the fluid understanding of *dispositio.* According to the Salernitan master Johannes Platearius, "to tell the truth, all the kinds of *lepra* are incurable, particularly *leonina* and *alopicia;* however, when there is a disposition to *lepra,* the patient can be saved [*preservari*] from a future greater evil." Platearius apparently understood "disposition" as a morbid condition, or even as a phase in what became commonly designated as the "inception" of the disease. Compounding the confusion, he moved directly from "preservative" measures, such as bloodletting and purges, to palliation with lotions, plasters, and the like. Predisposition and actual disease, as well as the corresponding objectives of prevention, remediation, and even palliation, were still confused in the early fourteenth century. In a manuscript from the University of Montpellier, a "proven recipe" (*experimentum*) was prescribed "for the palliation and treatment of those disposed to *lepra,* which cured someone who had all cracked hands and who appeared predisposed to *lepra."* In this same academic milieu, Jordanus de Turre blurred the sequence of therapeutic objectives in his *Notes on Lepra.* His "general rules for treatment" began (items numbered 1 through 6) with *canones* for heading off the causes of the disease, primarily by adjustments in diet and lifestyle; then, the rules skipped rather abruptly (9–12) to the palliation of effects such as hoarseness and fetid skin; the rest (13–21) prescribed, in random order, remedies for the inception, advance, and "status or confirmation." A similar lack of structure, and of distinction between predisposition and onset, marked Jordanus's particular instructions that followed his general rules.[23]

Guy de Chauliac learned the distinction between *lepra* in actuality and in potentiality or predisposition from Jordanus de Turre. Once more bringing order into his teacher's sketchy propositions, Guy stated.

> [T]he preservative regimen has three objectives. The first is that the [offending] matter not be generated; the second, that what has been generated be expelled; but third, that the liver and the complexion of the entire body be rectified.

The first and most genuinely preventive goal was "achieved by the due administration of the six non-naturals," that is, by the proper diet and lifestyle. Specific recommendations addressed suspected "dietary" causes of leprosy, which ranged from salty meats to intense emotions (Chap. 6). In Marseille in 1464, Guy's first two preservative objectives were combined when the examiners ordered Bartholomeus Guiraudus, whom they found *dispositus*, "to abstain, for one year, from certain foods, which we have identified to him, and to have himself purged by expert physicians, lest he fall into the said grave disease." In Nuremberg around 1470, the Schedels advised Conrad to observe "a good diet," to avoid extreme cold and heat, and to be "purged of bad humors by an expert physician." With a similar combination of dietary and medicinal considerations, the Cologne faculty in 1566 judged Johann Behem clean while "admonishing him henceforth to avoid errors in lifestyle, and to pay attention to the treatment and evacuation of the body by means of the proper drugs." A few months later, the faculty "absolved" Christus of Langen, whom they had found, a year earlier, "not entirely clean but not infected with *lepra* to the extent that he should be separated." At that initial visit the doctors had applied Chauliac's third approach: they had ordered treatment by a physician "who should bring the body out of its humoral disorder [*cacochymia*] and, after removing the imbalance [*intemperie*] of the liver, correct the body's cutaneous infection with the appropriate topical medicines."[24]

Therapeutics
Problems of Methodology

Archival documents, while quite informative on "preservative" governance during a reprieve, are mostly silent on the medical treatment of "those who are actually sick." As a result, our assessment of actual therapeutics depends almost entirely on academic propositions, with limited confirmation that they were applied or even considered in practice. Other methodological pitfalls, more insidious and more complex, concern the efficacy or, at least, the validity of the prescribed treatments. If we judge the early choice and applications of medicines by current scientific and aesthetic standards rather than on their own terms, we falsify history. Such anachronism results both in Whiggish caricatures of early remedies, as products of ignorant superstition and bizarre fantasy and, inversely, in their romantic glorification, as models of alternative medicine or as harbingers of later discoveries. In the framework of premodern medicine, learned and popular pharmacopoeias flowed from different sources, although they interacted constantly. Scholastic therapeutics revolved around humoral and qualitative pathology, depended more on rational principles than on methodical observation, and corresponded to a multifarious semiology. Folk cures relied on perceptions and assumptions, with little room for explanations and classifications. There were several links, however, between learned and popular medicine. Both evinced faith in *experimenta*, that is, recipes that were inexplicable but "proven" by experience. Another common feature was the notion that certain "specifics" possessed an efficacy that, while "occult," or hidden from understanding, was inherent in some aspect of their nature, such as a similarity to a sign or effect of leprosy. Doctors and empirics also shared a social awareness that expressed itself, for example, in the distinction between exotic drugs for rich patients and domestic substances for the poor.[25]

It is tempting but fruitless, at least with regard to leprosy, to speculate about apparent parallels between modern therapeutics and premodern prescriptions, whether from academic sources or vernacular lore. The greatest temptation, I believe (without ignoring the value and serendipitous results of current pharmacological searches), is to interpret coincidences as evidence of prescientific intuition. In general, any assessment needs to take into account the number and diversity of remedies that were recommended for leprosy and which surpassed those for any other disease. Touati has tallied 270 medicinal simples prescribed since antiquity; he also notes that, in contradiction to stereotypes about the slavish derivativeness of medieval authors, fewer than half of these were adopted by Montpellier authors (90 by Bernard de Gordon, and 97 by Henri de Mondeville). A valid comparison between this profusion of substances (polypharmacy) and today's multidrug therapy for Hansen's disease is precluded by the fact that both sprang from entirely different concerns. In particular, the historian's chief challenge is to understand the *intended* effect of individual substances, rather than to verify their *actual* efficacy. This does not diminish the interest of occasional coincidences—for example, between the modern concern with personal hygiene and medieval recommendations of baths and soap. It is equally intriguing to note the coincidence between the late nineteenth-century discovery that a primary infection affects the nasal mucous membrane and fourteenth-century prescriptions for "head purges" (*caputpurgia*). These head purges, many of which were sternutatories (sneeze-inducing chemical substances), were to be administered, "one drop at a time, through a nasal funnel," or with a wick of "fresh wool, silk, or coarse hemp," to alleviate catarrh.[26]

When old remedies for *lepra* seem to anticipate recent treatments of Hansen's disease, we should keep in mind that the contexts and objectives were entirely different. This will keep in check speculation about parallels between current laboratory experiments with the armadillo and scattered recipes that included distantly analogous animals, such as the turtle and porcupine. Moses Maimonides (1135–1204) believed that the meat of a porcupine, "if dried and some of it imbibed in oxymel, is beneficial for mutilating leprosy." Albertus Magnus noted that "a hedgehog's skin, burned together with its quills, returns hair to scar tissue," and Henri de Mondeville applied this idea to a recipe for palliative care. In 1483, after Eustache de la Fosse claimed that "*lepre* would be cured by the blood of turtles in the Cape Verde Islands," King Louis XI of France, who was rumored to be infected with leprosy, sent a search expedition to the islands. The later royal physician Ambroise Paré indeed asserted that the blood of giant turtles is good "for those who are afflicted with leprosy." There can be little doubt that these premodern recipes were based on traditional premises such as "the doctrine of signatures," that is, the belief that the Creator or Nature designed certain animals or plants as remedies for ailments with similar appearances or complementary qualities.[27]

There are other differences between premodern and modern premises for the choice of medicinal substances. If the medieval popularity of myrobalans (five kinds of South Asian plum- or prunelike fruits) seems to foreshadow the early twentieth-century recourse to chaulmoogra oil from India, we are brought back to reality by the expressly laxative objective of the former. One recipe from fifteenth-century France, "for dissolving the humors in the leprous head and body," included "Indian pills," presumably compounded from myrobalans. The inclusion of sulfur in old prescriptions can evoke the modern use of sulfones only if we overlook the enormous difference between the element and the compounds, or if we fail to note that the sulfur was administered in baths, and primarily for palliative rather than curative purposes.[28]

Whatever the value of speculating about the correspondence between past and present pharmacopoeias, numerous variables virtually preclude the exact and certain interpretation of the recipes we find in the sources. The medium of transmission, language, was often also a barrier, as it was in responses to leprosy in general (Chap. 3). The medicinal nomenclature of ingredients, preparations, and applications left ample room for misunderstanding until it became standardized. As a trivial but cautionary instance, I have discovered that historians occasionally confuse two meanings of the term *epithimum*, which occurs repeatedly in prescriptions, namely, "flower of thyme" (from *epi-*

thymum) and "poultice" (from *epithemum*, "application"). The names of vegetable ingredients were—and are—particularly susceptible to mistranslations.
Generic designations of plants, without identification of the species, left
room for vastly differing allocations in the modern Linnaean system. Moreover, northern European varieties were often substituted for the Mediterranean specimens of the sources. The greatest obstacle to "translating" many
recipes is the lack of precise information on one or all of the following: the
part of the plant to be prepared (seed, flower, leaves, stem, roots), the method
of preparation (boiling, steeping, distilling, drying, and so on), the manner
of administration (for example, in syrup, pills, or lotions), and the dosage.

Treatment, Step 1: Regimen

The intended effects of simples and compounds can readily be inferred from
the choice of medicines for "those who are actually sick." The administration of medicines, however, was to be preceded by dietary adjustments, even
in treating the truly sick. Guy de Chauliac outlined this treatment as having
four objectives:

> The first, in order to temper the affected humor, a good regimen is achieved by
> the proper management of the six non-naturals. . . . The second, evacuation of
> the burned humors, by means of phlebotomy and drugs. . . . The third, rectifica-
> tion of the inflicted damage, by the proper administration of snakes, potions,
> and confections. . . . And the fourth, correction of the symptoms, by means suit-
> able to the nature of those that occur.

The treatment of actual leprosy overlapped considerably with responses to
the incipient disease, especially in the dietary recommendations. We are already
familiar with most of these recommendations and their correspondence to
the causes of leprosy. In general, the affected or "imbalanced" (*lapsus*) humor
needed to be tempered by steering the six non-naturals "toward the cold and
the humid," in the words of Guy de Chauliac. A basic rule, expressly observed
by the Schedels but reiterated through the centuries, was to keep the patient
from salty, acidic, and spicy foods, and from extremes of cold and heat. Many
therapeutic guidelines, however, showed a particular and possibly intensifying preoccupation with one factor in the regimen: sexual activity.[29]

The authors unanimously adopted Avicenna's teaching that the leprous
patient should "be kept away entirely from coitus." A few were prepared to

allow sex when the disease was still "in the incipient stage." Most, however, agreed with Bernard de Gordon's stern warning that "coitus is not suitable in *lepra*, even though it is the popular and erroneous opinion that it not only helps but cures." This thesis endured, though with varying rationales and applications. The principal argument was that sex dried and dissolved the natural moisture, overheated the body, and stirred the poisonous matter. Some of the justifications seem inconsistent with the therapeutic emphasis on purging (to which we will turn shortly), and thus they raise suspicions about underlying fears and less rational motives. The male perspective surfaced in various expressions and concerns. Thomas Scellinck urged patients to avoid, literally, "playing with women" (*vrouwenspele*). Varanda (almost) corrected the gender imbalance when he advised that "women should say farewell to venereal business, because the disturbance of the humors and warming of the body will make the bad humors rush outward, more readily causing tuberosities that are followed by ugliness of the face."[30]

Nothing eliminated sex more definitively—or suggested latent fears more shockingly—than castration. It is not clear when or where this drastic remedy was first recommended against leprosy, or even whether it was ever applied as part of premodern medical practice. One can, of course, neither ignore the horrendous possibility that the suggestion may have been part of popular responses nor deny the fact that castration was practiced in "civilized" leprosaria well into the twentieth century (and eventually "emended" to vasectomy). In any event, the harsh measure may have been mentioned first by Johannes Platearius at Salerno. Theodoric Borgognoni was probably the first to state that, in *leonina*, "removal of the testicles is a help." Theodoric expressed a reservation, "Yet it is harmful in the cold types," which would have included all but the advanced lepromatous forms. Another surgeon, Jan Yperman, erroneously credited "Avicenna and many other masters" with the advice "that the testicles of a leprous man should be removed, in order to prolong his life." The fourteenth-century masters at Montpellier were silent on the procedure, but an alumnus, Velasco de Tharanta, raised the formal question of "whether the removal of the testicles helps in the treatment of *lepra*, since the testicles send warmth into the whole body but the bodies of *leprosi* are cold and dry." His answer, while premised on the complexional characterization of gender, was somewhat contrived: "I say that it helps because, as there is excessive dryness in *lepra*, the entire body is humidified by the removal of the testicles and it becomes as if effeminated, and moisture is retained; also, the warmth, which was burning the blood, is removed."[31]

The convoluted logic in Velasco's rationalization makes me wonder whether it signified a slipping hold on a barely tenable but diehard thesis or a growing distance between discourse and actuality. The surgeon Pierre Bocellin conveyed the distinct impression that castration was more a topos than common practice. When mentioning the supposed benefits of the operation, he left no doubt that he was quoting someone else.

> Velasco de Tharanta says that by cutting off the genitals a man's complexion or, in other words, his behavior, is changed into the nature of a woman, and thereby acquires a moist complexion that is able to resist the dryness of *Lepra:* and by this unmanning or incision, the warmth of the liver is cooled; consequently, the humors are not burned which are the first cause of every kind of *Lepra.*

Ambroise Paré also cited Velasco, and, though adding, "I am of the same opinion," he gave no indication of any corresponding reality. Jean de Varanda showed ambivalence—and alluded to an underlying apprehension—when he opened his discussion of surgery with castration, for the purposes of "weakening [*enervandum*, unmanning] the sexual appetite, moistening the body, and retaining the beneficial seminal material. However, this kind of operation is not easily tolerated by patients, unless they are perhaps still quite young." Since it "cannot be performed on adults without grave danger to life," Varanda moved on to other surgical interventions. I mention in passing that the role of sex, in addition to the similarities in symptoms, accounted for the application of some identical treatments to leprosy and syphilis, even when the two diseases were not confused: this was the case for antimonials, mercury, and the "decoctions of woods"—by which a Louvain professor meant guaiacum in 1733.[32]

Treatment, Step 2: Evacuation

After tempering the imbalanced humor with dietary adjustments, the second objective of actual treatment, in Guy de Chauliac's scheme and in most itemized programs, was to evacuate the burned humors. The general and persistent preoccupation with purging in therapeutics was rooted in humoral physiology, and it reflected the prominence of "putrefaction" in the understanding of leprosy. A particular and somewhat paradoxical challenge lay in the pursuit of this preoccupation without drying or dissolving the "radical moisture." The challenge was inherent in most of the measures that were prescribed for evacuating the noxious humors. Phlebotomy, medicines, and baths were foremost among these measures, as we saw in the report of Doctor Hermann Schedel. Because of the theory that the retention and corruption of blood were the worst causes of leprosy—and paralleling the importance of blood in diagnostic practice—bloodletting consistently headed the evacuative protocol.

The principal method for drawing blood was by opening, or "cutting," a vein (venesection, from Latin, synonymous with phlebotomy, from Greek). Alternative but less common methods involved the application of leeches, or of cupping vessels in combination with scarification, "ventosacio et scarificacio" in Guy de Chauliac. The place where excess blood appeared to accumulate determined which vein was to be cut, although the authors seemed to prefer the basilic or the median vein. While the adverse effects of bloodletting were notoriously underestimated until at least the early nineteenth century, premodern authors did not recommend the treatment without misgivings. Avicenna cautioned that the patient needed "to rest for one week" after each bloodletting, and that "phlebotomy from the big veins is more harmful than helpful." It is surprising that Jordanus de Turre, right after citing Avicenna, disregarded his caution by recommending "an extreme phlebotomy, that is, a big one from the big veins, first from the hepatic vein," up to one *libra*, if a large excess of corrupt blood was indicated by the "fullness, distention, and color of the veins." This advice might not have been approved by one likely teacher of Jordanus, Bernard de Gordon, who warned that bloodletting "should be done properly, but we see those who operate commonly err, because they perform phlebotomies indiscriminately and far too frequently."[33]

We are likely to find even more indiscriminate excess in the bewildering polypharmacy—or multiplicity of medicinal substances—for purging the humors that caused leprosy. At first sight, the profusion seems to be a product of helplessness, fantasy, and pretense, rather than of rational therapeutics. This is a widespread perception of premodern materia medica, which combined the legacy inventoried by Dioscorides (fl. ca. 40–80 CE) with the storehouse from the Levantine trade and Arabic writings, and with home-grown western European curative lore. The profusion was particularly pronounced in the treatment of leprosy, creating the impression that authors and practitioners indiscriminately piled up remedies in a desperate attempt to hit an elusive mark—just as they kept adding signs in diagnostics. A closer look, however, reveals logical structure, deliberate purpose, and sensible nuance. Steadily focused on the evacuation of the offending humors, the *pharmacia* was not randomly diverse but precisely differentiated according to the efficacy of the simples, the complexion of the targeted humor, and the circumstances of the patient. In a preparatory step, the offending material was to be "digested" by means of a syrup that was conventionally based on fumitory, and in which the juice of borage, flowers of violets, roseate honey, and several other mild simples complemented the consistency, palatability, and digestive action of the liquid. Predominantly warm humors called for the addition of cooling agents, such as lettuce; cold humors, for warming agents such as dill and lavender. Bernard de Gordon specified further adjustments, according to the season and other particulars, including for "a woman with retention of her menses." For the poor patient he prescribed simple additions, of vinegar mixed with honey (oxymel) or sugar (*oxyzuccara*); for "the delicate and rich," more precious electuaries, that is, paste-like sugary compounds, based on amber, pearls, and the like.[34]

Evacuation of the digested material began with light, or "easy," laxatives, in which the ubiquitous fumitory and borage were reinforced with pulp from the pods of *Cassia fistula;* these were to be given daily, or at least twice a week. More forceful purging was achieved by means of pills or a decoction of myrobalans, the *kebuli* ("from Kabul") for phlegmatic matter, the green for a warm humor. The dosages of still more potent cathartics, including the bitter *yera rufini* and *yeralogodion,* were to be varied because "some patients are strong, others weak, and some live in a warm, others in a cold climate." Avicenna prescribed two drastic cathartics, black hellebore and pulp of colocynth, surprisingly without noting their toxicity; in fact, in asserting that the pulp of colocynth was "not harmful," he ignored the danger of intestinal hemorrhage. The Latin authors, whatever their allegiance to Avicenna, evinced greater caution. Bernard de Gordon warned that pulp of colocynth was to be used "in a very small quantity," and that "all the drugs derived from hellebore and from colocynth are dangerous." Concerned that these time-honored cathartics should be an available option for treating leprosy, he devised a curious recipe for reducing their toxicity.

In order that these may hold no danger, go to a garden with a lot of radish, fenugreek, and parsley. Pull them away from the roots, though not entirely. Pierce with an awl, and stuff with black hellebore. Cover properly and let stand on the ground for forty days. Then, when we wish to use this, all the stuffing should be

removed, and an oxymel be prepared with the rest. It will be a marvelous digestive and evacuant.

Rather than settling for this tempered recipe, subsequent authors moved
away entirely from the most powerful purgatives. Jordanus de Turre taught
that "strong medicine is harmful" and "all hellebore should be avoided." Guy
de Chauliac, who agreed with Bernard de Gordon on most other points, simply omitted hellebore from his treatment, as did the later members of the
Montpellier school, including Bocellin, Du Laurens, and Varanda. Du Laurens, as well as his academic contemporaries and successors, dropped
polypharmacy from the treatment by evacuation. Jean de Varanda explained,
"[P]urgative medicines greatly weaken the body before they are able to work
on the disease with their full strength." There were still advocates of polypharmacy in the seventeenth century, such as Samuel Hafenreffer, but they
became the exception rather than the rule.[35]

After bloodletting and cathartics, a third way of evacuating bad humors
was by bathing. Regular freshwater baths cleansed only the skin, while the
"deeper matter" of leprosy required the medicated vapor bath (*stupha*), followed by rubs, ointments, and rinses. After citing the usual prescriptions for
medicating the water, Jean de Varanda added an alternative that—though
barely enticing—might serve the purging objective. He recommended, "[I]f
there are no means for taking a bath, let the patients swim regularly in a
stream with pure water; and if it contains leeches, it is very advisable to move
them to various parts of the body for drawing a sufficient amount of blood."
While balneal treatments were often intended for palliation rather than cure,
a continued interest in curative benefits was evident. From at least 1532 to the
early eighteenth century, the Antwerp examiners ordered or advised leprous
patients to seek treatment at the spa in Aachen (Aix-les-Bains). In 1702,
when Mattheus van Braeckel returned from Aachen, with a doctor's affidavit
that he had taken the cure, the examiners found him unimproved: this convinced them that he was definitively incurable and therefore leprous, after
annual judgments at the *examen leprosorum* had been indecisive for thirty years.
Opening a new window onto balneotherapy, Uddman reported, "[F]or these
past years, Londoners have communicated to the learned world the manner
in which they have felicitously attacked *lepra*, with seawater." Their treatment,
"by means of which they have several times vanquished an evil otherwise
deemed immedicable," concentrated on the ingestion of seawater "in a large

dose," together with the application of warm, sharp, and penetrating oils. After detailing the recipe, Uddman concluded rather impetuously that the success of this therapy "wonderfully" confirmed his hypothesis about *lepra* being caused by minuscule spiral worms, which were expelled and repelled by this treatment.[36]

Treatment, Step 3: Rectification

The same etiological hypothesis led Uddman to reject treatment with mercury. This remedy, which was most often used in combination with vapor baths, was applied more notoriously to syphilis than to leprosy, and it held a special place among the followers of Paracelsus. On the premise that quicksilver does not adequately kill worms (and that Baglivi proved this experimentally), Uddman concluded, "[I]t is no wonder that the treatment of *lepra* has been tried in vain with unctions and salivations of mercury." He placed his hope, rather, in the application of seawater, which for him was not only a purgative but also an antidote. In contrast with purgatives, which adhered closely to therapeutic conventions, the category of antidotes inspired the most innovative attempts to treat leprosy. Their lasting appeal lay in supposedly "specific" qualities for "rectifying the inflicted damage" (*impressionis illate rectificatio*), in Guy de Chauliac's phrase for his third objective of the *cura*. A baffling and mysterious disease called for extraordinary and exotic agents. This quest was—and remains—timeless, although it often followed different paths and focused on different agents. Uddman called seawater a "wonderful" remedy. A few years later, in 1778, Godfried Schilling (see Plate 4 in Chap. 3) devoted several pages of his book on *lepra* to "a secret [*arcana*] method of treatment which a certain Surinam freedwoman applies." The antidotes included gold, snake meat, blood, and other substances that have received the modern label "bizarre and fantastic." It is not necessary here to contest this simplification, and it would be equally distracting to assemble the sensational elements into a *Wunderkammer*, or "cabinet of curiosities." Rather, a brief survey will suffice for placing the antidotes and specifics in perspective.[37]

Avicenna listed gold, rather casually, among the simples for treating leprosy, and he provided no details about the preparation. For three centuries, it received little attention from the authors, even from those who generally adopted Avicenna's prescriptions. Then, the interest waxed and waned, largely along with the rising and declining place of alchemy in medicine. Jan Yperman included "filings of pure gold" in a prescription that he called "a treasure for hiding *laserie*"; he placed this prescription ahead of the presumably more conventional remedies, for which he cited Avicenna and Theodoric. Early alchemists, including Johannes de Rupescissa and Guillelmus Sedacerii (Sedacer, Sedacinus, Sedaciano) later in the fourteenth century, alluded to the benefits for leprosy, without precise prescriptions. They also entertained the tangential but intriguing notion that lead is leprous gold. Joannes de Vesalia (fl. 1430–1476), great-grandfather of Andreas Vesalius, provided more detail on the (inferentially recent) method of preparation. He observed, "The sages alchemically compound potable water of gold," which protects against pestilence and is "helpful for *lepra*, because it strengthens the heart, spirit, and natural warmth." De Vesalia dismissed the warning by Averroës that calcined gold is poisonous. He argued that the toxicity was "resolved by artifice" in liquefaction, and that there must have been earlier methods for making it safe, since it was already "part of recipes and antidotes for *lepra* according to Avicenna." Lieven Lemnius (1505–1568), a compatriot of de Vesalia's and friend of his great-grandson's, sang the praises of the restorative value of gold, "beyond the look of gold coins and rings that delight the eyes." In particular, he continued, "ignited and hot gold, immersed in wine, strengthens the innate faculties" when taken internally; "also, when this liquid is applied externally, it controls *lepra*" and several skin diseases. Gold remained attractive, at least for listing in recipes, into the seventeenth century. Among a dozen items that he cited, without comment, from *Liber de secretis naturae* (better known as *Coelum philosophorum*, 1525, a compilation on alchemy by Philippus Ulstadius), Hafenreffer included "potable gold, which is prepared from gold filings and honey reduced into a water, and into which is mixed some portion of *aqua vitae*."[38]

By this time, however, Jean de Varanda gave the impression that skepticism had eclipsed the confidence of de Vesalia, Lemnius, and others in *aurum potabile*. In a critique of the Paracelsians, to which we will return shortly, Varanda rejected their assertion that "potable gold is a sure antidote to *elephantia*, for it restores the natural moisture that was destroyed by the disease." He scoffed,

> [W]hich conceivable remedy could drive out so grave an evil that is not only seated deep inside but also enfolds the outside of a person? None, to be sure, unless perhaps some *Chymista* extracts the patient's liver, purifies it with all the contained blood, and then instills the gold liquid to replace the lost natural moisture.

These reservations about the use of gold were founded on philosophy and on the practical issue of toxicity, rather than on the socioeconomic implications that might be the first to come to mind.[39]

Alberto de Bologna preferred gold needles for applying cautery to leprous patients, but he admitted, "[T]he patient's indigence or avarice sometimes forces us to use iron." The economic differentiation was turned into a straight class distinction—and the medical efficacy of gold was dismissed—by a Flemish wag, who grumbled,

> It is commonly claimed that gold is a proven remedy for *lepra*. This is both true and false. True, because the rich are allowed to continue living in their own houses; even when they are so leprous that their nose is falling off and that they feel no pain when they cut or burn themselves, it is still not called *"lepra."* But when poor people have the smallest spot on their toe, they have to be put away immediately. This is where gold is helpful, but not in any other way.

The complaint is borne out by some documents. In 1382, Charles VI of France, overruling strenuous objections from the residents of the leprosarium of Saint Cloud, by royal pardon allowed leprous Jean Julam and his healthy wife to live in one of their "several homes" and "to visit their estates." In sum, gold as a remedy triggered enough questions to justify skepticism about its actual administration to patients, unless they were gullible as well as rich.[40]

Unlike gold, snake meat as an antidote was ubiquitous in the sources and widely available on apothecaries' shelves, in widely varying adaptations. The best-known use was as the principal substance in theriac, an antidote to snake venom and other poisons. A syrupy compound of dozens of ingredients, this antidote was first promoted in the second century BCE by the Greek poet Nicander of Colophon, and it was the main panacea for two millennia. The meat of snakes—specifically vipers—was introduced and associated with leprosy by Galen, in his book on the virtues of simples. In personal reminiscences, which have fascinated readers through the ages, he related the cases of five patients, each of whom was "healed" (*sanatus*) from leprosy—here, as throughout his works. Galen used the terms *elephantia* and *lepra* interchangeably. Two men shed their skin after drinking wine that accidentally contained a dead viper (*echidna*). Another man, who willfully drank wine "that was poisoned in this manner," first became *leprosus* but then recovered. A fourth patient was regularly engaged in the catching of vipers; he ate their meat, "prepared in a casserole in the manner of eels," at Galen's recommendation, and "his disease vanished." The fifth individual, a rich man, was commanded by the gods in a dream to come to Pergamum and to follow a daily cure of drinking a viper potion and rubbing his body with a viper lotion: "in a few days, his disease turned into *lepra*, and then it was healed by the medicine that God prescribed."[41]

Divine sanction, promotion by the Prince of Physicians, and anecdotal concreteness: What more could be needed for the perfect advertisement of a miracle drug? As for the appeal inherent in the drug itself, we can only speculate. It seems to have sprung from a combination of sources: a survival of mithridatism, that is, the ancient notion of building resistance to a poison by measured but repeated exposure; the doctrine of signatures, with the analogy between scaliness and skin diseases; the archetypal symbolism of the cycle of life, death, and rebirth; and the principle that one poison repels another. It is neither possible nor necessary to pursue all these ramifications here. A few focused observations should enable us to evaluate their significance for the premodern therapy of leprosy. Avicenna reinforced the recommendations of Galen with his own authority, and he expanded them to the entire therapeutic spectrum. Jordanus de Turre cited Avicenna as teaching that, in treating leprosy, "it is necessary to make use of vipers in food, drink, electuaries, ointments (made of the black viper), and grand theriac."[42]

It is remarkable that neither Galen nor the later authors explained the specific efficacy of snake meat for leprosy while, on the contrary, they devoted detailed instructions to its preparation. Galen recommended cooking it à la eel, as we saw. Avicenna added various frills, including that it should be "boiled with scallions, dill, chickpeas, and a little salt" and that the juice might be enriched, "perhaps with some squab, until it is good." One of Bernard de Gordon's preparations was by steeping snakes in "the best wine." Alberto de Bologna suggested that the dish could be made "more delectable" and "exotic" with the addition of partridges. Velasco de Tharanta reported that some people placed the "lightly cooked" snake meat "in a pastry" and added a small amount of aromatic spices, and that this was "the style of some Lombards, as I have heard." Such culinary details may have appealed to taste buds, but they seem marginal to the serious treatment of advancing leprosy. On the other hand, there are suggestions that despair about efficacious remedies underlay the recourse to this wonder drug. Bernard de Gordon encouraged his colleagues to "use these serpents in every way we could possibly consider because, briefly said, we have no other method for leprous patients after cleansing the body."[43]

The first doubts about snake meat seem to have been voiced in the sixteenth century. André Du Laurens distanced himself from "those who try to convince us that they presently have an antidote against *lepra* in the flesh of vipers" while the use of the remedy has "proven the contrary until now." Two centuries later, Uddman more or less closed the saga of this wonder drug. "For repelling *lepra*," he reported, "the ancients applied a concoction of snake juice, which has had little success, however, in our time." Nevertheless, he left open the possibility that the original recipe had merit. "We know that dried genuine vipers were once transported by the shipload from Egypt to Venice: but now the pharmacists have substituted our common *Berum* [for the Near Eastern *colubrum*] in order to have something to give, as long as it is snake."[44]

Antiquity also provided the source for the therapeutic application of blood to patients with leprosy. Pliny reported that the disease began in Egypt, where it was "deadly for the people when it afflicted the kings, for their baths were usually prepared with human blood for treating it." The topic fueled feverish fantasies. Most directly in line with Pliny's report was the story that Emperor Constantine, infected with leprosy, was advised "to have three thousand children slaughtered and to bathe in their warm blood"; however, he did not receive the advice from medical attendants but from pagan priests. In some poems, with equal lack of verisimilitude, physicians proffered similar remedies in their role as healers. A late eleventh-century Latin version of *Ami et Amile* had physicians tell the leprous Ami that he could recover his health by bathing in the blood of the two infant children of his friend Amile; in all the later French versions, however, it was an angel who proposed this remedy to Ami. In the twelfth-century tale *Der arme Heinrich*, which introduced our first chapter, a Salernitan master assured "poor Henry" that he would be cured by "the blood of a virgin's heart." Michael Scot (d. ca. 1235), a translator of several works on natural philosophy who became more renowned as a magician, claimed that a hot bath medicated with the blood of two-year-old toddlers "undoubtedly cures leprosy."[45]

While all these instances obviously belonged to the realm of imagination, we approach the milieu of learned medicine—though by an idiosyncratic path—with Hildegard of Bingen. Abbess Hildegard recommended the addition of "menstrual blood, as much as it is possible to have," to a medicinal mixture for a bath against *lepra* caused by sexual "incontinence." Academic physicians considered human blood primarily for palliative purposes, to which we will turn shortly. They included animal blood, however, in their curative prescriptions. A few, for example, prescribed "the blood of a hare [*leporis*]," curiously combining name association (*lepus-lepra*) and the common belief that parts of swift animals boosted the body's vital faculties. In a more unusual substitution, Hildegard instructed "the man, who contracts *lepra* from anger," to seek out a place "where a little blood from horses runs into the earth when they undergo bloodletting." The man should fold this blood-soaked earth into a poultice and apply the poultice while taking a bath. Hildegard explained that "the warmth of this blood, with the temperateness of the earth, resists the *lepra*, like an enemy vanquishes an enemy."[46]

For the person "who becomes leprous from gluttony and drunkenness," Hildegard prescribed "the droppings of swallows." This prescription evokes a gamut of "strange" substances that ranged from mouse dung and pigeon manure, in the much maligned *Dreckapotheke*, to ground pearls and lapis lazuli, in the exotic apothecary for the rich. Some of these substances were acknowledged at the time as marginal—for example, by Guglielmo da Varignana (d. 1339) at the University of Bologna. In his medical manual, or *practica*, Guglielmo prefaced his compilation of remedies with the admission, "We do not intend to record in this chapter the common treatment of *lepra*, but only those things that the ancients praised." His inventory of uncommon items, while indiscriminate, was fascinating enough to make an excerpt irresistible.

> Rasis said that, if a hoopoe eye is hung on a leprous patient, the *lepra* ceases. He also said that, if the skin that is cut in the circumcision of a child is dried, ground up, mixed in a potion with a little musk, and given to someone who has begun to suffer from *lepra*, the disease stabilizes and does not increase. He further said that, if fresh semen is applied to the black macules that are a sign of *lepra*, it cures. . . . Avicenna said that there is a sort of antidote [*bezoar*] for *lepra*, namely the flesh of a toad.

In several instances, physicians pointed to the origin of toad and frog remedies in folk lore, while accepting their use, at least implicitly. Lemnius wrote in his book *De miraculis occultis naturae,*

I know that in people who are infected with *elephantiasis* or *lepra*, called *melaetsch* in the [Flemish] vernacular, this disease is tamed by the frequent use of swamp frogs, because this little aquatic animal mitigates the heat of the blood and the burning of the black bile.

Jean de Varanda most likely had popular empirics in mind, rather than learned practitioners, when he stated that *"practici* praise the eating of frogs and forest turtles for people with advanced *elephantia."* We may also recall that Godfried Schilling recorded the empirical practice of a South American healer.[47]

One group of physicians stood out as claiming kinship with the *practici* but also supremacy among theoreticians. Paracelsus and his followers, who endeavored to illuminate the occult while extending its range, advertised revolutionary remedies for leprosy. Two succinct and relatively early responses to Paracelsus, one from an admirer and the other from a critic, save us from having to collate his dispersed and disparate propositions. In his *Nosologiae harmonicae dogmata hermatica,* Heinrich Petraeus, doctor of medicine and philosophy and professor ordinary of anatomy and surgery in Marburg, professed his admiration for the "leader of sacred rites" (*mysteriarcha*). He praised "Paracelsus and the other Hermetics" for seeking the most beneficial treatment for *elephantiasis.* Petraeus valued this treatment for

> relying on a great range of indications and on the power of the most efficacious medicaments. For, since this disease is star-related [*astralis*] and humoral [*balsamicus*], it can be cured only by the restoration of the innate heat and vital moisture and the resolution of the radical impurities.

There is some irony in the prominence of two Galenic archprinciples, namely, innate heat and vital moisture, recast as *astrum innatum* and *balsamum vitale,* in the therapeutics of the most outspoken anti-Galenist. Petraeus summarized the doctrine on which Paracelsian therapeutics were based: *"lepra* and its kinds, all desperate chronic diseases, are removed by a philosophical transfer [*per translationem philosophicam*]," which was also called a "regeneration." This doctrine was formulated in the command of Paracelsus "to investigate, with the utmost zeal and diligence," the role of the sun and planets "in the regeneration and transmutation of the leprous metals in the human body, into the gold and silver of health." Samuel Hafenreffer credited Paracelsus with a recipe of "the quintessences of antimony, gold" and other substances, for "radically [*radicitus*] removing all leprous signs."[48]

In direct contrast with his contemporary in Marburg and his younger counterpart in Tübingen, Jean de Varanda in Montpellier was very skeptical. He observed that there was "great controversy" between the Paracelsian and Dogmatist physicians on the treatment of *elephantia*. As he saw it, the Paracelsians "audaciously promise the complete removal and therapy of this affliction, following the example and authority of the master who boasts of having restored twelve *elephanticos* to health in one year." Their claims would make one believe that, "by the action of an occult property and power, some magical antidote [*alexipharmacon*] could expel this ugly disease just as well as guaiacum expels the venereal pox [syphilis]." Varanda unmistakably, albeit cautiously, favored the *Dogmatici*, who, "guided by reason and correct demonstration, contend that complete *elephantiasis* cannot be brought to total cure." After reiterating the arguments for this thesis, he speculated, with a mixture of hope and defeatism,

> Allowing for the possibility that there may be antidotes for *elephantiasis*, they will not always restore health, just as taking *bezoar* antidotes does not always rid one of poison, either because they were administered too tardily or the poison acted on the heart too quickly; nor does guaiacum always cure the venereal pox, unless it is applied at the proper time.

With similarly nuanced logic, Varanda concluded,

> We do not doubt that *elephantici* have occasionally been cured by some remedy. These were not in the confirmed stage, however, or surely they were more affected on the outside. This is how Paracelsus deceived the people, by holding all *elephantiasis* patients for confirmed, and then boasting about their cure.[49]

PALLIATION

Cosmetics

Confirmed leprosy was supposed to fall outside the range of *cura* and to be open only to palliative treatment. There were overlaps, however, between the two phases. Guy de Chauliac identified the fourth objective of the curative regimen as the "correction of symptoms," which more commonly coincided with a general understanding of palliation. In more specific terms, the palliative regimen had three objectives. The triple aim, congruent with three meanings of the word *palliation*, was the cloaking, extenuation, and mitigation of symptoms. Efforts to cloak or disguise disfigurement, while extended to exposed skin and general appearance, were concentrated on the face. This concentration, especially in a part of medicine which was geared primarily to men, compounded "the difficulty of reconciling (or, on the other hand, differentiating) aesthetic and reconstructive surgery." Michael McVaugh draws our attention to this difficulty in a pathbreaking study on rational surgery in the Middle Ages, and he points out that Galen already drew a line between medical and "purely aesthetic treatments." While Galen objected to the "depraved cosmetics" for beautifying a natural appearance, "as women do," he justified the decorative medicine that corrected "impetigo and scabies and leprosy, conditions that are not natural but unnatural." McVaugh also indicates that Henri de Mondeville addressed the tension with a sharpened sensitivity.[50]

According to Henri, wax was "commonly added to ointments for decorating the face of leprous patients and others," but it was not recommended for regular use because "women who anoint their faces with it, wind up with terribly wrinkled faces." Cosmetic concerns, which no doubt intensified with the advance of urbanization, inspired a variety of remedies for patients with advanced leprosy, men as well as women. As an added irony, physicians and surgeons prescribed remedies that suspected patients might employ in attempts to hide their symptoms and cheat the examiners, as we saw in the accusations of Grette Swynnen Thielens of Diedenhofen and Felix Platter's "young man from Colmar." An ingredient, no doubt less "commonly added" to the remedies, was liquid gold. Lemnius claimed that this, mixed with some tartar, "removes all blemishes, no matter how ugly, and beautifies the reddish and warty nose and the chin, cheeks, and forehead (the parts in which such eruptions present themselves embarrassingly, indecorously, and with little respect for civility)."[51]

The most extreme cosmetic prescription—at least to modern sensitivities—called for mixing the patient's own blood in a facial ointment. It was mentioned by several authors, including Theodoric Borgognoni, Bernard de Gordon, and Henri de Mondeville. Velasco de Tharanta questioned this recipe, not on aesthetic grounds, as some might expect, but "because the blood of a leprous patient is corrupt"; therefore, he believed "that it would be better if this ointment were made with the blood of a young and healthy person, or of a hare." Earlier, however, Alberto de Bologna had distanced himself from this alternative by reporting that "*some* use the fresh blood of children as a palliative, especially for the face." In reality, the alleged application of this sort of remedy caused Perrette de Rouen to be convicted in Paris in 1408. A sworn midwife, Perrette had procured a stillborn child whose fat would serve to anoint the face of a leprous lord. After seven weeks in prison, she was sentenced to the pillory and the loss of her license; King Charles VI granted a pardon for the latter part of her severe punishment. Although some seventeenth-century authors, such as Hafenreffer, still mentioned "*quinta essentia* of human blood" (again citing Philippus Ulstadius), only animal blood seems to have retained a place in cosmetics for leprosy. Varanda offered a rationale for the use of animal blood when he prescribed remedies "for the redness and deformity of the face" caused by leprosy: hardened matter would be resolved by "the warm blood of a hare, or of a dove." Bad humors, on the other hand, would be repelled by "virginal milk" (probably the plant *Pulmonaria officinalis*, rather than the "water of mercury" favored by alchemists).[52]

A second aim of palliation was the extenuation or the correction, insofar as possible, of certain severe effects. Some of these, such as hair loss, coincided with cosmetic concerns. For other symptoms, ranging from hoarseness and nasal blockage to ulcerations, the authors referred to the chapters that they devoted to these subjects elsewhere in their manuals. When corrective palliation was detailed in discussions of leprosy, the focus was most often on nodules and "tuberosities." Since baths, ointments, and internal remedies were mild but minimally efficacious, there was a tendency to prefer surgical intervention. Bernard de Gordon pointed out that there were

> many methods of removing growths: some cut a cross into them, others destroy them with corrosives, and others cauterize; this is the best method for lifting them: lift the node with the fingernail, and then cut it off at the base with a razor; do this for each [nodule], if the patient is able to tolerate it.

While the sensitive modern reader will understandably recoil at these methods, we should keep in mind that, even in those "barbaric" times, practitioners evinced a wide range of sensitivities.[53]

Degrees of difference between "aggressive" and "conservative" medicine were reflected in references to the standard procedure of cautery. The procedure, in which skin or tissue was seared with a hot iron (actual cautery) or corrosive agents (potential cautery), aimed either at destroying unwanted growths or edges or at stimulating humoral flow. Cautery was naturally more traumatic when used for the former purpose, whereas for the latter purpose, it was analogous to cupping (and, in a broad sense, to acupuncture). After describing fifteen different methods, without differentiating their respective purposes, Henri de Mondeville suggested, "[I]f these are not sufficient for the patient, let the surgeon turn to Albucasis, who teaches some seventy and says that the more cauteries the better for the patient." Guy de Chauliac, another surgeon, did not share this enthusiasm. He cautioned that cauteries "ought not to be administered except after all the other treatments," and he confided, "[E]ven though Albucasis proposes ten cauteries, these are not my usual procedures." The administration of cautery was contingent not only on the preference but also on the skill of the practitioner and, above all, on the tolerance of the patient. Potential cautery, with caustics, was more complicated and less precise, but it was recommended for patients who were either too delicate or too afraid of the red-hot iron. On the other hand, someone with advanced leprosy might not need to fear actual cautery because, as Henri de Mondeville learned from Albucasis, "the patient, having lost sensation, does not feel fire as much as a healthy person." This, in itself, might lead to mortal danger. Alberto de Bologna, expanding on an observation by Avicenna, warned, "[D]umb physicians [*stolidi medici*] sometimes kill the patient when they cauterize on the head, because with too heavy a hand they reach to the substance of the brain." It is not clear whether this danger was among the reasons that cautery largely disappeared from the treatment of leprosy after the fourteenth century.[54]

Ultimate Medical Treatments: For Life or Death?

The startling allegation that practitioners "sometimes kill the patient," when taken together with other seemingly casual allusions, may raise a sinister suspicion that a leprous patient was occasionally hastened to his or her death—in quite an egregious interpretation of "palliation." Bernard de Gordon expressed alarm at the implicitly mortal risk of one particular treatment. He charged that "people err shamefully, everyday, in the bloodletting of *leprosi*. Many times I have seen that physicians aimed for nothing else than drawing all the blood." From Galen's original reminiscences on viper meat, to Avicenna's interpretation, and then to the commentators' exegeses, an insinuation of intentional killing persisted, while another, of suicide, faded. In Galen's first story, harvesters gave a flask of wine containing a viper "to the *elephantia* patient, in the belief that such a person would rather die than live." In his third story, a sick philosopher, crushed by anxiety and depression, drank toxic wine after asserting "that it was better to die than to be in such a pitiful condition."[55]

While Galen's third story disappeared, the first underwent subtle changes. Avicenna understood that the viper-infested wine was given "by accident or with the intention of killing, so that this leprous patient would die and have rest or be quiet." Alberto de Bologna clarified tersely that the purpose may have been so that the *leprosus* "would no longer circulate, or would be relieved [*quiescat*] of his disease." It is doubtful that we will find evidence of the actual elimination, mercy killing, or suicide of leprosy sufferers, or even clear expressions of such intentions. Moreover, toward the end of the period under consideration, societies were searching for a less direct, albeit equally terminal, substitute, in the removal of patients to "colonies" in isolated places— unlike most of the medieval leprosaria—and then in unseen and virtually inaccessible peninsulas on remote islands (a visit to Kalaupapa or Culion leaves an unforgettable perspective on history). A step in this direction may have been the foundation, in 1631, of a place to receive the sick (*nosodochium*) in Kronoby (Ostrobotnia, Finland): the purpose, according to Isaac Uddman, was "to keep the sufferers of this disease away from the daily commerce of people."[56]

While we can only speculate about harsh measures by society, and the possible role of medicine in these, there is ample evidence that medical practitioners intended to prolong life, to mitigate the deterioration of advanced leprosy, and, at least, not to make the patient suffer unnecessarily. The Hippocratic precept "to do no harm" was unmistakable in the advice of Jordanus de Turre, that strong medicine should be avoided because "it is harmful in confirmed *lepra*." Sometimes, emergencies might call for extreme measures in an attempt to save a life. Jean de Varanda reported that, when the disease had harmed the throat so much that there was "great fear of suffocation, practitioners cut the jugular veins in the neck." While one may question the odds for survival with such procedures, Varanda left no doubt that the paramount objective of palliation was "to prevent confirmed *elephantici* from racing too fast to their deaths." This prevention was achieved by keeping the body moist, resisting putrefaction, and fortifying the heart and liver.[57]

Two dramatically different yet equally concrete statements sum up the range of palliative objectives and, more generally, of medical responses to leprosy. Ambroise Paré concluded his chapter on *ladrerie* with the following thought:

There is no cure for confirmed *lepre*, even though one still gives snakes to drink and eat, bleeds, cups, and bathes the patients, or uses many other remedies. It is true that by these means one can palliate and repel the humor inward, so that the patients are not recognized as such. But I would not recommend this to be done, out of fear that they might abuse the women and have dealings with the healthy. On the other hand, to make them live longer, I always advise them to be castrated, for the reasons stated earlier, and also in order to prevent offspring.

This bluntness, which does not fairly represent Paré's overall merits, is counterbalanced by the moral sensibility of another statement. Felix Platter recounted that "a respectable man" had come to him with clear signs of full-blown *elephantiasis.*

Although I had doubts about the success of any treatment, we undertook one, in part in order to enable the man to keep the disease secret for a time, and in part to spare him from the dreaded expulsion from his home as soon as it would become known, for the patient promised never to leave his own residence. The man lived for another year and a half, and he died of another disease.

Platter's elaborate treatment included cooling potions, gentle laxatives, an "aperitive" electuary, tablets of viper flesh, moisturizing lotions, herbal baths, and even an ointment with mercury (which he belittled in his theoretical discourse).[58]

The subtle ambivalence in both of these statements exemplifies the chiaroscuro of medical—and, indeed, human—responses to leprosy. While Paré's conclusion leaves an impression of callousness, it shows that prolonging the patient's life was a normal concern. While Platter's anecdote epitomizes exemplary care, it raises a suspicion that the patient's "respectable" status may have been a factor. In contrast with the stark finality of Paré's last clause, continued caring echoes in Platter's afterword, that the man died of another disease. His remark breaks through the two-layered and widespread assumption that leprous patients died of leprosy and that, well before death approached, they were deserted by the medical practitioner. A paradoxical contribution of medicine, in this case a combination of one practitioner's mercy and another's error, was poignantly recaptured in a recollection of François Raymond. In 1746 the thirty-two-year-old wife of Etienne Menager had consulted him "on the horrible state in which she had been for many years." After describing her symptoms, Raymond reported with perceptible empathy, "[A]ntimonies and emollients did not have any happy effect: the unfortunate woman died two years later; her death was hastened by the frictions with mercury that a surgeon had given her." No one could more fittingly bring this historical survey to a conclusion than François Raymond, who observed, in his *Histoire de l'Élephantiasis* (1767), that the symptoms of the disease add up to "un cancer universel," a cancer of the whole body.[59]

Conclusion

In conversations about leprosy (hardly an everyday item on the menu), people routinely suggest comparisons with AIDS, although parallels are illusory—even with regard to social stigma—and differences are too vast for fruitful contrasting. A far more logical counterpart, and a suitable foil for recapitulating our examination, is the epidemic that became known as the "Black Death" and which swept across Europe in recurrent waves from 1347 to the eighteenth century. These waves and the presence of leprosy have generated more extensive documentation, in academic writings and archival records, than any other premodern health-related issue. The name and identity of both "leprosy" and "bubonic plague" have been and remain subjects of change and debate. That leprosy has been given a greater number of names reflects a longer history, more complex responses, and a greater diversity of contexts. A major contrast, obvious yet worth emphasizing, is that the "Great Mortality," in the words of Guy de Chauliac, who lived through the first two waves, "besieged almost the entire world," whereas leprosy affected relatively small numbers; the former was a communal catastrophe, the latter an intensely personal experience, as several sources prove poignantly. Nevertheless, we have also discerned collective dimensions, not only in community and faculty responses to patients, but also in events such as the *Schau* and in associations in leprosaria and wandering fraternities.[1]

By definition, leprosy was a chronic disease, a slow disintegration and corruption, while plague brought swift death and caused immediate terror. The fear of contagion by sharing the air with certain patients is put in relief by Guy's remark that the plague "was so contagious that one person caught it from another not only by being close but even by looking." The sequestration of leprous patients, however inhuman, pales in comparison with the panic that, in 1348, "finally reached the point that guards were posted in cities and towns, and these let no one enter unless well known." While medical writers explained both diseases by natural causes, conventional humoral physiology proved useless and intellectually unsatisfactory for the plague, which required swift responses and cosmic explanations. Classifications and correlates assigned leprosy a precise place in nosology, with logical consistency, if not practical benefit; "pestilential fevers" occupied a separate class, with a secondary and vague assignment to the category of apostemes.[2]

The complicated deliberations about the signs of leprosy make the diagnosis by buboes look simple and intuitive. The *iudicium leprosorum*, notwithstanding the doubts and challenges, represented a high point in the medicalization of disease and in the social standing of learned practitioners. In sharp contrast, the plague "was a cause of uselessness and shame for the medical men," according to Guy, because, when they were brave enough to visit the sick, "they did little and gained nothing." Their only recourse was prevention—although Guy paid lip service to "actual treatment" (*cura in casu*). On the other hand, physicians professed the incurability of leprosy, yet they recommended measures for prolonging life and palliating the ultimate effects. Neither leprosy nor plague was conquered by premodern medicines, but both boosted the development of public health policies and private hygiene.[3]

A concluding thought is inspired by an eminent historian, who observes, "More than a disease-entity, the Black Death was for [the university] physicians a social disaster, one which they scarcely characterized from a clinical point of view." The same physicians, on the contrary, viewed leprosy as a somatic disaster, as an illness of the whole body. They characterized it, within the "clinical" framework of the time—and with little recourse to metaphor—as affecting the skin and flesh, face and extremities, heart and liver. They understood the "disease-entity" in the light of traditional humoral physiology, rather than in moral or biblical terms. Even when susceptible to societal anxieties, the medical practitioners cared primarily about the physical and emotional state of their patients—leaving the care of souls to others. In sum, Saul Brody's portrayal of the "disease of the soul" finds a counterpoise in premodern medicine, which treated leprosy as a "malady of the whole body."[4]

Notes

CHAPTER ONE. The Sources

1. The quotation that opens this chapter is from Hartmann von Aue, *Der arme Heinrich*, ed. and trans. (into modern German) Helmut de Boor (Frankfurt: Fischer, 1971), p. 32, vv. 371–377.

Belatedly, in the enchanting Gorges du Tarn, I came across a case that, if explored further, might illustrate the change. When Saint Enimie, Merovingian princess and sister of Dagobert, was stricken with leprosy, she found no help in human medicine, according to the thirteenth-century Provençal troubadour Bertrand de Marseille; in the early hagiography, the saint seems to have pleaded directly with God. I have not seen Bertrand de Marseille, *La vie de sainte Énimie*, trans. (from Romance into modern French) Félix Buffière (La Canourgue, France: Éd. la Confrérie, 2001). There is reason to look forward to Cristina Álvares, *La peau de la Pierre: Études sur la Vie de Sainte Énimie, de Bertrand de Marseille* (Braga, Portugal: CEHUM, 2006).

2. Hartmann von Aue, *Der arme Heinrich*, p. 14, vv. 119–123; "und wart nû als unmaere, / daz in niemand gerne ane sach," ibid., vv. 126–127; ibid., p. 18, vv. 165–178.

3. William of Tyre, *Historia rerum in partibus transmarinis gestarum*, XXI, 1–2, trans. James Brundage, *The Crusades: A Documentary History* (Milwaukee, Wis.: Marquette University Press, 1962), p. 142.

4. Hartmann von Aue, *Der arme Heinrich*, p. 18, vv. 173–187. Gilles de Corbeil, *Egidii Corboliensis Viaticus de signis et symptomatibus aegritudinum*, ed. Valentin Rose (Leipzig: Teubner, 1907), pp. 74, 78. See Enrique Montero Cartelle and Ana Isabel Martín Ferreira, "Le *De elephancia* de Constantin l'Africain et ses rapports avec le *Pantegni*," in Charles Burnett and Danielle Jacquart, eds., *Constantine the African and ʿAlī ibn al-ʿAbbās al-Maǧūsī. The "Pantegni" and Related Texts* (Leiden: E. J. Brill, 1994), pp. 233–246. Gariopontus, *Passionarius*, chap. 5, "De elephantia," in Rome, Bibliotheca Angelica MS 1496, eleventh to twelfth century, fol. 77r–v. I am grateful to Florence Eliza Glaze, who supplied me not only with a citation but also with a transcript of this text. Johannes Platearius, *Practica brevis*, capitulum 1, "De egritudinibus cutis et primo de lepra" (printed with *Practica Joannis Serapionis*, Lyon: Jacobus Myt, 1525), fols. 221v–222r. Johannes de Sancto Paulo, *Breviarium de signis, causis, et curis morborum*, Bethesda, Maryland, National Library of Medicine (hereafter cited as NLM) MS 27.

5. "November 1220, Statuut van het Godshuis der Hoge Zieken te Ieper," *De Leprozerij, genaamd het Godshuis der Hoge Zieken te Ieper. Oorkonden. I. 1176–1300*, ed. O. Mus, in *Bijdragen tot de*

geschiedenis van de liefdadigheisinstellingen te Ieper 2 (1950), pp. 74–75. "Het statuut van het Godshuis der Hoge Zieken," *De Leprozerij, genaamd het Godshuis der Hoge Zieken te Ieper. Oorkonden. II. 1300–1400*, ed. O. Mus, in *Bijdragen tot de geschiedenis van de liefdadigheisinstellingen te Ieper* 3 (1951), pp. 5–10. A persuasive caution against this inference from the "défaut des sources" is sounded by François-Olivier Touati, *Maladie et société au Moyen Âge: La lèpre, les lépreux et les léproseries dans la province ecclésiastique de Sens jusqu'au milieu du XIVe siècle*, Bibliothèque du Moyen Âge 11 (Brussels: De Boeck Université, 1998), pp. 455–457. For evidence, albeit somewhat later, see Griet Maréchal, "Lepra-onderzoek in Vlaanderen (XIVde–XVIde eeuw)," *Annales de la Société belge d'histoire des hôpitaux* 14 (1976), pp. 28–63, esp. pp. 30–31.

6. Malcolm Barber, "Lepers, Jews and Moslems: The Plot to Overthrow Christendom in 1321," *History* 66 (1981), pp. 1–17. Michael R. McVaugh, *Medicine before the Plague: Practitioners and Their Patients in the Crown of Aragon, 1285–1345* (Cambridge: Cambridge University Press, 1993), p. 220.

7. Julius Pagel, ed., *Die Chirurgia des Heinrich von Mondeville (Hermondaville) nach Berliner, Erfurter, und Pariser Codices* (Berlin: Hirschwald, 1892), p. 422.

8. The entire report, "d'une lecture difficile," was first transcribed and published by Ernest Wickersheimer, "Lèpre et Juifs au moyen âge," *Janus* 36 (1932), pp. 43–48.

9. Karl Sudhoff, "Dokumente zur Ausübung der Lepraschau in Frankfurt a./Main im XV. Jahrhundert," *Lepra* (Leipzig) 13, fasc. 3 (1913), pp. 164–165.

10. Ibid., p. 151.

11. Ibid., p. 153.

12. At the sentence ending "that neither we nor anyone else would have noticed anything," Losser added in the margin, "and this way, we might well have avoided a groundless complaint."

13. Sudhoff, "Dokumente zur Ausübung der Lepraschau," pp. 154–155.

14. Ibid., p. 152.

15. For a convenient synopsis, see R. J. Schaefer, "Die Lepraatteste in den Akten der medizinischen Fakultät an der alten Universität Köln," *Lepra* 13, fasc. 4 (1913), pp. 249–252.

16. Sudhoff, "Dokumente zur Ausübung der Lepraschau," p. 156.

17. Ibid., pp. 157–159.

18. Ibid., pp. 161–163.

19. Guenter Risse, *Mending Bodies, Saving Souls: A History of Hospitals* (New York: Oxford University Press, 1999), pp. 167–190. Before his book was published, Professor Risse graciously shared with me not only his references to this case but even his copies of the documents. Ted Gugelyk and Milton Bloombaum, *The Separating Sickness: Ma'i Ho'oka'awale*, 3rd ed. (Honolulu: Separating Sickness Foundation, 1996). I am grateful to John Wilhelm and Sally Squires for keeping me abreast of their progress in preparing a documentary for public television on Carville.

20. Archives are the "normal" repositories of the certificates, even of those recorded in the Deanery Books (*Dekanatsbücher*) of the University of Cologne, which are now in the Cologne Stadtarchiv (under Univ. V 137, 138). A few certificates are scattered across European libraries, in codices such as Erfurt Universitätsbibliothek Amplonianum MS Q 225, fol. 78r; Paris, Bibliothèque Nationale (hereafter cited as Paris BN) MS lat. 5237; Paris BN MS lat. 7138; and Nuremberg Stadtbibliothek MS Cent. V, 42.

21. Sudhoff, "Dokumente zur Ausübung der Lepraschau," p. 158.

22. For the introduction of the concept of a "new Galen," see Luis García-Ballester, "*Artifex factivus sanitatis:* Health and Medical Care in Medieval Latin Galenism," in Jon Arrizabalaga, Montserrat Cabré, Lluís Cifuentes, and Fernando Salmón, eds., *Galen and Galenism: Theory and Medical Practice from Antiquity to the European Renaissance* (Aldershot, England: Ashgate Publishing, 2002), VI, p. 144. García-Ballester's seminal article, originally published in Spanish in 1982, is updated in "The *New Galen:* A Challenge to Latin Galenism in Thirteenth-Century Montpellier," in Arrizabalaga et al., *Galen and Galenism*, V, pp. 55–83. See further below, Chap. 4.

23. Bernard de Gordon, *Tabula de decem ingeniis curandi morbos*, in Basel Universitätsbibliothek MS D.I.ii, fol. 120rb. Bernard's influence is assessed in Luke Demaitre, "Bernard de Gordon et son influence sur la pensée médicale aux XIVᵉ et XVᵉ siècles," in Daniel Le Blévec and Thomas Granier, eds., *L'Université de Médecine de Montpellier et son rayonnement (XIIIᵉ–XVᵉ siècles)*, Actes du colloque international de Montpellier (17–19 mai 2001) (Turnhout, Belgium: Brepols, 2004), pp. 113–114. Bernard's career and writings were explored in Luke Demaitre, *Doctor Bernard de Gordon, Professor and Practitioner* (Toronto: Pontifical Institute of Medieval Studies, 1980).

24. *Guigonis de Caulhiaco (Guy de Chauliac) Inventarium sive Chirurgia magna*, ed. Michael R. McVaugh (Leiden: E. J. Brill, 1997), vol. 1, p. 282 (hereafter cited as Guy de Chauliac, *Chirurgia*).

25. Bernard de Gordon, *Lilium medicine*, 1.21, in Paris BN MS lat. 16189, fol. 19va (references hereafter are to this manuscript); ibid., fol. 19va–b.

26. Ibid., fol. 18ra.

27. Ibid., fol. 20ra.

28. Translated from H. Keussen, "Urkunden und Aktenstücke zur Geschichte der Leprauntersuchung in Köln und Umgegend, 1357–1712," *Lepra* 14, fasc. 2 (1914), p. 111. More context for this letter is given in Chap. 2.

29. See Chap. 2.

30. Keussen, "Urkunden," pp. 111–112.

31. Different tallying methods (in particular, the inclusion of clarificatory sections) yield results that diverge somewhat from the partial tabulation in the seminal article by François-Olivier Touati, "Les traités sur la lèpre des médecins Montpelliérains: Bernard de Gordon, Henri de Mondeville, Arnaud de Villeneuve, Jourdain de Turre et Guy de Chauliac," in Le Blévec and Granier, *L'Université de Médecine de Montpellier*, p. 224.

32. Johann Jakob Huggelin, *Von dem Aussatz durch was Zeichen ein jeder Mensch so mit dem Aussatz behafftet erkant möge werden*.... (Frankfurt: Peter Schmid, 1566), sig. b iᵛ. Jodocus Lommius Buranus, *Medicinalium observationum libri tres. Quibus notae morborum et quae de his possint haberi praesagia, iudiciumque, proponuntur* (Antwerp: Gulielmus Sylvius, 1560), fols. 23r–24r.

33. Theodoric, *Cyrurgia*, printed in *Cyrurgia Guidonis de Cauliaco* (Venice: Gregorius de Gregoriis, 1513), fols. 130r–134v. *The Surgery of Theodoric*, trans. (from the Latin) Eldridge Campbell and James Colton (New York: Appleton-Century-Crofts, 1955), vol. 2, pp. 167–179.

34. Pierre Bocellin, *Practique sur la matiere de la contagieuse et infective maladie de lepre* (Lyon: Masse/Macé Bonhomme, 1540), pp. 2r, 19r. See Ferdinand Brunot, *Histoire de la langue française des origines à 1900*, 2nd ed. (Paris: A. Colin, 1927), tome 2, p. 49.

35. Jean de Varanda, *Tractatus de elephantiasi seu lepra; Item, de lue venerea, et hepatitude, seu hepatis*

ατονια (Montpellier: François Chouet, 1620), pp. 6–9. Henricus Petraeus, "*Diss[ertatio] harmonica XLV. De elephantiasi, seu lepra Arabum;* Respondente Dn. Henrico Nisio Tremonia-Westphalo, 1615," in Petraeus, *Nos[ologiae] harm[onicae] dogm[ata] herm[etica],* vol. 2 (Marburg: Paulus Egenolphus, 1616), pp. 400–416.

36. Samuel Hafenreffer, *Nosodochium, in quo cutis, eique adhaerentium partium, affectus omnes . . . fidelissime traduntur . . . Renovatum & plurimis in locis auctum* (Ulm, Germany: Balthasar Kühnen, 1660).

37. Isaac Uddman, *Lepra, quam, dissertatione medica, venia exper[tae] Facultat[is] Med[icinae] ad Reg[iam] Acad[emiam] Upsal[ensem] Praeside viro generosissimo Carolo von Linne . . . pro gradu doctoris submittit* (Uppsala, 1765), p. 1.

38. Ibid., pp. 2, 9–14. More on definitions below, in Chap. 4. Lisbet Koerner, *Linnaeus: Nature and Nation* (Cambridge, Mass.: Harvard University Press, 1999), p. 233 n. 183.

39. François Raymond, *Histoire de l'Éléphantiasis: Contenant aussi l'origine du scorbut, du Feu St. Antoine, De la Variole, etc.* (Lausanne, 1767), pp. 3, 14.

CHAPTER TWO. *Iudicium leprosorum*

1. Guillaume des Innocens, *Examen des éléphantiques ou lépreux* (Lyon: Thomas Soubron, 1595), p. 3.

2. Guy de Chauliac, *Chirurgia,* p. 285. *Examinandi leprosorum ordo,* in Vatican, Biblioteca Apostolica Vaticana, Palatine (hereafter cited as Vat. Pal.) MS Latin 1225, early fifteenth century, fol. 17r.

3. De La Fons, Baron de Mélicocq, "Documents inédits sur la maladrerie et les lépreux de Lille. Épreuves de ladres," *Revue des Sociétés Savantes des Départements,* 3ᵉ série, 4 (1864), p. 459. Ignaz Schwarz-Wien, "Zur Geschichte der Lepraschau," *Archiv für Geschichte der Medizin* 4 (1911), p. 384. H. Keussen, "Beiträge zur Geschichte der Kölner Lepra-Untersuchungen," *Lepra* 14, fasc. 2 (1914), pp. 80–98. A doctor's "surgery" seems envisioned in the instruction that the examiners "should retire with the subject to a bright, high, well-aired and secure room, away from noise—if possible—and from the traffic of passersby in the house" (des Innocens, *Examen,* p. 98).

4. A. Garosi, "Perizie e periti medico legali in alcuni capitoli di legislazione statutaria medioevale," *Rivista di Storia delle Scienze Mediche e Naturali* 20 (1938), pp. 157–167. Karl Sudhoff, "Lepraschaubriefe aus Italien," *Archiv für Geschichte der Medizin* 5 (1912), pp. 434–435. See below, Chaps. 4, 5, and 7. The secondary role of medical practitioners was still evident at Brugge in 1309, where the three lay examiners of the Magdalen leprosarium were "assisted" by a physician and a surgeon; from the title "medecyn," it is not certain whether the physician was university-educated: see A. Van den Bon, *Het achthonderd jaar oud Sint-Janshospitaal van de stad Brugge* (Brugge: Kommissie van Openbare Onderstand, 1974), p. 81.

5. Marsilius of Padua, *The Defender of Peace: The "Defensor pacis,"* trans. (with an intro.) Alan Gewirth (New York: Columbia University Press, 1956), p. 145.

6. Ibid., pp. 148–149.

7. *Cartulaire de l'Université de Montpellier,* ed. Commission du Cartulaire, with intro. by Alexandre Germain (Montpellier: Ricard, 1890), vol. 1, p. 188. McVaugh, *Medicine before the Plague,* pp. 218–219.

8. See Chap. 7. Pierre Pansier, "Les procès en suspicion de lèpre dans la région d'Avignon aux XIVe et XVe siècles," *La France Médicale* 58 (1911), pp. 281–284, 348–350.

9. *Cartulaire de l'Université de Montpellier*, vol. 1, p. 344. Guy de Chauliac, *Chirurgia*, p. 284.

10. Ernest Wickersheimer, "Les maladies épidémiques ou contagieuses (peste, lèpre, syphilis) et la Faculté de Médecine de Paris, de 1399 à 1511," *Bulletin de la Société française d'Histoire de la médecine* 13 (1914), p. 27. Ernest Wickersheimer, *Dictionnaire biographique des médecins en France au moyen âge* (1936; reprint, Geneva: Librairie Droz, 1979), pp. 723–724, and Danielle Jacquart, *Supplément* (Geneva: Librairie Droz, 1979), p. 264. Pierre Delattre, "Cas de lèpre à Antoing: Mœurs médiévales," *Annales de l'Est* 5 (1909), p. 434.

11. Karl Sudhoff, "Weitere Lepraschaubriefe aus dem 14.–17. Jahrhundert," *Archiv für Geschichte der Medizin* 5 (1912), pp. 154–155. See below, Chap. 4. Keussen, "Urkunden," p. 108; compare below, Chap. 3.

12. O. Schell, "Zur Geschichte des Aussatzes am Niederrhein," *Archiv für Geschichte der Medizin* 3 (1910), p. 342. The *Protokolle* are abstracted and placed in context in Keussen, "Beiträge," pp. 80–112; I am grateful to Guenter Risse for setting me on my way to this veritable gold mine. Keussen, "Beiträge," pp. 92–93, 98.

13. Keussen, "Beiträge," pp. 93–94.

14. Ibid., pp. 94–95. Above, Chap. 1; see also below, Chap. 4. Keussen, "Beiträge," pp. 102–103, 106–107, 112.

15. Sudhoff, "Dokumente zur Ausübung der Lepraschau," pp. 142–148, 169. See above, Chap. 1.

16. Tübingen Universitätsarchiv MS 20/10, "Gutachten der medizinischen Fakultät." I found my way to this valuable source thanks to Mitchell Hammond. A dossier rather than a single codex, 20/10 contains requests to the Faculty of Medicine: Nos. 1–8, 10–11, 13, and following; Nos. 9 and 12; Nos. 40, 43, and 50; Nos. 127 and 133. See below, Plate 4 (Chap. 3).

17. Tübingen Universitätsarchiv MS 20/10, No. 119 (the report was filed at the time with an intriguingly erroneous caption on the cover, "Testimonium obstetricis").

18. Robert Herrlinger, "Die Nürnberger Leprösenschau im 16. Jahrhundert," *Ärztliche Praxis* 3, no. 13 (March 1951), p. 16.

19. Varanda, *Tractatus de elephantiasi*, p. 25. Van den Bon, *Het achthonderd jaar oud Sint-Janshospitaal*, p. 81.

20. Herrlinger, "Die Nürnberger Leprösenschau," p. 17.

21. Nuremberg Stadtbibliothek MS Cent. V, 42, pp. 153v–155v.

22. Ibid., pp. 149v–150v, 151r–v, 152r–153r, 156r–v, 140v–142r. Palma also compiled, from a wide range of authors, a Latin tract on diagnosis and judgment, *Collectanea de Lepra*, which is extant only in this same manuscript, pp. 75r–85r.

23. Guenter Risse generously provided me with a photocopy, a transcript, and a modern German translation of this dossier (my English translation is based on the original text). The documents are abstracted in Keussen, "Beiträge," pp. 109–110. Grette's story is thoughtfully retold and placed in context by Risse, *Mending Bodies*, pp. 167–172.

24. Guy de Chauliac, *Chirurgia*, p. 285. Keussen, "Beiträge," p. 110.

25. Petrus Forestus, *Observationum et curationum chirurgicarum libri quinque* (Frankfurt: Zacharias Palthenius, 1610), p. 124.

26. Robert Jütte, "Lepra-Simulanten," in Martin Dinges and Thomas Schlich, eds., *Neue*

Wege in der Seuchengeschichte, Medizin, Gesellschaft und Geschichte, Beiheft 6 (Stuttgart: Franz Steiner Verlag, 1995), pp. 25–42, esp. p. 34.

27. Ambroise Paré, *On Monsters and Marvels,* trans. (with intro. and notes) Janis L. Pallister (Chicago: University of Chicago Press, 1982), pp. 76–77.

28. Maréchal, "Lepra-onderzoek in Vlaanderen," p. 51. A. F. C. Van Schevensteen, *La Lèpre dans le Marquisat d'Anvers aux temps passés,* extrait du *Recueil des Mémoires couronnés et autres mémoires de l'Académie royale de Médecine de Belgique,* tome 24 (Brussels: L'Avenir, 1930), p. 29.

29. Van den Bon, *Het achthonderd jaar oud Sint-Janshospitaal,* p. 81. There was a "gilde der akkerzieken" in the region outside Brugge which was independent of the Brugge municipal authority (the "Brugse Vrije"): Maréchal, "Lepra-onderzoek in Vlaanderen," p. 42.

30. Sudhoff, "Dokumente zur Ausübung der Lepraschau," p. 157. Ambroise Paré, *Les oeuures d'Ambroise Paré . . . : Diuisees en vingt huict liures, auec les figures & portraicts, tant de l'anatomie que des instruments de chirurgie, & de plusieurs monstres,* reueuës & augmentees par l'autheur, 4th ed. (Paris: Gabriel Buon, 1585), p. 757. Des Innocens, *Examen,* p. 19. Varanda, *Tractatus de elephantiasi,* p. 7. Schell, "Zur Geschichte des Aussatzes," pp. 343–344.

31. Jütte, "Lepra-Simulanten," p. 34. Henri Stein, "Quatre lépreux faussaires en 1627," *Annales de la Société historique et archéologique du Gâtinais* 15 (1897), pp. 186–187.

32. Nuremberg Stadtbibliothek MS Cent. V, 42, p. 154r. Karl Sudhoff, "Die Clever Leprosenordnung vom Jahre 1560," *Archiv für Geschichte der Medizin* 4 (1911), pp. 386–388.

33. Philippus Schopff, *Kurzer aber doch aussfürlicher Bericht von dem Aussatz . . .* (Strasbourg: B. Jobin, 1582), sig. A2. Friedrich Bühler, *Der Aussatz in der Schweiz: Medicinisch-historische Studien,* Inaugural-Dissertation, medizinische Fakultät Bern (Zurich: Polygrapisches Institut, 1902), p. 64.

34. O.C.M.W.-Archief, Fonds Magdalena Leprozerie, Brugge, reg. nr. 4, fol. 4v.

35. Ibid., lias 80, stuk 14; additional references in Maréchal, "Lepra-onderzoek in Vlaanderen," p. 48 n. 98. W. L. Braekman, "Tekenen en behandeling van melaatsheid: Een Gents traktaatje uit de vijftiende eeuw," *Volkskunde* 81 (1980), pp. 269–279; comment in Ria Jansen-Sieben, "Ziektebeeld en behandeling in de Middelnederlanse medische literatuur," in Marleen Forrier, Walter De Keyzer, and Michel Van der Eycken, eds., *Lepra in de Nederlanden (12de–18de eeuw)* (Brussels: Algemeen Rijksarchief, 1989), pp. 37–38. On the connection between *pakerie* or *pakers,* the French *pouacre,* and the Latin *podagra* (gout), see Jansen-Sieben, "De Middelnederlandse terminologie," in Forrier, De Keyzer, and Van der Eycken, *Lepra in de Nederlanden,* p. 31; the term is thought to refer to scabies. I have been unable to find persuasive clues to the meaning of "green" leprosy. A. Dewitte, *De Geneeskunde te Brugge in de Middeleeuwen* (Brugge: Heemkundige Kring M. van Coppenolle, 1973), pp. 37–39. Maréchal, "Lepra-onderzoek in Vlaanderen," p. 31.

36. Piers D. Mitchell, "Pre-Columbian Treponemal Disease from 14th Century AD Safed, Israel, and Implications for the Medieval Eastern Mediterranean," *American Journal of Physical Anthropology* 121 (2003), p. 123. I have benefited in various ways from Dr. Mitchell's wide-ranging insights and from his personal advice to "keep an open mind."

37. Van Schevensteen, *La Lèpre dans le Marquisat d'Anvers,* pp. 79–80. Maréchal, "Lepra-onderzoek in Vlaanderen," p. 54.

38. Des Innocens, *Examen,* p. 132. Hafenreffer, *Nosodochium,* p. 116.

39. Archives départmentales de l'Yonne, H 2398 (second half of liasse), pp. 1–6 (emphasis added).

40. See above, Chap. 1. Nuremberg Stadtbibliothek MS Cent. V, 42, p. 155r.

41. Fred Bergman, "Hoping against Hope? A Marital Dispute about the Medical Treatment of Leprosy in the Fifteenth-Century Hanseatic Town of Kampen," in Hillary Marland and Margaret Pelling, eds., *The Task of Healing: Medicine, Religion and Gender in England and the Netherlands, 1450–1800* (Rotterdam: Erasmus, 1996), p. 31. Sudhoff, "Dokumente zur Ausübung der Lepraschau," p. 156. Garosi, "Perizie e periti," p. 162 n. 3.

42. Albert Bourgeois, *Lépreux et maladreries du Pas-de-Calais (X^e–XVIIIe siècles)*, Mémoires de la Commission Départementale des Monuments Historiques du Pas-de-Calais, tome 14.2 (Arras, 1972), p. 216.

43. Ibid., p. 218. The text is edited in Bruno Tabuteau, "Une léproserie normande au Moyen Age: Le prieuré de Saint-Nicolas d'Evreux du XIIe au XVIe siècle. Histoire et corpus des sources, thèse de doctorat d'histoire inédite," Université de Rouen, 1996, tome 4, n° 230. I am grateful to Bruno Tabuteau for kindly sharing this interesting document. In 1492, the city of Manosque set the honoraria of a physician and surgeon at 2½ florins if the subject was declared healthy, and 5 florins if he or she was found infected: Michel Hébert, "Un diagnostic de lèpre aux Baux-de-Provence à la fin du XVe siècle," *Provence historique* 45, fasc. 183 (January–February 1996), p. 132 n. 3, pp. 135–136.

44. Archives municipales, Nevers, CC 120, 1545–1546, pp. 11v–12.

45. Keussen, "Beiträge," p. 88 (1513) and passim. Ibid., pp. 110–111.

46. Pansier, "Les procès en suspicion de lèpre," p. 349. Sudhoff, "Dokumente zur Ausübung der Lepraschau," p. 159.

47. Foreest, *Observationum et curationum chirurgicarum libri quinque*, pp. 124–125. Andreas Vesalius, preface to *On the Fabric of the Human Body. Book I. The Bones and Cartilages*, translation of *De humani Corporis Fabrica Libri Septem* by William Frank Richardson in collaboration with John Burd Carman (San Francisco: Norman, 1998), p. xlviii.

48. Vesalius, preface to *On the Fabric of the Human Body*, p. xlvii. J. Pourrière, *Les hôpitaux d'Aix-en-Provence au Moyen Age, XIIIe, XIVe et XVe siècles* (Aix-en-Provence: Paul Roubaud, 1969), p. 141 nn. 5, 6.

49. Felix Platter, *Observationes*, quoted (in German) in Bühler, *Der Aussatz in der Schweiz*, pp. 64–66.

50. Des Innocens, *Examen*, pp. 96, 3. Most of the judicial terms in this paragraph are found concentrated in a template that was drawn up, in the 1440s, by the dean of the medical faculty of Vienna and the other "public physicians of the lords dukes of Austria": see Karl Sudhoff, "Vier Schemata für Lepraschau-Atteste der Wiener medizinische Fakultät," *Archiv für Geschichte der Medizin* 6 (1912–1913), p. 392. Keussen, "Beiträge," p. 91 (1527).

51. Varanda, *Tractatus de elephantiasi*, p. 28. Paré, *Les oeuures*, pp. 1203–1204. The template is extant only in this edition of the French version of Paré's works; Paré, *Les oeuures*, p. 1194.

52. Sudhoff, "Vier Schemata," pp. 392–393. Guy de Chauliac, *Chirurgia*, pp. 285–286. *The Cyrurgie of Guy de Chauliac*, ed. Margaret S. Ogden (London: published for the Early English Text Society by the Oxford University Press, 1971), p. 383. Bocellin, *Practique*, fol. 16r.

53. Bocellin, *Practique*, fol. 16r. Varanda, *Tractatus de elephantiasi*, p. 31.

54. André Du Laurens, *Operum . . . tomus alter, continens scripta therapeutica X, nimirum tractatum*

De crisibus . . . De senectute, De morbo articulari, De lepra, De lue venerea . . . (Frankfurt: Gulielmus Fitzerus, 1628), p. 39.

CHAPTER THREE. The Many Labels of Leprosy

1. Wickersheimer, "Les maladies épidémiques," p. 27 (emphases added). Compare Du Laurens, above, Chap. 2.

2. Thomas de Saint-Pierre, deacon of the diocese of Coutances, Master of Medicine, Paris 1371; Jean le Lièvre, cleric of the diocese of Autun, Master of Medicine regens, Paris, 1392; Robert de Saint-Germain, priest of the diocese of Rouen, Master of Medicine, Paris 1396. Entries in Wickersheimer, *Dictionnaire biographique.*

3. See above, Chap. 2. Auxerre, Archives départementales de l'Yonne, G 1633, item 3.

4. "The Long Road," in Gugelyk and Bloombaum, *Separating Sickness*, pp. 91–92. Latin text in the First Book of Letters of the University of Cologne, in Paris BN MS lat. 5237, fol. 261v; transcribed by Keussen, "Urkunden," p. 108.

5. Paris BN MS lat. 7138, fol. 4, published by Ernest Wickersheimer, "Beiträge zur Geschichte des Aussatzes in Frankreich und in den benachbarten Ländern," *Archiv für Geschichte der Medizin* 5 (1912), p. 146. Karl Sudhoff, "Lepraschaubriefe aus dem 15. Jahrhundert," *Archiv für Geschichte der Medizin* 4 (1911), p. 371. Sudhoff, "Dokumente zur Ausübung der Lepraschau," pp. 156, 161.

6. Sudhoff, "Dokumente zur Ausübung der Lepraschau," p. 160. Keussen, "Beiträge," p. 94. Sudhoff, "Dokumente zur Ausübung der Lepraschau," p. 147. Archives Départmentales de l'Hérault, II E 35/42 (acte 1, daté du 14 juin 1570), transcribed by Jean-Claude Toureille (jctou@arisitum.org). I am grateful to Bruno Tabuteau for bringing this document to my attention.

7. Schwarz-Wien, "Zur Geschichte der Lepraschau," p. 383. Enric Moliné, "Un diagnòstic mèdic del 1372 a La Seu," *Església d'Urgell* 183 (July–August 1989), p. 6. Michael McVaugh kindly sent me a copy of this brief and obscure but very valuable article. Sudhoff, "Dokumente zur Ausübung der Lepraschau," pp. 170, 147, 163, and passim. Piera Borradori, *Mourir au Monde: Les lépreux dans le Pays de Vaud (XIIIᵉ–XVIIᵉ siècle)*, Cahiers lausannois d'Histoire médiévale, 7 (Lausanne: Université de Lausanne, 1992), p. 188. Keussen, "Urkunden," p. 110. Transcribed by Jean-Claude Tourelle, "Un lépreux à Ganges en 1570," http://hypo.ge-dip .etat-ge.ch/www/cliotexte/sites/Arisitum/cdf/lepre.html. I thank Bruno Tabuteau for alerting me to this site. Schwarz-Wien, "Zur Geschichte der Lepraschau," p. 384.

8. Des Innocens, *Examen*, p. 96.

9. Touati, *Maladie et société*, pp. 381–388. Guy de Chauliac, *Chirurgia*, p. 285.

10. Touati, *Maladie et société*, p. 382 n. 15. Des Innocens, *Examen*, p. 1. O. Mus, ed., *De Leprozerij, genaamd het Godshuis der Hoge Zieken te Ieper. Oorkonden. IV. 1546–1581*, in *Bijdragen tot de geschiedenis van de liefdadigheidsinstellingen te Ieper* 5-A (1953), 30 (1559) and p. 38 (1560) (hereafter cited as Mus, *Oorkonden IV*). Sudhoff, "Die Clever Leprosenordnung" p. 387. Varanda, *Tractatus de elephantiasi*, pp. 6–7.

11. Friedrich Kluge, *Etymologisches Wörterbuch der deutschen Sprache*, 22nd ed. (Berlin: de Gruyter, 1989), s.v. "Miselsucht." The first documented use of the word *misellus* in France was in Melun, in 1165, according to Jan Frederik Niermeyer, *Mediae Latinitatis Lexicon minus*, vol. 1

(London: Brill, 1976). Tony Hunt, ed., *Anglo-Norman Medicine*, vol. 1, *Roger Frugard's "Chirurgia"; the "Practica brevis" of Platearius* (Cambridge, England: D. S. Brewer, 1994), p. 248. Danielle Jacquart and Claude Thomasset, *Lexique de la langue scientifique (Astrologie, Mathématiques, Médecine . . .): Matériaux pour le Dictionnaire du Moyen Français* (Paris: Institut National de la Langue Française, 1997), p. 182.

12. There is, however, a very similar name in Hawaii, *Maʻi Hoʻokaʻawale,* or "the separating sickness": see Gugelyk and Bloombaum, *Separating Sickness.* Petraeus, *Diss[ertatio] harmonica XLV,* p. 401. Leviticus 13:3; also, "habitabit extra castra," Leviticus 13:46, which was interpreted as "he shall live outside the town," although it was translated in the King James Bible as "without the camp his habitation shall be." Sudhoff, "Dokumente zur Ausübung der Lepraschau," p. 158. For more on Losser, see above, Chap. 1.

13. Sudhoff, "Dokumente zur Ausübung der Lepraschau," p. 164 (emphasis added); also pp. 151, 152, 162. Keussen, "Urkunden," p. 109 (emphasis added). Sudhoff, "Dokumente zur Ausübung der Lepraschau," p. 147 (February 3, 1439). Brugge Stadsarchief, Klerken van de Vierschaar, quoted by A. Dewitte, "Te Magdalenen, 1542–1545," *Biekorf, Westvlaams Archief voor geschiedenis, archeologie, taal-en volkskunde* 103 (2003), p. 305. Huggelin, *Von dem Aussatz,* sig a iiii.

14. See, for example, Jansen-Sieben, "De Middelnederlandse terminologie," p. 27. "In spring occur melancholia, madness, epilepsy, bloody flux, angina, colds, sore throats, coughs, skin eruptions and diseases [λεπραι], eruptions turning generally to ulcers, tumours and affections of the joints," *Aphorisms,* 3.20, in *Hippocrates, with an English Translation by W. H. S. Jones,* vol. 4 (Cambridge, Mass.: Harvard University Press, 1979), pp. 128–129. The translator notes, "It is not possible to translate the Greek terms for the various skin diseases, as the modern classification is so different from the ancient. We may be sure, however, that λεπρα included many diseases besides leprosy" (p. 129). Mirko D. Grmek, *Diseases in the Ancient Greek World,* trans. Mireille Muellner and Leonard Muellner (Baltimore: Johns Hopkins University Press, 1989), p. 157; see the entire chap. 6, pp. 152–176. Grmek refers to the Aristotelian Theophrastus.

15. Leviticus 13:2 and 3, 11 and 15, 29, and 13.12 and 13, respectively; further, passim. Petraeus, *Diss[ertatio] harmonica XLV,* p. 406.

16. Bocellin, *Practique,* p. [3v]. Mus, *Oorkonden* IV, p. 57; note: Waelraem had been examined, found contagious, and recommended for sequestration on May 14, 1570: Mus, *Oorkonden* IV, p. 51. The citation is Leviticus 13:29–31.

17. Aulus Cornelius Celsus, *De medicina,* Liber Tertius, Caput 25, 1 (Intratext Edition CT, copyright *Èulogos* 2001). I have based much of this paragraph on the unsurpassed synopsis in Grmek, *Diseases,* pp. 168–171.

18. Galen, *De methodo medendi ad Glauconem,* Kühn XI.142. For a useful perspective, see Robert James Hankinson, "Galen on the Use and Abuse of Language," in Stephen Everson, ed., *Language* (Cambridge: Cambridge University Press, 1994), pp. 166–187. *Galeno ascripta Introductio seu Medicus,* Kühn XIV.757.

19. Galen, *Definitiones medicae,* Kühn XIX.427. Galen, *Commentarium III in Hippocratis de humoribus,* Kühn XVI.442. Galen, *De tumoribus praeter naturam,* Kühn VII.727; also, in the Latin translation by Niccolò da Reggio (ca. 1330), in *Galieni Opera* (Venice, 1490), vol. 1, fol. 155vb. Galen, *De accidenti et morbo liber VI,* in *Galieni Opera,* vol. 2, fol. 149va.

20. "Ad elefantiosos medicamen," "Ad elefantiasim," and similar phrases, Ulrich Stoll, *Das*

"Lorscher Arzneibuch": Ein medizinisches Kompendium des 8. Jahrhunderts (Codex Bambergensis Medicinalis 1); text, Übersetzung und Fachglossar [*Sudhoffs Archiv. Beihefte.* Heft 28] (Stuttgart: Franz Steiner Verlag, 1992), pp. 80–83, 114, 128–130. "Curatio elephantiosorum," NLM MS 8, fol. 179. "Ad maculas corporales et ad lepram," Stoll, *Das "Lorscher Arzneibuch,"* pp. 84, 202. "V. De elephantia. Elephantiosos apprehendimus," Gariopontus, *Passionarius,* fol. 77r–v. I am indebted to Eliza Glaze, who furnished me with a transcription of the chapter from this manuscript.

21. Ana Isabel Martín Ferreira, ed., *Tratado médico de Constantino el Africano: Constantini Liber de Elephancia* (Valladololid: Universidad de Valladolid, 1996), esp. pp. 23–28 on the questions of authorship. Variant titles, ibid., p. 74. Constantinus Africanus, *Viaticum,* in *Opera parva Abubetri* (Lyon, 1510), fol. [98]v. Bartholomeus Salernitanus, *Practica,* in Salvatore de Renzi, ed., *Collectio Salernitana,* vol. 4 (Naples, 1856), p. 362. Johannes Platearius, *Practica brevis,* fol. 221v. Gilles de Corbeil, *Viaticus,* pp. 74–78. In the eighth century, for example, "Elefantiosum ita apprehendimus: primum ex maculis, que initio inpediginibus similes nascutur," Stoll, *Das "Lorscher Arzneibuch,"* p. 128. Gariopontus, *Passionarius,* fol. 77r. Johannes de Sancto Paulo, *Breviarium,* sig. aᵛ.

22. The term *al-baras,* while rather indistinctive, was routinely applied to leprosy in Arabic. Among "infectiones cutis maculose," the white ones "are called *albaras,*" according to Guy de Chauliac: *Chirurgia,* p. 290.

23. Paulus Iuliarius, *De lepra et eius curatione* (Verona: Antonius Putelletus, 1545), pp. 1v–2r.

24. Girolamo Fracastoro, *De contagione et contagiosis morbis,* ed. and trans. Wilmer Cave Wright (New York: Putnam, 1930), p. 159. Mus, *Oorkonden IV,* p. 11. Jansen-Sieben, "De Middelnederlandse terminologie," p. 30. Lommius, *Medicinalium observationum libri tres,* pp. 23–24. Forestus, *Observationum et curationum chirurgicarum libri quinque,* pp. 350, 391–397.

25. Du Laurens, *Operum,* pp. 36–37. Varanda, *Tractatus de elephantia,* p. 6. Raymond, *Histoire de l'Élephantiasis,* p. 6.

26. Van Schevensteen, *La Lèpre dans le Marquisat d'Anvers,* p. 63.

27. Giovanni Boccaccio, *Decameron,* I.10, trans. G. H. McWilliam (Hammondsworth, England: Penguin Books, 1973), pp. 108–110. Albertus de Zanchariis, *Glose super tractatum Avicenne de cura lepre,* in Paris BN MS lat. 7148, fol. 28r. Galen, *De ingenio sanitatis,* 2.2, in *Galieni Opera,* vol. 2, fol. [171]ra.

28. See Ernst Robert Curtius, *European Literature and the Latin Middle Ages,* trans. (from the German) Willard R. Trask (New York: Harper and Row, 1953), pp. 43–44. For Isidore's etymology of "the disease elefantiacus," *Etymologie,* IV.8, para. 12, trans. William Sharpe, reproduced in Ed Grant, ed., *A Source Book in Medieval Science* (Cambridge, Mass.: Harvard University Press, 1974), p. 704. Bocellin, *Practique,* fols. 2r–3ra. Suffice it to cite a few of the "ancient" etymologies available to Bocellin. "[I]nterpretatur lepra quasi lesio prava," Albertus de Zanchariis, *Glose,* fol. 24r. "Interpretatur quasi lesio petrosa," in *De lepra et primo de interpretatione,* in Basel Universitätsbibliothek MS D.III.10, fol. 6r. "[V]idetur dici a leporia greco quod est erunna," Simon de Cordo Januensis (of Genoa, fl. 1285–1295), *Clavis sanationis* (Venice, 1507), fol. [XLVra], adding the information, "in psalterio greco ubi est leporia in latino est erumna"; the loci in the Vulgate are Ecclesiastes 5:16 and 2 Corinthians 11:27. Guy de Chauliac, *Chirurgia,* p. 283. In allocating the sign to the "tip" of the nose I am following Juhani Norri, *Names of Sicknesses in English, 1400–1550: An Exploration of the Lexical Field* (Helsinki: Suomalainen

Tiedeakatemia, 1992), p. 238; however, a translation as "the bridge of the nose" seems equally plausible if one considers the Middle English elucidation, "i. of þe coppe of þe nose"; moreover, since I am unable to find a Latin term *lepus* or *lepor* with the supposed meaning of "tip" or "head," I suspect that Guy, while appearing to reconnect etymology and observation, may at the same time have adopted a scribal error, namely, the substitution of "a lepore nasi" for an original statement that leprosy could be recognized by the patient's "hare nose," "a leporis naso." Peter Richards, *The Medieval Leper and His Northern Heirs* (Cambridge, England: D. S. Brewer, 1977), p. 118, also pp. 116–117 and illustrations.

29. Aretaeus, *Aretaei Cappadocis medici libri VIII*, Iunio Paulo Crasso Patauino interprete (Paris: Guilielmus Morelius et Iacobus Putaneus, 1554), pp. 203–207. Du Laurens, *Operum*, p. 36. Gilles de Corbeil, *Viaticus*, p. 77.

30. Aretaeus, *Aretaei Cappadocis*, p. 207. Grmek, *Diseases*, p. 168. Rolandus [Parmensis], *Methodus medendi certa, clara et brevis* (Basel: Henricus Petrus, 1541), fol. 303r; same in the Anglo-Norman translation of Roger Frugardi, *Chirurgia*, in Hunt, *Anglo-Norman Medicine*, p. 86. *Galeno ascripta Introductio seu Medicus*, Kühn XIV, p. 757. Aretaeus, *Aretaei Cappadocis*, p. 211. Avicenna, *Canon medicinae*, Latin translation by Gerard of Cremona (Venice: Petrus Maufer, 1486), sig. FF 7ra. Gilles de Corbeil, *Viaticus*, p. 76. "Vocatur leoninia quia leo est animal furiosum, sic lepra leonina est furiosa id est velocissima," in *De lepra et primo de interpretatione*, fol. 6v. "Dicitur leonina et est incurabilis quia cito occidit," Johannes de Sancto Paulo, *Breviarium*, sig. aᵛ. Aretaeus, *Aretaei Cappadocis*, p. 212.

31. *De generatione animalium*, 768b, trans. A. Platt, in *The Complete Works of Aristotle: The Revised Oxford Translation*, ed. Jonathan Barnes, vol. 1, Bollingen Series LXXI.2 (Princeton: Princeton University Press, 1984), p. 1190. For a different translation and interpretation, see Aristotle, *Generation of Animals*, trans. A. L. Peck, in the Loeb Classical Library, no. 366 (Cambridge, Mass.: Harvard University Press, 1963), p. 413 and note a. See also Grmek, *Diseases*, p. 400 n. 42. Aretaeus, *Aretaei Cappadocis*, p. 207. Rufus of Ephesus, trans. Grmek, *Diseases*, p. 168.

32. Galen, *De accidenti et morbo* II, cap. 7, in *Galieni Opera*, vol. 2, fol. 141ra. Compare *De morborum causis*, Kühn VII, 33. *De tumoribus*, in *Galieni Opera*, vol. 1, fol. 155vb. Compare Kühn VII.728. Guy de Chauliac, *Chirurgia*, p. 284. Velascus de Tharanta, *Practica quae Philonium dicitur* (Lyon: Matthias Huss, 1490), fol. 256rb. Huggelin, *Von dem Aussatz*, sig a iiii. Varanda, *Tractatus de elephantiasi*, pp. 5–6. Du Laurens, *Operum*, p. 37.

33. Saul Nathaniel Brody, *The Disease of the Soul: Leprosy in Medieval Literature* (Ithaca, N.Y.: Cornell University Press, 1974), pp. 51–53 and passim. Varanda, *Tractatus de elephantiasi*, p. 8.

34. See above, Chap. 1.

35. Leviticus 13:6, 11, 15, 17, 23. The terms *immundus* and *mundus* alternate constantly, for example, throughout certificates issued by the medical faculty of Cologne, Keussen, "Beiträge," esp. pp. 92–98. Hildegard of Bingen, *Causae et curae*, ed. Paulus Kaiser (Leipzig: Teubner, 1903), p. 161. Karl Sudhoff, "Wurzacher Lepraschaubriefe aus den Jahren 1674–1807," *Archiv für Geschichte der Medizin* 5 (1912), p. 428; see also p. 429; Sudhoff, "Dokumente zur Ausübung der Lepraschau," p. 159. Keussen, "Beiträge," p. 92. Galen, *Commentarium III in Hippocratis de humoribus*, Kühn XVI.442.

36. For an excellent introduction to the early development of the notion, see Jacques Jouanna, "Air, miasme et contagion à l'époque d'Hippocrate et survivance des miasmes dans la médecine posthippocratique (Rufus d'Éphèse, Galien et Palladios)," in Sylvie Bazin-

Tachella, Danielle Quéruel, and Évelyne Samama, eds., *Air, miasmes et contagion: Les épidémies dans l'Antiquité et au Moyen Age*, Hommes et Textes en Champagne (Langres, France: Dominique Guéniot, 2001), pp. 9–28. Leviticus 13:44–46.

37. Gariopontus, *Passionarius*, fol. 77r. Sudhoff, "Lepraschaubriefe aus dem 15. Jahrhundert," pp. 371–372. Borradori, *Mourir au Monde*, pp. 188–189 (emphasis added). Sudhoff, "Lepraschaubriefe aus Italien," pp. 434–435.

38. Moliné, "Un diagnòstic mèdic," p. 6. Mus, Oorkonden IV, p. 31. Sudhoff, "Lepraschaubriefe aus dem 15. Jahrhundert," pp. 373–374.

39. Aretaeus, *Aretaei Cappadocis*, p. 207. Avicenna, *Canon medicinae*, sig. FF 7rb. "Lepra est vilis membrorum corrupcio," in *De lepra et eius curis*, MS Sloane 282, fol. 129r, at the British Library. "Lepra est morbus turpis," Pagel, *Die Chirurgia des Heinrich von Mondeville*, p. 422. Keussen, "Urkunden," pp. 81, 82, 84, 94. Stoll, *Das "Lorscher Arzneibuch,"* p. 128. Petrarch, *De remediis utriusque fortunae*, liber II, Dialogus CXIII, in *Opera omnia* (Basel, 1554; facsimile ed., 1965), p. 228. Adam Smith, *The Wealth of Nations*, bk. 5, chap. 1, pt. 3, art. 1: Online at http://socserv2.socsci.mcmaster.ca/~econ/ugcm/3113/smith/wealth/wealbk05.

40. Aretaeus, *Aretaei Cappadocis*, pp. 209, 212. Mus, *Oorkonden* IV, p. 27. Galen, *De tumoribus*, in *Galieni Opera*, vol. 1, fol. 155vb. Bocellin, *Practique*, p. [3r–v]. Iuliarius, *De lepra*, p. 3r. Raymond, *Histoire de l'Élephantiasis*, pp. 6–7.

41. Aretaeus, *Aretaei Cappadocis*, p. 207. Gilles de Corbeil, *Viaticus*, p. 78. Avicenna, *Canon medicinae*, sig. FF 7ra. Lommius, *Medicinalium observationum libri tres*, p. 23v. *Galeno ascripta Introductio seu Medicus*, Kühn XIV.757. Bocellin, *Practique*, p. 3r. Velascus de Tharanta, *Practica*, fol. 256ra. Keussen, "Urkunden," pp. 81, 82, 84, 87, 89, 94.

42. Avicenna, *Canon medicinae*, sig. FF 6vb. See Richard J. Durling, *Galenus Latinus* II, vol. B, e.g., φαυλός. "Malignior omnibus," Gariopontus, *Passionarius*, fol. 77; "molesta," Albertus de Zanchariis, *Glose*, fol. 24v; "lesio prava," Bocellin, *Practique*, p. 3r; "nullus est tetrior," Du Laurens, *Operum*, p. 36. "Sed tu dices, Cum omnis infirmitas sit mala cum egrum esse sit male vivere sicut sanum bene vivere, eciam cum causa finalis omnis egritudinis sit lesio operacionum egritudinem sequens sicut umbra corporis ut Galienus 3 de accidenti et morbo, ad quid Avicenna dixit quod lepra est mala infirmitas," Albertus de Zanchariis, *Glose*, fol. 24v.

43. "Dicitur mala . . . ratione sue essencie," Albertus de Zanchariis, *Glose*, fol. 24v. "Est autem mala quia omne genus egritudinum amplectatur ut malam complexionem malam composicionem et solutam continuitatem," Bartolomeo da Montagnana, *De lepra et aliis cutis defedationibus*, in Erfurt Amplonianum MS 269, fol. 1va. Also Theodoric, *Cyrurgia*, fol. 130r; Bernard de Gordon, *Lilium*, fol. 17vb; Niccolò Bertucci, *Nusquam antea impressum collectorium totius fere medicine* (Lyon: Claudius Davost, 1509), fol. ccxliiij^va. Bocellin, *Practique*, p. [3v]. "Racione sue essencie que . . . alterans membra ad malam complexionem et presertim contrariam vite scilicet ad frigidam et siccam," Albertus de Zanchariis, *Glose*, fol. 24v. Also, Theodoric, *Cyrurgia*, fol. 130r. "Quoniam generatur ex omnibus causis ex quibus generantur omnes morbi sicut ex parentela, et est morbus hereditarius. Item generatur in matrice menstruis immundis repleta, licet semina sint munda. Item ex corrupto regimine matricis. Item potest generari ex coitu leprosi cum pregnante, quia inficitur fetus, licet non inficiatur matrix. Item propter malum regimen, sicut per nimium usum carnis vaccine buballine lentium et omnium leguminum . . . Et est mala malicia materie que venenositate non caret," Montagnana, *De lepra*, fol. 1va. Bocellin, *Practique*, p. [3v]. Du Laurens, *Operum*, p. 38. Jean Fernel,

Pathologiae libri VII, in his *Universa medicina* (Paris: Andreas Wechelus, 1567), p. 335. "Malicia in visceribus nobilibus presertim in epate et splene," Albertus de Zanchariis, *Glose*, fol. 24v. "Tercio mala est malicia modi eveniendi cum sepe generetur a corrupcione complicata," Montagnana, *De lepra*, fol. 1va. "Est in ea molestia racione virtutis dispensantis toti cum ipsa virtute vehementer dehiciat," Albertus de Zanchariis, *Glose*, fol. 24v. Bertucci, *Nusquam antea*, fol. ccxliiij^va. "Quarto iterum mala est malicia effectus … adeo ut complete prohibitum sit tales ab hominum conversacionibus exules fieri propter verendam venenositatem," Montagnana, *De lepra*, fol. 1va. Compare Du Laurens, *Operum*, p. 36. "Interpretatur lepra quasi lesio prava ultimo pacientem ducens ad mortem … ultimo ad mortem deducens que advenit ex completo dominio morbi super virtutem," Albertus de Zanchariis, *Glose*, fol. 24v. Theodoric, *Cyrurgia*, fol. 130r.

44. Du Laurens, *Operum*, p. 37. Compare Aretaeus, *Aretaei Cappadocis*, p. 207. Fernel, *Pathologiae libri VII*, p. 335. Simon de Cordo Januensis, *Clavis sanationis*, fol. XLVra. Leviticus 13:2, 9, 20, and so on. In one exceptional case, a town physician of Ieper in 1572 cited Leviticus in the same paragraph in which he used the word *plaga*; ironically, he evidently understood it as "lesion" rather than in the biblical sense of "plague." Sudhoff, "Wurzacher Lepraschaubriefe," p. 431.

CHAPTER FOUR. Definitions and Explanations

1. Moliné, "Un diagnòstic mèdic," p. 6.
2. See, for example, Figure 1.1.
3. Avicenna, *Canon medicinae*, sig. FF 6vb.
4. Iuliarius, *De lepra*, p. 1v.
5. Phlegm was the humor with the least relevance to *lepra*; aside from a relatively unspecified type of "white phlegm" (*leukophlegma*), two kinds were normally differentiated, one normal (a sweet constituent of the blood, analogous to serum), and the other pathological; the latter was further subdivided into acid or salty (*salsu-*) phlegm and gelatinous or glassy (*vitreosum*) phlegm, from cold mucus discharged by the brain. The most lucid introduction to Galen's system of physiology is Margaret Tallmadge May, *Galen on the Usefulness of the Parts of the Body*, Περί χρείας μορίων, *De usu partium*, trans. (from the Greek) with an intro. and commentary (Ithaca, N.Y.: Cornell University Press, 1968), vol. 1, pp. 44–64, esp. pp. 53–54.
6. "Octavo advertendum quod error secunde digestive in epate potest esse causa lepre remota, sed error virtutis immutative in carne est causa immediata," Bernard de Gordon, *Lilium*, fol. 21ra.
7. Guy de Chauliac, *Chirurgia*, p. 282. Hans von Gersdorff, *Feldtbuch der Wundtartzney* (Strasbourg: Joannes Schott, 1517), sig. M iijvb. Gersdorff's abridged citation of Galen, "den worten Galieni in vj" (and Chauliac's "in 6°"), actually referred to the sixth book of Galen's work known as *De accidenti et morbo*.
8. Mus, *Oorkonden* IV, pp. 37, 38.
9. See above, Chap. 3.
10. Bernard de Gordon, *Lilium*, fol. 17vb.
11. Iuliarius, *De lepra*, pp. 2v–3.
12. Stoll, *Das "Lorscher Arzneibuch,"* p. 128; "Elephantiosos apprehendimus ex maculis," Gari-

opontus, *Passionarius*, fol. 77r; "Leprosos apprehendimus et primum cognoscimus ex maculis," Johannes de Sancto Paulo, *Breviarium*, sig. aᵛ. Hildegard of Bingen, *Causae et Curae*, pp. 18, 37, 160–161, 207, 211, 213. Huggelin, *Von dem Aussatz*, sig. a iiij.

13. Sudhoff, "Lepraschaubriefe aus Italien," p. 434.

14. Theodoric, *Cyrurgia*, fol. 130r. García-Ballester, "*Artifex factivus sanitatis*," p. 144.

15. Text of the three sermons in Nicole Bériou and François-Olivier Touati, *Voluntate Dei Leprosus: Les lépreux entre conversion et exclusion aux XIIème et XIIIè siècles,* Testi, Studi, Strumenti, 4 (Spoleto: Centro Italiano di Studi sull'Alto Medioevo, 1991), pp. 128–155.

16. Danielle Jacquart, "De *crasis à complexio:* Note sur le vocabulaire du tempérament en latin médiéval," in Guy Sabbah, ed., *Textes médicaux latins antiques* (Saint-Etienne: Université de Saint-Etienne, 1984), pp. 71–76. The twelfth-century Latin translation from the Greek is available in *Galenus Latinus:* Richard J. Durling, ed., *Burgundio of Pisa's Translation of Galen's ΠΕΡΙ ΚΡΑΣΕΩΝ "De complexionibus,"* Ars medica II, Abt. Bd. 6.1 (Berlin: Walter de Gruyter, 1976).

17. *Galieni Opera*, vol. 2, fol. 21rb. See Luke Demaitre, "Medieval Notions of Cancer: Malignancy and Metaphor," *Bulletin of the History of Medicine* 72 (1998), 609–637. Sudhoff, "Dokumente zur Ausübung der Lepraschau," p. 151.

18. See Nancy G. Siraisi, *Taddeo Alderotti and His Pupils: Two Generations of Italian Medical Learning* (Princeton: Princeton University Press, 1981), pp. 284–285, 288–289, and passim. García-Ballester, "The *New Galen*," pp. 55–83.

19. Bernard de Hangarra (Engarra) was cited in 1294 as physician to Rossolin, lord of Lunel: Wickersheimer, *Dictionnaire biographique*, p. 74. "Utrum lepra sit mala complexio tocius corporis. Dixit Cancellarius quod non, quia ordo nature est ut membris nobilibus detur nutrimentum nobilius unde extremitatibus cedit malum ex quo primo inficiuntur," *Supra libellum de mala complexione diversa hec sunt dubitata quorum positiones secundum magistrum bernardum de hangarra montis pessulani quondam Cancellarium recitantur,* in Munich, Codex latinus monacensis (hereafter cited as CLM) 534, fol. 43r. I am grateful to Michael McVaugh for providing me with a photocopy of fols. 42v–45r of this manuscript.

20. Arnau de Vilanova, *Commentum supra tractatum Galieni de malicia complexionis diverse*, ed. Luis García Ballester and Eustaquio Sanchez Salor, Arnaldi de Vilanova Opera Medica Omnia XV (Barcelona: Universidad de Barcelona, 1985), p. 155.

21. Ibid., p. 156.

22. Bernard de Gordon, *Lilium*, fol. 20rb.

23. Ibid., fol. 20va.

24. As formulated in the basic textbook, Joannitius (Hunain ibn Ishāq, 809–877), *Isagoge.* An annotated English translation, with additional notes by Michael McVaugh, is conveniently available in Grant, *Source Book*, pp. 709–710. Bernard de Gordon, *Lilium*, fol. 17vb.

25. Keussen, "Urkunden," p. 112.

26. "At the request and the command of the scientific man, Master Bernard de Gordon, eminent professor of the science of medicine at the famous school of Montpellier," in Paris BN MS lat. 7131. Translated from the French quotation in E. Nicaise, trans. and ed., *La Chirurgie de maître Henri de Mondeville* (Paris: Alcan, 1893), p. 1. Pagel, *Die Chirurgia des Heinrich von Mondeville*, p. 423.

27. Luke Demaitre, "The Relevance of Futility: Jordanus de Turre (fl. 1313–1335) on the

Treatment of Leprosy," *Bulletin of the History of Medicine* 70 (1996), pp. 25–61. Jordanus de Turre, *De lepra nota,* edited in Demaitre, "Relevance of Futility," pp. 54–56.

28. Guy de Chauliac, *Chirurgia,* p. 282.

29. Velascus de Tharanta, *Practica,* fol. 256ra.

30. Bernard de Gordon, *Lilium,* fol. 17vb. "Non potest hanc materiam corruptam et ineptam assimilare carni," ibid. Velascus de Tharanta, *Practica,* fol. 256va.

31. For some aspects of this vernacularization, see Luke Demaitre, "Medical Writing in Transition: Between *Ars* and *Vulgus,*" *Early Science and Medicine* 3, no. 2 (1998), pp. 88–102. Bocellin, *Practique,* p. 3v (while the book was published in 1540, Bocellin dated his preface in 1539).

32. Nancy G. Siraisi, "Giovanni Argenterio and Sixteenth-Century Medical Innovation: Between Princely Patronage and Academic Controversy," in Michael R. McVaugh and Nancy G. Siraisi, eds., *Renaissance Medical Learning: Evolution of a Tradition* (in *Osiris,* 2nd ser., 6 [1990]), p. 162; the entire essay, pp. 161–180, with copious references, is most instructive. For a development in Italy which paralleled the Montpellier "New Galen," see Tiziana Pesenti, "Galenismo e 'novatio': La scuola medica vicentina e lo Studio do Padova durante il periodo veneto di Galileo (1592–1610)," in Lino Conti, ed., *Medicina e biologia nella revoluzione scientifica* (Assisi: Edizioni Porziuncula, 1990), pp. 107–147.

33. Du Laurens, *Operum,* p. 36. Compare Guy de Chauliac, *Chirurgia,* pp. 5–8.

34. Du Laurens, *Operum,* pp. 36–37.

35. Ibid., pp. 37–38.

36. Ibid., p. 38.

37. Varanda, *Tractatus de elephantiasi,* p. 7.

38. Ibid.

39. Ibid., p. 8.

40. Ibid., pp. 9–12.

41. Bertucci, *Nusquam antea,* fol. 264vb.

42. Uddman, *Lepra,* pp. 2–4. Raymond, *Histoire de l'Éléphantiasis,* p. 13.

43. Johann David Ruland, *De Lue sive Lepra Venerea Disputatio . . . sub Praesidio Gothofredi Weidneri* (Frankfurt-an-der-Oder, 1636), sig. A2ᵛ. Van Schevensteen, *La Lèpre dans le Marquisat d'Anvers,* p. 85. Sudhoff, "Wurzacher Lepraschaubriefe," pp. 427, 433, 426.

CHAPTER FIVE. "Une maladie contagieuse et héréditaire"

1. The quotation that opens this chapter is from Sudhoff, "Wurzacher Lepraschaubriefe," pp. 431–432.

2. Owsei Temkin, "An Historical Analysis of the Concept of Infection," in Owsei Temkin, *The Double Face of Janus and Other Essays in the History of Medicine* (Baltimore: Johns Hopkins University Press, 1977), pp. 465–471. Mirko D. Grmek, "Les vicissitudes des notions d'infection, de contagion et de germe dans la médecine antique," *Mémoires du Centre Jean Palerne* (1984), pp. 53–70. Above all, Vivian Nutton, "Seeds of Disease: An Explanation of Contagion and Infection from the Greeks to the Renaissance," *Medical History* 27 (1983), pp. 1–34; and Vivian Nutton, "The Reception of Fracastoro's Theory of Contagion: The Seed That Fell among Thorns?" in McVaugh and Siraisi, *Renaissance Medical Learning,* pp. 196–234. Archives Départmentales de l'Hérault, II E 35/42.

3. Varanda, *Tractatus de elephantiasi*, pp. 11–12. See Nutton, "Seeds of Disease," p. 8. Girolamo Fracastoro, *De sympathia et antipathia rerum liber unus. De contagione et contagiosis morbis et curatione libri iii* (Venice: Heirs of Lucasantonius Junta, 1546). Edited and translated into English by Wilmer Cave Wright (New York: Putnam's, 1930), p. 60. For more context, see Nutton, "Reception of Fracastoro's Theory of Contagion," esp. pp. 200–201, with "seedlets" as the English translation for *seminaria*. Fernel, *Pathologiae libri VII*, I.4. While Fernel was correct in noting that medical literature agreed on the contagious nature of ophthalmia, he overstated the "consensus" by ignoring the authors who believed in a transmission by rays rather than flow: see Nutton, "Seeds of disease," p. 9. Santorio Santorio, *Galeni Ars medica . . . Commentaria in artem medicinalem Galeni . . .* (Venice: Marcus Antonius Brogiollus, 1630), p. 123. It is highly more likely that Varanda was directly influenced by Fracastoro, Fernel, and even Santorio than by similar ideas that had percolated in the thirteenth century. These earlier ideas are surveyed in François-Olivier Touati, "Historiciser la notion de contagion: L'exemple de la lèpre dans les sociétés médiévales," in Bazin-Tachella, Quéruel, and Samama, *Air, miasmes et contagion*, pp. 179–180 and n. 63.

4. Arnau de Vilanova, *Commentum supra tractatum Galieni de malicia complexionis diverse*, pp. 248–249.

5. McVaugh, *Medicine before the Plague*, pp. 221–222. Velascus de Tharanta, *Practica*, fol. 258vb.

6. Varanda, *Tractatus de elephantiasi*, p. 26. Maréchal, "Lepra-onderzoek in Vlaanderen," p. 44. E. C. Van Leersum, ed., *De "Cyrurgie" van Meester Jan Yperman: Naar de Handschriften van Brussel, Cambridge, Gent en Londen* (Leiden: A. W. Sijthoff's, [1912]), p. 177.

7. "Cavendo sibi semper de anhelitu examinandi ne medicus inficiatur," *Signa lepre*, in Vat. Pal. MS Latin 1207, fol. 18r. Also, "caveant ergo medici ne ab infirmo suscipiant anhelitum," *De lepra*, in Basel Universitätsbibliothek MS D.III.10, fol. 11r. Further, Foreest, *Observationum et curationum chirurgicarum libri quinque*, p. 124.

8. This paragraph, and the following three, owe much to the incisive *mise au point*, with copious references, by Touati, "Historiciser," pp. 157–187. Aretaeus, *Aretaei Cappadocis*, p. 213. Hartmann von Aue, *Der arme Heinrich*, p. 14, vv. 126–127.

9. Cicero, *De oratore* 3, 41, 164; Pliny, *Historia naturalis* 23, 8, 80, § 157 and 26, 3. Gilles de Corbeil, *Viaticus*, p. 74. Johannes Platearius, *Practica brevis*, fol. 221vb. "Fit aut ex corrupcione aeris . . . Fit eciam hoc modo qui assistunt leprosis et eis confabulantur," Johannes de Sancto Paulo, *Breviarium*, sig. aᵛ. See also Ferreira, *Tratado médico*, p. 74. "Quandoque adiuvat illud totum corruptio aeris in seipso: aut propter vicinitatem leprosorum quia egritudo est invadens," Avicenna, *Canon medicinae*, sig. FF 7ra. Galen, *De differentiis febrium*, I.3 and 4, in *Galieni Opera*, vol. 2, fols. 222ra, 222rb.

10. Touati, "Historiciser," p. 182. Quoted in ibid., p. 182 n. 70. The *Breviarium* was long attributed to Arnau de Vilanova. Bernard de Gordon, *Lilium*, fol. 17vb.

11. Pagel, *Die Chirurgia des Heinrich von Mondeville*, p. 424. Barber, "Lepers, Jews and Moslems," pp. 1–2, 7–8. Jordanus de Turre in Demaitre, "Relevance of Futility," p. 56 (in view of the context, however, Jordan's expression, "frequenter mutari de loco ad locum," may have referred to moving the patients around within their domicile). Bertucci, *Nusquam antea*, fols. 264va, 264vb.

12. Guy de Chauliac, *Chirurgia*, p. 283. Sudhoff, "Lepraschaubriefe aus Italien," p. 435; also

Garosi, "Perizie e periti," p. 162. See Touati, "Historiciser," pp. 183–185. A pathbreaking and compelling defense of this thesis is Touati, *Maladie et société*, pp. 270–275 and passim.

13. Garosi, "Perizie e periti," p. 162. Sudhoff, "Lepraschaubriefe aus Italien," p. 434. Sudhoff, "Weitere Lepraschaubriefe," p. 156. E. C. Van Leersum, ed., *Het "Boeck van Surgien" van Meester Thomas Scellinck van Thienen: Naar de handschriften van de Koninklijke Bibliotheek te 's-Gravenhage en het British Museum te Londen uitgegeven*, Opuscula selecta Neerlandicorum de Arte Medica, Fasciculus Septimus (Amsterdam: F. van Rossen, 1928), p. 198.

14. Schwarz-Wien, "Zur Geschichte der Lepraschau," pp. 383–384.

15. Ernest Wickersheimer, "Eine kölnische Lepraschau vom Jahre 1357," *Archiv für die Geschichte der Medizin* 2 (1908–1909), p. 434. See above, Chap. 3. Karl Sudhoff, "Was geschah mit den (nach erneuter Schau) als 'leprafrei' Erklärten und aus den Leprosorien wieder Entlassenen, von behördlicher und ärztlicher Seite? Zwei Aktenstücke, zugleich als Fortsetzung seiner Studien zur Lepraschau," *Archiv für Geschichte der Medizin* 6 (1912–1913), p. 153. Keussen, "Urkunden," p. 90.

16. Quoted by Edouard Jeanselme, "Comment l'Europe, au moyen âge, se protégea contre la lèpre," *Bulletin de la Société française d'histoire de la médecine* 25 (1931), p. 112.

17. Translated from Borradori, *Mourir au Monde*, pp. 185–186.

18. Ibid., pp. 188–189.

19. "Contra duos presbyteros leprosos ex opido sancte sirie oriundos. Non extra predictum opidum in eorum bordis sed inter populum in ipso oppido commorantes. Et quod deterius est secum tenent concubinas que publice vestimenta ecclesiastica ad ecclesiam predicti loci defferunt ad missas per ipsos leprosos in predicta sancte sirie ecclesia celebrandas in scandalum et periculum tocius populi predicti sancte sirie opidi. Et quod ad hoc sibi placeat remedium apponere, alias domini ut domini temporales predicti loci remedium apponent secundum quod juri fuerit," Troyes, Archives départmentales de l'Aube, G 1281, fol. 62r, Wednesday, July 9, 1516 (emphases added).

20. Erfurt Amplonianum MS F 288, late fourteenth century, fol. 84v.

21. Velascus de Tharanta, *Practica*, fol. 256ra. It is likely that Velasco adopted these conditions from an earlier source, but I have not been able to find it in Johannes Platearius, to whom it was attributed by Bocellin, *Practique*, p. 3v. Avicenna, *Canon medicinae*, sig. FF 6vb. Velascus de Tharanta, *Practica*, fols. 258r, 259ra.

22. Bocellin, *Practique*, p. 3v (emphasis added). A marginal note refers not only to Leviticus 13 but also to Canon Law, "de poenitentia distinctio prima, canone, Voluissent iniqui," ibid. "The whole herd in the fields falls by one scabies," Juvenal, *Satirae*, 2.79. "Bad infections [*mala contagia*] of a nearby herd will be harmful," Virgil, *Bucolica*, 1.51. Touati, *Maladie et société*, pp. 709–710 and nn. 74–75.

23. Paré, *Les oeuures*, pp. 752, 756.

24. Ibid., p. 757.

25. Des Innocens, *Examen*, pp. 15–19.

26. "Ceterorum dicti loci habitancium maximum periculum cedere posset, maxime cum dictus morbus lepre contagiosus existat, nisi per nos super hoc provideretur et remedia con[stituerentur]," Auxerre, Archives départementales de l'Yonne, G 1633, item 3, March 21, 1416. See also Ganges in 1570, at n. 2 above. Sudhoff, "Dokumente zur Ausübung der Lepraschau," p. 151. Sudhoff, "Lepraschaubriefe aus dem 15. Jahrhundert," pp. 373–374.

27. "Invenimus dictum Thionotum non esse leprosus in actu perfecto sed pro tuciore detur eidem terminus in mense septembri et interim sibi dicatur vel moneatur ipsum ut non vadat in pressuris gencium et magnis communionibus," Auxerre, Archives départementales de l'Yonne, G 1633, item 2 (emphasis added). See also the preceding note. Sudhoff, "Dokumente zur Ausübung der Lepraschau," p. 151. Mus, *Oorkonden* IV, p. 31.

28. M. Barthélemy, "Procès-verbal de visite d'un lépreux en 1464 et relation juridique d'une autopsie en 1499," *Actes du comité médical des Bouches-du-Rhône* 20 (1881–1882), pp. 118–119; 119–120.

29. Wickersheimer, "Beiträge," p. 146. Mus, *Oorkonden* IV, p. 27. "zoo dat niemandt danof last ende bestavynghe omme de grooten stanck vande zeerheyt haers lichgaems meer nemen wil, want danof infectie ende bedervenesse een ighelyck commen zoude," Mus, *Oorkonden* IV, pp. 37, 51, 52. Sudhoff, "Wurzacher Lepraschaubriefe," p. 430.

30. Fracastoro, *De sympathia.* Du Laurens, *Operum,* pp. 37–39 (emphasis added). Petraeus, "Diss[ertatio] harmonica XLV," pp. 402, 401.

31. Uddman, *Lepra.* I first became aware of this thesis thanks to Richards, *Medieval Leper and His Northern Heirs,* p. 101. I am preparing an English translation of Uddman's interesting work, which is insufficiently accessible because it was written in Latin and copies are scarce.

32. Quotations from the brilliant summation of the search until Fracastoro in Nutton, "Seeds of Disease," p. 2.

33. Ibid., p. 11. Uddman, *Lepra,* pp. 9–10.

34. Richards, *Medieval Leper and His Northern Heirs,* pp. 89–90.

35. Uddman, *Lepra,* pp. 9, 2.

36. Mus, *Oorkonden* IV, p. 31. Sudhoff, "Wurzacher Lepraschaubriefe," pp. 426, 427.

37. Bernard de Gordon, *Lilium,* fol. 17vb. Henri de Mondeville, *Chirurgia,* p. 422.

38. One of the few exceptions was Avicenna, who differentiated between causes of leprosy in "the embryo's complexion itself" and in "the disposition of the womb," *Canon medicinae,* sig. FF 6vb.

39. "Et quandoque accidit propter hereditatem," Avicenna, *Canon medicinae,* sig. FF 6vb (emphasis added). Bocellin, *Practique,* p. 4r (emphasis added). Johannes Platearius, *Practica brevis,* fol. 221vb (emphasis added). Theodoric, *Cyrurgia,* fol. 130ra (emphasis added). "Materia lepre tante est pernecabilitatis quod in matrice corrigi non potest, et ideo est morbus heredetarius, et podagra similiter," Bernard de Gordon, *Lilium,* fol. 21ra. Bertucci, *Nusquam antea,* p. 244vb, citing Averroës, *Colliget,* bk. 3. Lommius, *Medicinalium observationum libri tres,* p. 24r. Paré, *Les oeuures,* p. 751. Du Laurens, *Operum,* p. 39.

40. Bertucci, *Nusquam antea,* p. 244vb. Similarly, "apparenter aliquando sanantur; sicut est de lepra aut ptisi, que in semine hereditantur: nam in principio non sunt leprosi, sed postquam incurrerunt durant plurimo tempore vite," Giovanni Santasofia, *Super Tegni,* in Seville, Biblioteca Columbina, MS 7-7-18, fol. 17va; I owe this text to the kindness of Tiziana Pesenti (pers. comm., April 1998). Du Laurens, *Operum,* p. 39. Platter, *Observationes,* quoted in Bühler, *Der Aussatz in der Schweiz,* pp. 64–66.

41. Des Innocens, *Examen,* pp. 100–101.

42. Bernard de Gordon, *Lilium,* fol. 17vb.

43. The coincidental mention by Aretaeus in the chapter on *elephantiasis* of the potential that is "in the seed" (*in semine*) had no bearing on the heredity of leprosy. Aretaeus, *Aretaei Cap-*

padocis, p. 214. Du Laurens, *Operum*, p. 39. Guy de Chauliac, *Chirurgia*, p. 283. Sudhoff, "Dokumente zur Ausübung der Lepraschau," p. 159. "Quomodo est hoc possibile quod leprosus generet, et si generet, quomodo est hoc quod masculum cum sperma sit corruptum et distemperatum in substancia et complexione, et talia impediunt generacionem et potissime filiorum? Dico quod licet ista impediant, plus tamen impeditur generacio propter peccatum complexionis quam substancie spermatis," Bernard de Gordon, *Lilium*, fol. 21ra. Paré, *Les oeuures*, pp. 751, 756.

44. Velascus de Tharanta, *Practica*, fols. 256rb, 258rb.

CHAPTER SIX. Causes, Categories, and Correlations

1. The quotation that opens this chapter is from Sudhoff, "Dokumente zur Ausübung der Lepraschau," p. 158.

2. Dino del Garbo, or Dinus Florentinus, *Expositio super tertia, quarta, et parte quintae fen IV, libri Avicennae* (Venice: Johannes Hamman, for Andreas Torresanus, 1499).

3. See above, Chap. 4.

4. Du Laurens, *Operum*, p. 39. The "regimen of health" is discussed most comprehensively by Pedro Gil-Sotres, with the collaboration of J. A. Paniagua and L. García-Ballester, "La higiene medieval," introduction to Arnau de Villanova, *Regimen sanitatis ad regem Aragonum*, ed. Luis García-Ballester and Michael R. McVaugh, in *Arnaldi de Villanova Opera Medica Omnia* X.1 (Barcelona: Universitat de Barcelona, 1996), pp. 471–861.

5. Bernard de Gordon, *Lilium*, fol. 17vb. Velascus de Tharanta, *Practica*, fol. 256ra. "Caliditas aeris," Avicenna, *Canon medicinae*, sig. FF 7ra. Bertucci, *Nusquam antea*, p. 244va. Cologne 1548, in Keussen, "Urkunden," p. 94. Paré, *Les oeuures*, p. 752; also Petraeus, "Diss[ertatio] harmonica XLV," p. 403. Godfried Guglielmus [Wilhelm] Schilling, *De lepra commentationes*, ed. J. D. Hahn (Leiden: Sam[uel] and Joan[nes] Luchtmans, 1778), p. 19. Uddman, *Lepra*, p. 5.

6. Bertucci, *Nusquam antea*, p. 244va–b. Paré, *Les oeuures*, pp. 752, 751. Petraeus, "Diss[ertatio] harmonica XLV," p. 403 (citing Forestus).

7. Avicenna, *Canon medicinae*, sig. FF 7ra. Bernard de Gordon, *Lilium*, fol. 18ra; Velascus de Tharanta, *Practica*, fol. 256rb; and so on. Bernard de Gordon, *Lilium*, fol. 18ra. Theodoric, *Cyrurgia*, fol. 130ra. Bernard de Gordon, *Lilium*, fol. 17vb. Uddman, *Lepra*, p. 6. Johannes Platearius, *Practica brevis*, fol. 221vb. Richards, *Medieval Leper and His Northern Heirs*, pp. 95–96.

8. Galen, *Commentarium in III Hippocratis de humoribus*, Kühn XVI.442. Iuliarius, *De lepra*, p. 2v. Avicenna, *Canon medicinae*, sig. FF 7ra. Galen, *De accidente et morbo VI*, in *Galieni Opera*, vol. 2, p. 149va.

9. Sudhoff, "Dokumente zur Ausübung der Lepraschau," p. 151. Petraeus, "Diss[ertatio] harmonica XLV," p. 404, citing Paracelsus, *Paramirum* 2.8.

10. Velascus de Tharanta, *Practica*, fol. 256rb. We can only speculate about possible connections between these "glands" and those in scrofula, the tuberculosis of the lymphatic glands in the neck, whose name is derived from "sow" (*scrofa*).

11. Hildegard of Bingen, *Causae et curae*, pp. 207, 160. Huggelin, *Von dem Aussatz*, sig a viʳ.

12. Bertucci, *Nusquam antea*, p. 244va. See Luke Demaitre, "The Medical Notion of 'Withering' from Galen to the Fourteenth Ventury," *Traditio* 34 (1992), pp. 259–307. Avicenna, *Canon*

medicinae, sig. FF 7ra. Hildegard of Bingen, *Causae et curae*, p. 18. Also, on *gula* and *lepra*, Hildegard of Bingen, *Causae et curae*, p. 160.

13. Iuliarius, *De lepra*, p. 2v. Avicenna, *Canon medicinae*, sig. FF 7ra. Galen, *Commentarium in III Hippocratis de humoribus*, Kühn XVI.442. Celsus, *De medicina*, Liber Tertius, Caput 25, 3. Bernard de Gordon, *Lilium*, fol. 20vb. Sudhoff, "Dokumente zur Ausübung der Lepraschau," p. 151. Cologne 1548, in Keussen, "Urkunden," p. 95.

14. Avicenna, *Canon medicinae*, sig. FF 7ra. Guy de Chauliac, *Chirurgia*, p. 283; Du Laurens, *Operum*, p. 39; Iuliarius, *De lepra*, p. 2v. Sudhoff, "Wurzacher Lepraschaubriefe," p. 427. *Albertus Magnus on Animals: A Medieval "Summa Zoologica,"* trans. and annotated by Kenneth F. Kitchell and Irven Michael Resnick (Baltimore: Johns Hopkins University Press, 1999), vol. 2, p. 1330. I thank Irv Resnick for bringing this statement to my attention.

15. General background in Janice Delaney, Mary Jane Lupton, and Emily Toth, *The Curse: A Cultural History of Menstruation*, rev. ed. (Urbana: University of Illinois Press, 1988). Migne, *Patrologia Latina*, 25, col. 173–174. Quoted by Marie-Hélène Congourdeau, "Sang féminin et génération chez les auteurs byzantins," in Marcel Faure, ed., *Le sang au Moyen Âge*, Actes du quatrième colloque international de Montpellier, Université Paul-Valéry, 27–29 novembre 1997, Les Cahiers du CRISIMA, n° 4 (Montpellier: Université Paul-Valéry, 1999), p. 22.

16. Avicenna, *Canon medicinae*, sig. FF 7ra (emphasis added). Johannes Platearius, *Practica brevis*, fol. 222ra. I cannot exclude the possibility that the mention was an interpolation in the printed edition of 1525, which I consulted; the earliest manuscript copies would need to be checked in order to be certain. Brian Lawn, ed., *The Prose Salernitan Questions*. Edited from a Bodleian Manuscript (Auct. F. 3. 10). An Anonymous Collection dealing with Science and Medicine written by an Englishman c. 1200. With an Appendix of ten related Collections, Auctores Britannici Medii Aevi, V (London: Oxford University Press, 1979), esp. Questions B 14, 33, 187; Ba 92; and P 116. Ferreira, *Tratado médico*, pp. 74–80. Enrique Montero Cartelle, *Constantini Liber de Coitu. El Tratado de andrologia de Constantino el Africano*, Estudio y edicion critica. (Santiago de Compostela: Universidad de Santiago de Compostela, 1983). Gilles de Corbeil, *Viaticus*, p. 75.

17. Theodoric, *Cyrurgia*, fol. 130ra. Bernard de Gordon, *Lilium*, fol. 17vb. Pagel, *Die Chirurgia des Heinrich von Mondeville*, p. 422. Bertucci, *Nusquam antea*, p. 244vb.

18. "Item nota, istis temporibus pueri leprosi maxime generantur in conceptione menstruantis quia tunc mulier tempore menstruorum plus delectatur in coitu quia sanguis eius mordicat et sic venit titillacio. Et sic cum puer concipitur in fluxu menstruorum sine dubio incurrerit lepram vel scabiem," "De lepra et primo de interpretatione," in Basel Universitätsbibliothek MS D.III.10, fol. 6a-v. Du Laurens, *Operum*, p. 39 (emphasis added). Petraeus, "Diss[ertatio] harmonica XLV," p. 402.

19. Touati, *Maladie et société*, pp. 114–119.

20. Hildegard of Bingen, *Causae et curae*, p. 161. Bertucci, *Nusquam antea*, p. 244vb.

21. Lawn, *Prose Salernitan Questions*, p. 185. Erfurt Amplonianum MS F 288, fol. 85r. Donald J. Ortner, "Male/Female Immune Reactivity and Its Implications for Interpreting Evidence in Human Skeletal Pathology," in A. L. Grauer and P. Stuart-Macadam, eds., *Sex and Gender in Paleopathological Perspective* (Cambridge: Cambridge University Press, 1998), pp. 79–82. For caveats about the implications, see Charlotte A. Roberts, "The Antiquity of Leprosy in Britain: The Skeletal Evidence," in Charlotte A. Roberts, Mary E. Lewis, and K. Manches-

ter, eds. *The Past and Present of Leprosy: Archaeological, Historical, Palaeopathological and Clinical Approaches*, Proceedings of the International Congress on the Evolution and Palaeoepidemiology of the Infectious Diseases 3, University of Bradford, July 26–31, 1999 (Oxford: Archaeopress, 2002), p. 215. My ongoing and very tentative tally of three hundred premodern certificates yields ratios that hover around 65 percent male. Danielle Jacquart and Claude Thomasset, *Sexualité et savoir médical au Moyen Age* (Paris: Presses Universitaires de France, 1985), pp. 247–257. The question "Dubitatur quare si mulier coeat cum leproso non fit lepra" was entertained by one of the most "scholastic" physicians, Gentile da Foligno (d. 1348), in his treatise on *lepra*, printed in *Dinus in chirurgia cum tractatu eiusdem de ponderibus et mensuris . . . additi sunt insuper Gentilis de Fulgineo super tractatu de lepra* (Venice: Junta, 1536), fols. 54vb–55ra.

22. Above, Chap. 1, at n. 25. Bernard de Gordon, *Lilium*, fol. 18ra. My thinking about these paradoxes has been stimulated, above all, by Touati, "Les traités sur la lèpre," esp. p. 221.

23. Bernard de Gordon, *Lilium*, fol. 18ra. Pagel, *Die Chirurgia des Heinrich von Mondeville*, p. 429. Françoise Bériac, *Histoire des lépreux au moyen âge: Une société d'exclus* (Paris: Imago, 1988), pp. 69–70, without reference.

24. Johannes Platearius, *Practica brevis*, fol. 222ra. Gilles de Corbeil, *Viaticus*, p. 74.

25. Bernard de Gordon, *Lilium*, fols. 17vb–18ra. Lawn, *Prose Salernitan Questions*, pp. 9, 18, 101, 249. Paré, *Les oeuures*, p. 752. Petraeus, "Diss[ertatio] harmonica XLV," p. 402.

26. Bertucci, *Nusquam antea*, p. 244vb. Keussen, "Urkunden," p. 94. Velascus de Tharanta, *Practica*, fol. 256rb, citing Hippocratic *Aphorism* 6.23. Hildegard of Bingen, *Causae et curae*, p. 161. Also Petraeus, "Diss[ertatio] harmonica XLV," p. 403. Du Laurens, *Operum*, p. 39.

27. See Luke Demaitre, "The Art and Science of Prognostication in Early University Medicine," *Bulletin of the History of Medicine* 77 (2003), p. 768. For an illustrative instance of this simplification, see Demaitre, "Relevance of Futility," p. 55, lines 39–42.

28. Sudhoff, "Dokumente zur Ausübung der Lepraschau," p. 158. These stages were adumbrated in Avicenna, *Canon medicinae*, sig. FF 7ra. Sudhoff, "Weitere Lepraschaubriefe," pp. 154–155.

29. Petraeus, "Diss[ertatio] harmonica XLV," p. 401. Also Gentile da Foligno, *Super tractatu [Avicennae] de lepra*, printed with *Dinus in chirurgia cum tractatu eiusdem de ponderibus et mensuris . . .* (Venice: Junta, 1536), fol. 55rb. And Velascus de Tharanta, *Practica*, fol. 256va, citing *De proprietatibus rerum* of Bartholomeus Anglicus. Ferreira, *Tratado médico*, p. 76. Johannes Platearius, *Practica brevis*, fol. 221vb. Johannes de Sancto Paulo, *Breviarium*, fol. 1v. Gilles de Corbeil, *Viaticus*, pp. 76–78. Van Leersum, *De "Cyrurgie" van Meester Jan Yperman*, p. 175. Simon de Cordo Januensis, *Clavis sanationis*, fol. 45ra. Guy de Chauliac, *Chirurgia*, p. 283. Bocellin, *Practique*, p. 9v.

30. It should be noted, however, that Gilbertus Anglicus, or Gilbert the Englishman (fl. 1250), already developed the division in the *Compendium medicine*, which circulated widely. See the annotated English translation by Michael McVaugh, in Grant, *Source Book*, pp. 753–754.

31. Barthélemy, "Procès-verbal," p. 118 (Marseille). Also at Constance in 1541: Borradori, *Mourir au Monde*, p. 223. Guy de Chauliac, *Chirurgia*, p. 283. Similarly, Gilbertus Anglicus, in Grant, *Source Book*, pp. 753–754. See F. O. Touati, "Facies leprosorum: Réflexions sur le diagnostic facial de la lèpre au Moyen Age," *Histoire des sciences médicales* 20 (1986), p. 64. Du Laurens, *Operum*, p. 38.

32. Petraeus, "Diss[ertatio] harmonica XLV," p. 407 (citing Paracelsus, *Paramirum* 6.1) and p. 408 (emphasis added).

33. Uddman, *Lepra*, pp. 3–4.

34. Ibid., p. 2.

35. Galen, *Comm. III in Hipp. de humoribus*, Kühn XVI, p. 442.

36. See above, Chap. 2, for context on juries. Bourgeois, *Lépreux et maladreries*, pp. 214, 215, 217, and passim. Compare the categories of Sister Barbara Sagers, above, Chap. 2.

37. Bernard de Gordon, *Liber pronosticorum*, ed. in Alberto Alonso Guardo, *Los pronósticos médicos en la medicina medieval: El "Tractatus de crisi et de diebus creticis" de Bernardo de Gordonio* (Valladolid: Universidad de Valladolid, 2003), pp. 132, 138, 146.

38. The complete tree is schematized, from Vat. Pal. MS Latin 1083, in Demaitre, "Medieval Notions of Cancer," p. 615.

39. Bernard de Gordon, *Lilium*, fol. 20vb.

40. For a survey of cancer attributes, see Demaitre, "Medieval Notions of Cancer," pp. 609–637. Rufus of Efesus, as transmitted by Oribasius, quoted in Grmek, *Diseases*, pp. 168–169. Raymond, *Histoire de l'Élephantiasis* pp. 5, 7–8. Avicenna, *Canon medicinae*, sig. FF 6vb–7ra.

41. Galen, *De tumoribus*, in *Galieni Opera*, vol. 1, fol. 155vb; compare Kühn 7.727. Gersdorff, *Feldtbuch der Wundtartzney*, fol. sig m iv^va. La Seu: see above, Chap. 3. Losser: see above, Chap. 1.

42. Sudhoff, "Dokumente zur Ausübung der Lepraschau," p. 158. "Dicit Galienus in fine tercie particule de morbo talia verba, Item mala digestio non generat maciem sed mutat speciem membri sicut videmus in morphea et lepra," Bernard de Gordon, *Lilium*, fol. 20ra. Arnau de Vilanova, *Commentum supra tractatum Galieni de malicia complexionis diverse*, p. 292.

43. Bourgeois, *Lépreux et maladreries*, pp. 238, 240. Gersdorff, *Feldtbuch der Wundtartzney*, sig ixxx^rb. Bourgeois, *Lépreux et maladreries*, p. 186.

44. Bourgeois, *Lépreux et maladreries*, pp. 240, 186. Mus, *Oorkonden IV*, p. 31.

CHAPTER SEVEN. Diagnosis

1. Bourgeois, *Lépreux et maladreries*, p. 218.

2. Keussen, "Urkunden," pp. 94–95 and passim. Gilles de Corbeil, *Viaticus*, p. 78.

3. Guy de Chauliac, *Chirurgia*, p. 285. Du Laurens, *Operum*, pp. 39–40. Hafenreffer, *Nosodochium*, p. 113.

4. Hafenreffer, *Nosodochium*, pp. 113–114.

5. Sudhoff, "Lepraschaubriefe aus Italien," p. 435. Sudhoff, "Lepraschaubriefe aus dem 15. Jahrhundert," p. 372. Sudhoff, "Wurzacher Lepraschaubriefe," p. 430. Keussen, "Urkunden," pp. 85, 92, and passim.

6. Keussen, "Urkunden," pp. 109 (1486) and 95 (1550). Bourgeois, *Lépreux et maladreries*, p. 186.

7. Wickersheimer, "Beiträge," p. 147. Sudhoff, "Lepraschaubriefe aus dem 15. Jahrhundert," p. 372. Keussen, "Urkunden," p. 95. Mus, *Oorkonden IV*, pp. 30–31. Bourgeois, *Lépreux et maladreries*, p. 215.

8. Guy de Chauliac, *Chirurgia*, p. 285. Du Laurens, *Operum*, pp. 37, 40. Hafenreffer, *Nosodochium*, pp. 115–116.

9. Paré, *Les Oeuures*, pp. 753–755.

10. Ibid., p. 753.

11. Guy de Chauliac, *Chirurgia*, p. 285.

12. Gersdorff, *Feldtbuch der Wundtartzney*, fol. 72v.

13. Uddman, *Lepra*, p. 5. Van Leersum, *De "Cyrurgie" van Meester Jan Yperman*, p. 177. Avicenna, *Canon medicinae*, sig. FF 6rb. Bertucci, *Nusquam antea*, p. 245ra. Galen, *Definitiones medicae*, Kühn XIX.427. Aretaeus, *Aretaei Cappadocis*, pp. 209–213.

14. Aretaeus, *Aretaei Cappadocis*, p. 212.

15. For example, in Hafenreffer, *Nosodochium*, p. 113. Avicenna, *Canon medicinae*, sig. FF 7rb. Hildegard of Bingen, *Causae et curae*, p. 161. Guy de Chauliac, *Chirurgia*, p. 284. Paré, *Les Oeuures*, p. 755.

16. Du Laurens, *Operum*, pp. 39–40.

17. Bernard de Gordon, *Lilium*, fol. 18rb.

18. Paré, *Les oeuures*, p. 756.

19. Sudhoff, "Dokumente zur Ausübung der Lepraschau," p. 158. Context in Chaps. 1 and 5.

20. Aretaeus, *Aretaei Cappadocis*, pp. 209–213. Avicenna, *Canon medicinae*, sig. FF 7rb. Campbell and Colton, *Surgery of Theodoric*, p. 168. Johannes de Sancto Paulo, *Breviarium*, sig. aᵛ. Johannes Platearius, *Practica brevis*, fol. 221v. Ferreira, *Tratado médico*, p. 76. Gariopontus, *Passionarius*, fol. 77.

21. "Leprosos apprehendimus et primum cognoscimus ex maculis similibus impetigini in inicio apparentibus," Johannes de Sancto Paulo, *Breviarium*, sig. aᵛ; "Elephantiosos apprehendimus ex maculis," Gariopontus, *Passionarius*, fol. 77; however, the eighth-century compendium from Lorsch already contained the expression "Elefantiosus ita apprehendimus: Primum ex maculis," Stoll, *Das "Lorscher Arzneibuch,"* p. 128. Bernard de Gordon, *Tabula de decem ingeniis*, fol. 120rb (see above, Chap. 1).

22. Bernard de Gordon, *Lilium*, fol. 18rb. Bourgeois, *Lépreux et maladreries*, pp. 185–186.

23. Bernard de Gordon, *Lilium*, fol. 18rb. Pagel, *Die Chirurgia des Heinrich von Mondeville*, p. 423.

24. "Ita quod illud examen debite fiat, accipe unam tabulam et scribas bona signa ad unam partem et mala ad aliam partem, et sic non deficies," *De examine leprosorum*, in Munich, CLM 660, fol. 46r. For a lucid English translation by Michael McVaugh, of a slightly variant version of this tract, see Grant, *Source Book*, pp. 754–755. Forestus, *Observationum et curationum chirurgicarum libri quinque*, p. 132. *Signa lepre*, in Vat. Pal. MS Latin 1207, fol. 16v.

25. Jordanus de Turre, *De lepra nota*, in Demaitre, "Relevance of Futility," pp. 54–56. This edition does not contain the diagram, which occurs in Basel Universitätsbibliothek MS D.I.11, fols. 103vb–104r and which, unfortunately, has proven difficult to reproduce photographically.

26. Bertucci, *Nusquam antea*, p. 245r. An intermediate version, between the *Nota* of Jordanus and the pages in the *Collectorium* of Bertucci, is extant in Vat. Pal. MS Latin 1225, fol. 17r–v, *incipit*, "Examinando [*sic*] leprosorum ordo hic observandus est."

27. Guy de Chauliac, *Chirurgia*, pp. 283–284.

28. Bertucci, *Nusquam antea*, p. 245r.

29. Bernard de Gordon, *Lilium*, fol. 18rb. Pagel, *Die Chirurgia des Heinrich von Mondeville*, p. 423.

30. Guy de Chauliac, *Chirurgia*, p. 284. Paré, *Les oeuures*, pp. 753–756.

31. Marseille: M. Barthélemy, "Procès-verbal," pp. 118–119; compare Chaps. 5 and 6.

32. Ganges: Archives Départmentales de l'Hérault, II E 35/42; compare Chaps. 3 and 5.

33. Bertucci, *Nusquam antea*, p. 245r. Bernard de Gordon, *Lilium*, fol. 18ra. Guy de Chauliac,

Chirurgia, p. 284, reading "fetiditas," which is no doubt a scribal error for *feditas*, or "hideous-ness": the proper and common Latin term for fetor is *fetor*; I have not encountered fetid lips listed among symptoms; and Avicenna and Bernard de Gordon suggested that lips became unsightly, "ingrossantur et denigrantur." For body fetor as a symptom, see Michael McVaugh, "Smells and the Medieval Surgeon," *Micrologus* 10 (2002), pp. 113–132, esp. pp. 115 and 131.

34. Barthélemy, "Procès-verbal," p. 120 (emphases added).

35. Paris BN MS lat. 6988A, fol. 9, quoted in Wickersheimer, "Beiträge," p. 147.

36. Stadtarchiv Nördlingen, R39, IV, F5, Nr. 37, "Ärztliche Atteste, 1503–1694," May 26, 1542. Compare the similarly casual, though later (1657), reference to syphilis by Cologne doctors who found Adelheidt Diercksens free of leprosy, even though she "presented large and wide ulcers scattered over her body: these ulcers were not judged leprous, for they were more suggestive [*redolebant*] of the French pox," Keussen, "Urkunden," p. 98.

37. Thus, "Signes généraulx ou équivocques" were contrasted with "signes espéciaulx uni-vocques" in Arras, 1506: Bourgeois, *Lépreux et maladreries*, p. 217. Touati, "Facies leprosorum," p. 62. Bernard de Gordon, *Lilium*, fol. 20ra.

38. Bernard de Gordon, *Lilium*, fols. 18ra, 19va–20ra. Bernard was consistent in his emphasis, for in his *Liber pronosticorum* (1295) he had declared, "[I]t should be understood that signs taken from the face are by far the most truthful [*veridica in ultimo*]": Guardo, *Los pronósticos médicos*, p. 288. "Observable pathognomonic lesions on the bones of the face" indeed reveal, as Grmek points out, "the initial phase of leprous infection," Grmek, *Diseases*, p. 153.

39. Avicenna, *Canon medicinae*, sig. FF 7rb. Bernard de Gordon, *Lilium*, fol. 20r.

40. Bernard de Gordon, *Lilium*, fol. 20ra.

41. Borradori, *Mourir au Monde*, pp. 223–224.

42. For example, "pustulae, excrescentiae, nodi, impetigines et morfeae ut plurimum in facie, et aliquando alibi," Pagel, *Die Chirurgia des Heinrich von Mondeville*, p. 423; "Lepers can be recognized by five signs: by the urine, by the pulse, by the blood, by the voice, and by different members . . . a leper's face is horrible to see; its natural expression being distorted, it is a terrible sight. These are the most obvious signs," translation of Jordanus de Turre by Michael McVaugh, in Grant, *Source Book*, p. 754. Guy de Chauliac, *Chirurgia*, p. 285. Bernard de Gordon, *Lilium*, fol. 18ra. Velascus de Tharanta, *Practica*, fol. 256rb.

43. Paré, *Les oeuvres*, pp. 753–756. Varanda, *Tractatus de elephantiasi*, p. 29.

44. "Veu l'intégrité de tous ses membres, sans aucune macule ny exténuation, ni déformité par tout son corps, combien que à la fache on y voit déformité du nez et de la bouche qui sont signes équivoques non suffisans pour le séquestrer," Bourgeois, *Lépreux et maladreries*, p. 218. Sudhoff, "Wurzacher Lepraschaubriefe," p. 425.

45. On Ramona Isern, see above, Chaps. 3 and 4. Keussen, "Urkunden," pp. 88 (1508) and 97 (1569).

46. Keussen, "Urkunden," pp. 94–95.

47. Mus, *Oorkonden IV*, pp. 38, 29.

48. Sudhoff, "Wurzacher Lepraschaubriefe," pp. 430–431.

49. Velascus de Tharanta, *Practica*, fol. 256rb (emphases added). Paré, *Les oeuvres*, p. 755 (emphases added). Bernard de Gordon, *Lilium*, fol. 18rb. Hafenreffer, *Nosodochium*, p.114.

50. See the stimulating recent assessment by Jacalyn Duffin, "Jodocus Lommius's Little

Golden Book and the History of Diagnostic Semeiology," *Journal of the History of Medicine and Allied Sciences* 61 (2006), pp. 249–287. Uddman, *Lepra*, p. 4. See above, Figure 1.1. Varanda, *Tractatus de elephantiasi*, pp. 29–30. Petraeus, "Diss[ertatio] harmonica XLV," p. 410.

51. Schlägl MS 102, fol. 144r (I am grateful to the Hill Museum & Manuscript Library, Saint John's University, Collegeville, Minnesota, for giving me access to this manuscript page). See Grmek, *Diseases*, p. 153.

52. Guy de Chauliac, *Chirurgia*, p. 284.

53. Forestus, *Observationum et curationum chirurgicarum libri quinque*, pp. 124, 132. Gersdorff, *Feldtbuch der Wundtartzney*, p. 73r.

54. Premodern hematological suppositions were evaluated positively, in the light of modern knowledge, by Stephen Ell, "Blood and Sexuality in Medieval Leprosy," *Janus* 71 (1984), pp. 153–164. Gilles de Corbeil, *Viaticus*, p. 78. Campbell and Colton, *Surgery of Theodoric*, p. 168. Translation of Jordanus by Michael McVaugh, in Grant, *Source Book*, p. 754. Karl Sudhoff, "Eine Blutprobe zur Erkennung der Lepra," *Archiv für Geschichte der Medizin* 6 (1912–1913), p. 159. Velascus de Tharanta, *Practica*, fol. 256v.

55. Guy de Chauliac, *Chirurgia*, p. 285. Danièle Iancu, *Etre Juif en Provence au temps du roi René* (Paris: Albin Michel, 1998), pp. 89, 91–92. Hébert, "Un diagnostic de lèpre," pp. 135, 134.

56. Wickersheimer, "Beiträge," pp. 147–150. Guy de Chauliac, *Chirurgia*, p. 285. Translation of Jordanus de Turre by Michael McVaugh, in Grant, *Source Book*, p. 755. Velascus de Tharanta, *Practica*, fol. 256r.

57. Huggelin, *Von dem Aussatz*, sig b iⱽ. Paré, *Les oeuures*, p. 755.

CHAPTER EIGHT. Prognosis, Prevention, and Treatment

1. Latin text in Sudhoff, "Was geschah," p. 153. For an initial exploration of this aspect, see Demaitre, "Relevance of Futility," pp. 25–61. Much of this chapter stands in the shadow of two recent and incisive contributions: François-Olivier Touati, "Pharmacopée et thérapeutique contre la lèpre au Moyen Age: Quelques réflexions méthodologiques," in *Questions d'histoire de la médecine: Actes du 13ᵉᵐᵉ Congrès National des Sociétés Savantes* (Paris: CTHS, 1991), pp. 17–26; and an elaboration for the medieval authors of one university, by the same author, "Les traités sur la lèpre," pp. 205–231, esp. pp. 211–231. Uddman pointed out that, in Europe south of Scandinavia, for the specific purpose of isolating incurable leprous patients, "constructa sunt *Nosocomia*, quae dicuntur *incurabilium* aut vulgo *Lazarea* (*Lazaretter*)," *Lepra*, pp. 4–5. Bériac, *Histoire des lépreux au moyen âge*, p. 262; other sweeping generalizations passim.

2. Touati, *Maladie et société*, pp. 331, 345, 353, 458. Three centuries later, the bishop of Sens entrusted Gauffridus de Rippeforti, licentiate in medicine, with the "governance and administration of the hospital of Saint James of Melun": Auxerre, Archives départmentales de l'Yonne, série G 187, fol. 3; I am not sure whether this hospital was a leprosarium, since it is not mentioned in the exhaustive inventory by François-Olivier Touati, *Archives de la lèpre: Atlas des léproseries entre Loire et Marne au Moyen Âge*, Mémoires et documents d'histoire médiévale et de philologie, no. 7 (Paris: Comité des Travaux Historiques et Scientifiques, 1996). Van Schevensteen, *La Lèpre dans le Marquisat d'Anvers*, p. 7. Bourgeois, *Lépreux et maladreries*, pp. 312 n. 95, 311–312, 314.

3. *Cartulaire de l'Université de Montpellier*, vol. 1, p. 188. See above, Chap. 1. Quoted in Nancy

Siraisi, *Medieval and Early Renaissance Medicine: An Introduction to Knowledge and Practice* (Chicago: University of Chicago Press, 1990), p. 41.

4. Pourrière, *Les hôpitaux d'Aix-en-Provence*, p. 154 n. 46. Wickersheimer, *Dictionnaire biographique*, p. 414. Danielle Jacquart, *Le milieu médical en France du XIIe au Xve siècle. En annexe 2ᵉ supplément au "Dictionnaire" d'Ernest Wickersheimer* (Geneva: Droz, 1981), pp. 183–184. Keussen, "Beiträge," p. 90.

5. Pagel, *Die Chirurgia des Heinrich von Mondeville*, p. 424. Guy de Chauliac, *Chirurgia*, p. 287. Des Innocens, *Examen*, p. 4.

6. Iuliarius, *De lepra*, p. 2r. For more background on the issue of incurability, see Plinio Prioreschi, "Did the Hippocratic Physician Treat Hopeless Cases?" *Gesnerus* 49 (1992), pp. 341–350.

7. Van Schevensteen, *La Lèpre dans le Marquisat d'Anvers*, pp. 50–54, 84.

8. Ibid., p. 60. Mus, *Oorkonden IV*, pp. 29, 52–53 (emphasis added).

9. Van Schevensteen, *La Lèpre dans le Marquisat d'Anvers*, pp. 63–64.

10. Aretaeus, *Aretaei Cappadocis*, pp. 207–208. English translation by Mireille Muellner and Leonard Muellner, in Grmek, *Diseases*, p. 169 (emphasis added). Jacques de Vitry, "Sermo ad leprosos et alios infirmos, thema sumptum ex epistula Iacobi, capitulo V," ed. Bériou and Touati, *Voluntate Dei Leprosus*, p. 115 (emphasis added).

11. Bernard de Gordon, *Lilium*, fols. 12v, 33v, 10v. Fernel, *Pathologiae libri VII*, p. 335; Hafenreffer, *Nosodochium*, p. 118. See Touati, *Maladie et société*, p. 243 n. 195. Bourgeois, *Lépreux et maladreries*, p. 218 n. 161. Quoted in Touati, *Maladie et société*, p. 707.

12. Jacobus de Voragine, *The Golden Legend: Readings on the Saints*, trans. William Granger Ryan (Princeton: Princeton University Press, 1993), vol. 2, p. 222. Saint Francis is depicted kissing the *leprosus* on the lips in an eighteenth-century print: see *The Life and Times of Saint Francis* (Philadelphia: Curtis, [1967]), p. 17. Ortrud Reber, *Elisabeth von Thüringen Landgräfin und Heilige. Eine Biografie* (Regensburg, Germany: Pustet Friedrich, 2006). Osvát Laskai (Osvaldus de Lasco), quoted (in Latin) in Maria Vida, "Die heilige Elisabeth und die Bedeutung der Aussätzigen in den Legenden und den ikonographischen Andenken des mittelalterlichen Ungarns," *Orvostörténeti Közlemények / Communicationes de Historia Artis Medicinae*, nos. 133–140 (1991–1992), p. 52. Christine Boeckl graciously provided me with a copy of this article, which is not readily accessible.

13. Walters Art Museum Catalogue, p. 94. Françoise Baron, "Le mausolée de Saint Elzéar à Apt," *Bulletin Monumental* 136, no. 3 (1978), p. 274. I am indebted to Griffith Mann of the Walters Art Museum for giving me access to this valuable article. See, for example, the sermon of Jacques de Vitry in Bériou and Touati, *Voluntate Dei Leprosus*, pp. 118–119.

14. Paré, *Les oeuures*, p. 756. Avicenna, *Canon medicinae*, sig. FF 7rb. Hafenreffer, *Nosodochium*, p. 127.

15. Quoted in Christa Habrich, "Die Arzneimitteltherapie des Aussatzes in abendländischen Medizin," in Richard Toellner, ed., *Lepra-Gestern und Heute. 15 wissenschaftliche Essays zur Geschichte und Gegenwart einer Menschheitsseuche. Gedenkschrift zum 650-jährigen Bestehen des Rektorats Münster-Kinderhaus* (Münster: Verlag Regensberg, 1992), p. 67.

16. Quoted in Bühler, *Der Aussatz in der Schweiz*, p. 66 (emphasis added).

17. Forestus, *Observationum et curationum chirurgicarum libri quinque*, p. 132. Du Laurens, *Operum*, p. 41 (emphasis added). Uddman, *Lepra*, p. 3 (emphasis added).

18. Bernard de Gordon, *Lilium*, fol. 18rb (emphasis added). Albertus de Zanchariis, *Glose*, fol. 28v, citing Haly Abbas (emphasis added). Guy de Chauliac, *Chirurgia*, p. 286 (emphasis added). See Demaitre, "Art and Science of Prognostication in Early University Medicine," p. 765–788.

19. Avicenna, *Canon medicinae*, sig. FF 7ra. Albertus de Zanchariis, *Glose*, fol. 29v.

20. On the *tempora morborum*, see above, Chap. 6. Guy de Chauliac, *Chirurgia*, p. 286.

21. See above, Chap. 2. Varanda, *Tractatus de elephantiasi*, p. 34.

22. Gilio di Portugallo, *Tractato di medicina*, in an Italian version written at Perpignan in 1463, NLM MS E 22, fol. 69r (emphasis added). I have not been able to track down the corresponding chapter in the *Practica*, which is, presumably, the Latin original (also extant as *Secreta medicine*). Guglielmo da Varignana, *Secreta medicinae* (Lyon: Jean Flajollet, 1539), fols. 74v–75r. Basel Universitätsbibliothek MS D.III.10, fol. 10v.

23. Johannes Platearius, *Practica brevis*, fol. 222ra. Vat. Pal. MS Latin 1174, fol. 6vb; the context is discussed by Michael McVaugh, "Two Montpellier Recipe Collections," in Nancy G. Siraisi and Luke Demaitre, eds., *Science, Medicine and the University, 1200–1550: Essays in Honor of Pearl Kibre*, pt. 2, *Manuscripta* 20 (1976), pp. 175–180. Jordanus de Turre, *Nota de lepra*, ed. in Demaitre, "Relevance of Futility," pp. 54–61.

24. Guy de Chauliac, *Chirurgia*, p. 286. Barthélemy, "Procès-verbal," p. 119. Keussen, "Beiträge," p. 97.

25. In addition to Touati, "Pharmacopée et thérapeutique," a perceptive introduction to the problems of methodology is Habrich, "Die Arzneimitteltherapie des Aussatzes," pp. 57–72, esp. pp. 59–60.

26. Touati, "Les traités sur la lèpre," pp. 214–216 (tabulations and list in an appendix, pp. 224–231). Guy de Chauliac, *Chirurgia*, p. 288. Bernard de Gordon, *Lilium*, fol. 18vb.

27. Fred Rosner and Suessman Muntner, trans. and eds., *The Medical Aphorisms of Moses Maimonides*, 2 vols. (New York: Yeshiva University Press, 1971), vol. 2, p. 123. Pagel, *Die Chirurgia des Heinrich von Mondeville*, p. 427. Lynn Thorndike, *A History of Magic and Experimental Science*, 8 vols. (New York: Columbia University Press, 1934–1958), vol. 4, p. 546. Ambroise Paré, *On Monsters and Marvels*, p. 121.

28. For myrobalans, see Bernard de Gordon, *Lilium*, fol. 18va; Guy de Chauliac, *Chirurgia*, p. 288, and virtually every author. London: British Library, Sloane MS 282, fol. 235r; a French provenance of the manuscript is suggested by the inclusion of a remedy *pro leproso* according to a prescription of Pierre d'Auxon, physician to Charles VI from 1405 to 1409, "magistri Petri de Avonia medici regis defuncti." Sulfur: Bernard de Gordon, *Lilium*, fol. 18vb; Guy de Chauliac, *Chirurgia*, p. 289; and so on.

29. Guy de Chauliac, *Chirurgia*, pp. 286–287.

30. Avicenna, *Canon medicinae*, sig. FF 7va. Vat. Pal. MS Latin 1174, fol. 8ra. Bernard de Gordon, *Lilium*, fol. 21rb. Van Leersum, *Het "Boeck van Surgien" van Meester Thomas Scellinck*, p. 199. Varanda, *Tractatus de elephantiasi*, p. 39.

31. Velasco de Tharanta claimed that "Platearius said that many *leprosi* are cured by removal of the testicles," Velascus de Tharanta, *Practica*, fol. 259ra; I have not found this in the Latin *Practica brevis* of Platearius, although it occurred in the Anglo-Norman version: see Hunt, *Anglo-Norman Medicine*, p. 249; indeed, Hunt notes that the phrase is "missing from the print" (p. 279 n. 188). Campbell and Colton, *Surgery of Theodoric*, p. 177. Guglielmo da Saliceto sug-

gested castration "en termes dubitatifs" according to Bériac, *Histoire des lépreux au moyen âge*, p. 261 (without reference). Van Leersum, *De "Cyrurgie" van Meester Jan Yperman*, p. 178. Velascus de Tharanta, *Practica*, fol. 259ra.

32. Bocellin, *Practique*, fol. 19v. For some of the overlaps between leprosy and syphilis, and their context, see Claudia Stein, *Die Behandlung der Franzosenkrankheit in der Frühen Neuzeit am Beispiel Augsburgs*, ed. Robert Jütte, Medizin, Gesellschaft und Geschichte, Beiheft 19 (Stuttgart: Franz Steiner Verlag, 2003), p. 112 and passim.

33. Guy de Chauliac, *Chirurgia*, p. 287. Avicenna, *Canon medicinae*, sig. FF 7rb, ed. in Demaitre, "Relevance of Futility," p. 56. Bernard de Gordon, *Lilium*, fol. 18va.

34. Bernard de Gordon, *Lilium*, fol. 18va. Varanda, *Tractatus de elephantiasi*, p. 33.

35. Bernard de Gordon, *Lilium*, fol. 18vb. Avicenna, *Canon medicinae*, sig. FF 7va. Bernard de Gordon, *Lilium*, fols. 18vb, 18va. Jordanus de Turre, *Nota de lepra*, p. 56. Velascus de Tharanta stated, "[S]trong medicines, *especially* those containing hellebore, should not be administered from the onset," *Practica*, fol. 257v (emphasis added).

36. Bernard de Gordon, *Lilium*, fol. 20vb. Varanda, *Tractatus de elephantiasi*, p. 53. Van Schevensteen, *La Lèpre dans le Marquisat d'Anvers*, p. 70. Uddman, *Lepra*, pp. 13–14. Jordanus de Turre noted, more than four centuries earlier, "Galen in the *Passionarius* advises to prepare a bath of seawater for drying out [the offending humor]: and this is true for *tyria* but, I believe, not for *allopicia* or *leonina*," *Nota de lepra*, p. 60.

37. According to Varanda, the use of mercury was "invented" for the treatment of leprosy and adopted to syphilis: "inunctiones ex hydrargiro quas nunc ad luis venereae expugnationem usurpare solemus primum as scabiem hanc foedam et elephantica ulcera inventa fuerunt," *Tractatus de elephantiasi*, p. 61. Uddman, *Lepra*, p. 13. Schilling, *De lepra*. pp. 57–61. Brody, *Disease of the Soul*, pp. 71–72. Bériac, *Histoire des lépreux au moyen âge*, p. 261.

38. Van Leersum, *De "Cyrurgie" van Meester Jan Yperman*, p. 179. Thorndike, *History of Magic and Experimental Science*, vol. 3, pp. 362, 630. *Sedacina*, I, V, 32, cited in Pascale Barthélemy, "Les liens entre alchimie et médecine: L'exemple de Guillaume Sedacer," in Chiara Crisciani and Agostino Paravicini Bagliani, eds., *Alchimia e medicina nel Medioevo*, Micrologus' Library 9 (Florence: Edizioni del Galluzzo, 2003), p. 119 and n. 34; Sedacer credited Aristotle with the notion, which he borrowed directly from pseudo–Albertus Magnus, *Semita recta*: Barthélemy, "Les liens," p. 120 and n. 37. *Facsimile van Joannes de Vesalia's pesttraktaat, Hs. Stresa*, Centro Internazionale di Studi Rosminiani, ASIC A2 21, Academia Regia Belgica Medicinae, Series Historica, no. 6 (Brussels: Koninklijke Academien voor Geneeskunde van België, 1998), p. 48r–v. Livinus Lemnius, *De miraculis occultis naturae Libri IIII* (Antwerp: Christophorus Plantinus, 1574), p. 310. Hafenreffer, *Nosodochium*, p. 128.

39. Varanda, *Tractatus de elephantiasi*, p. 32.

40. Albertus de Zanchariis, *Glose*, fol. 42r. Quoted from Amsterdam Universiteitsbibliotheek MS XV G 6 (and translated from Middle to Modern Dutch) by Jansen-Sieben, "Ziektebeeld en behandeling," p. 42. Paris, Archives Nationales, X¹ᵃ 30, fol. 257; I am indebted to Riccardo Famiglietti for bringing this document to my attention.

41. Galen, *De virtutibus simplicis medicine*, XI.1 (Latin translation by Gerard of Cremona) in *Galieni Opera*, vol. 2, fol. 80v.

42. For an incisive and balanced assessment, see Touati, "Pharmacopée," p. 21. Jordanus de Turre, *Nota de lepra*, p. 60.

43. Avicenna, *Canon medicinae*, sig. FF 7vb–8ra. Bernard de Gordon, *Lilium*, fol. 19ra. Albertus de Zanchariis, *Glose*, fol. 38r. Bernard de Gordon, *Lilium*, fol. 19ra.

44. Du Laurens, *Operum*, p. 41. Uddman, *Lepra*, pp. 12–13.

45. Pliny, *Historia naturalis*, 26, 5. Samuel Rosenberg, introduction to *Ami and Amile*, trans. Samuel Damon and Samuel Rosenberg (York, S.C.: French Literature Publications, 1981), pp. 3–5. For context, see Anne Berthelot, "Sang et lèpre, sang et feu," in Faure, *Le Sang au Moyen Âge*, pp. 25–37. Hartmann von Aue, *Der arme Heinrich*, p. 22, vv. 224–232. Thorndike, *History of Magic and Experimental Science*, vol. 2, p. 232.

46. Hildegard of Bingen, *Causae et curae*, p. 213. "Sanguis leporis": Bernard de Gordon, *Lilium*, fol. 19ra; Pagel, *Die Chirurgia des Heinrich von Mondeville*, p. 426; Guy de Chauliac, *Chirurgia*, p. 289. Hildegard of Bingen, *Causae et curae*, p. 212.

47. Hildegard of Bingen, *Causae et curae*, p. 211. Guglielmo da Varignana, *Secreta medicinae*, fols. 74r–75v. Lemnius, *De miraculis occultis naturae Libri IIII*, p. 313. Varanda, *Tractatus de elephantiasi*, p. 38.

48. Petraeus, "Diss[ertatio] harmonica XLV," p. 403. Hafenreffer, *Nosodochium*, pp. 127–128.

49. Varanda, *Tractatus de elephantiasi*, pp. 32–34.

50. Michael McVaugh, *The Rational Surgery of the Middle Ages*, Micrologus' Library 15 (Florence: Società Internazionale per lo Studio del Medioevo Latino, 2006), chap. 5, "Surgery between Alchemy and Cosmetics," including the translation of Galen and references; I thank the author for the privilege of reading his remarkable study in manuscript.

51. See above, Chap. 6; Chap. 2. Lemnius, *De miraculis occultis naturae Libri IIII*, p. 310. Pagel, *Die Chirurgia des Heinrich von Mondeville*, p. 430.

52. Campbell and Colton, *Surgery of Theodoric*, p. 177; Bernard de Gordon, *Lilium*, fol. 18rb; Pagel, *Die Chirurgia des Heinrich von Mondeville*, p. 426. Velascus de Tharanta, *Practica*, fol. 347vb. Albertus de Zanchariis, *Glose*, fol. 44v (emphasis added). Jacquart, *Supplément* to Wickersheimer, *Dictionnaire biographique*, p. 222: "Perrette de Rouen." Hafenreffer, *Nosodochium*, p. 128. Varanda, *Tractatus de elephantiasi*, p. 58.

53. Bernard de Gordon, *Lilium*, fol. 19ra.

54. Pagel, *Die Chirurgia des Heinrich von Mondeville*, p. 426. Guy de Chauliac, *Chirurgia*, p. 290. Pagel, *Die Chirurgia des Heinrich von Mondeville*, p. 426, and Albucasis, *On Surgery and Instruments: A Definitive Edition of the Arabic Text with English Translation and Commentary*, by M. S. Spink and G. L. Lewis (London: Wellcome Institute of the History of Medicine, 1973), p. 144. Albertus de Zanchariis, *Glose*, fols. 41v–42r, expanding on Avicenna, *Canon medicinae*, sig. FF 8rb.

55. Bernard de Gordon, *Lilium*, fol. 20vb.

56. Avicenna, *Canon medicinae*, sig. FF 7vb. Albertus de Zanchariis, *Glose*, fol. 47v. Uddman, *Lepra*, p. 2.

57. Jordanus de Turre, *Nota de lepra*, p. 56. Varanda, *Tractatus de elephantiasi*, pp. 61, 56.

58. Paré, *Les oeuures*, p. 757. Platter, quoted in Bühler, *Der Aussatz in der Schweiz*, pp. 62–63.

59. Raymond, *Histoire de l'Élephantiasis* pp. 14, 5.

Conclusion

1. Guy de Chauliac, *Chirurgia*, p. 118.

2. Ibid.

3. Ibid. A slightly freer and more lucid English translation is by Michael McVaugh in Grant, *Source Book*, pp. 773–774.

4. Jon Arrizabalaga, "Facing the Black Death: Perceptions and Reactions of University Medical Practitioners," in Luis García-Ballester, Roger French, Jon Arrizabalaga, and Andrew Cunningham, eds., *Practical Medicine from Salerno to the Black Death* (Cambridge: Cambridge University Press, 1994), p. 285; the entire essay, pp. 237–288, is the best assessment to date.

Index

leprosaria (*continued*)
  ity, 140–41, 145, 155; and contagion, 140–
  41, 143–44, 154–55; discharges from, 63;
  and examinations, 36, 41, 42–43, 58, 60–
  61, 244–45; financial arrangements in, 4,
  46, 60; and incurability, 244–45; vs. lepers' colonies, 275; pilgrimages to, 247;
  and professional physicians, 23–24; and
  saints, 81, 249
leprosy: academic vs. official sources on,
  22–26; attitudes toward, 62, 94–95, 98–
  102; attributes of, 95–98; description of,
  vii–viii, 31; as disease of soul vs. body,
  xii, 38; as disease of whole body, 118–20,
  180, 181, 277, 279; as divine punishment,
  102, 125; geography of, 31; loosening
  boundaries of, 130–31, 155, 230; mutations
  in, 224; pathology of, 104–15; prevalence
  of, viii, ix, 25, 43, 50, 54; remission of, 63,
  71, 252; sources on, ix, x, 1–33. See also
  Hansen's disease
le Roy, Martin, 65
L'Escripvain, Roland, 40–41
Leviticus. See Bible
Lille, 36, 109, 151, 155, 194–95
Linnaeus, Carolus (Carl von Linné), 30, 31,
  180
lions, 164, 208–9, 227. See also *leontiasis*
liver, 107, 135, 162, 175, 178; and definitions,
  108, 109, 122, 124, 126, 127, 128, 129; and
  treatment, 254, 255, 265
Lommius, Jodocus (Jodocus Lommius
  Buranus, Josse van Lom), 28, 89, 130,
  220, 233, 239; and definitions, 113, 127; on
  heredity, 156–57
"Lorscher Arzneibuch," 87, 110
Losser, Heinrich (Heinrich Glyperch of
  Koblenz), 10–17, 44, 58, 64, 149, 150, 200;
  on causes, 160, 161, 165, 167; and correlates, 192, 193; on heredity, 158; on localized vs. systemic disease, 115; and
  Maderus case, 13–15, 16, 41, 65, 68, 78, 82,
  160; on precursors, 194; on stages, 176;
  on symptoms, 211–12

Louis XI (king of France), 257
Louvain Faculty of Medicine, 245–46
Lucretius, 85–86
lupus, 187
Lynscydt, Peter, 167, 230–31

maculatus (spotted, soiled), 96–97
Maderus, Hans, 13–15, 16, 41, 65, 68, 78, 82,
  160
Magdalen leprosarium (Brugge), 47, 61
Magenc, Geoffroi, 243
Mahieu, Jacob (Jacques), 152, 232
Maimonides, Moses, 257
Mainz, Dietrich von, 82
Manchester, Keith, ix
Maréchal, Griet, ix
Marseille, 219, 221, 222
Marsiglio (Marsilius) of Padua, 37–38
Martí de Soler, 134–35
Martini, Petrus, 151
Martin of Tours, Saint, 1
Martin V, Pope, 40
Mary Magdalen, Saint (Saint Magdalen,
  Sainte Madeleine), 58, 81, 247
Maunesin, Maria Jacoba, 45, 116
McVaugh, Michael, 5, 39, 134, 272
medicine: academic, 5–6, 22–23, 26, 30–33,
  60, 80–85, 149; forensic, 72; vs. law, 10–11;
  premodern, vii, viii, ix, 16, 32; vernacular-
  ization of, 123, 208
menstruation, 11–12, 199, 200, 262, 268–69;
  as cause, 167–71; conception during, 156,
  162, 168–71
mercury, 264, 276, 277
Mesue, 123
Metz, 53
microscope, 31, 154
Minee, Jeanne, 63–64
Mitchell, Piers, ix, 62
mob violence, 5–6, 8, 57
Molino, Johannes de, 77
Montagnana, Bartholomeus de, 113
Montagu, Mother Marie Isabelle de, 245
Montpellier school, 3, 17, 29, 68, 133, 177;